Bipolar Disorder
Vulnerability

Bipolar Disorder Vulnerability
Perspectives from Pediatric and High-Risk Populations

Edited by

Jair C. Soares

Consuelo Walss-Bass

Paolo Brambilla

ACADEMIC PRESS
An imprint of Elsevier

Academic Press is an imprint of Elsevier
125 London Wall, London EC2Y 5AS, United Kingdom
525 B Street, Suite 1650, San Diego, CA 92101, United States
50 Hampshire Street, 5th Floor, Cambridge, MA 02139, United States
The Boulevard, Langford Lane, Kidlington, Oxford OX5 1GB, United Kingdom

Notices
Knowledge and best practice in this field are constantly changing. As new research and experience
broaden our understanding, changes in research methods, professional practices, or medical treatment
may become necessary.

Practitioners and researchers must always rely on their own experience and knowledge in evaluating and
using any information, methods, compounds, or experiments described herein. In using such information
or methods they should be mindful of their own safety and the safety of others, including parties for
whom they have a professional responsibility.

To the fullest extent of the law, neither the Publisher nor the authors, contributors, or editors, assume any
liability for any injury and/or damage to persons or property as a matter of products liability, negligence
or otherwise, or from any use or operation of any methods, products, instructions, or ideas contained in
the material herein.

Library of Congress Cataloging-in-Publication Data
A catalog record for this book is available from the Library of Congress

British Library Cataloguing-in-Publication Data
A catalogue record for this book is available from the British Library

ISBN 978-0-12-812347-8

For information on all Academic Press publications visit our
website at https://www.elsevier.com/books-and-journals

Working together
to grow libraries in
developing countries

www.elsevier.com • www.bookaid.org

Publisher: Nikki Levy
Acquisition Editor: Joslyn Chaiprasert-Paguio
Editorial Project Manager: Kathy Padilla
Production Project Manager: Mohana Natarajan
Cover Designer: Matthew Limbert

Typeset by SPi Global, India

Contents

Contributors

A.C. Altamura
Department of Neurosciences and Mental Health, Fondazione IRCCS Ca' Granda Ospedale Maggiore Policlinico, University of Milan, Milan, Italy

Isabelle E. Bauer
Center of Excellence on Mood Disorders, Department of Psychiatry and Behavioral Sciences, McGovern Medical School, University of Texas Health Science Center at Houston (UTHealth), Houston, TX, United States

Daniela V. Bavaresco
Neuronal Signaling and Psychopharmacology Laboratory; Neurosciences Laboratory, Graduate Program in Health Sciences, Health Sciences Unit, University of Southern Santa Catarina (UNESC), Criciúma, SC, Brazil

Boris Birmaher
Department of Psychiatry, School of Medicine, University of Pittsburgh, Pittsburgh, PA, United States

Paolo Brambilla
Department of Neurosciences and Mental Health, Fondazione IRCCS Ca' Granda Ospedale Maggiore Policlinico, University of Milan, Milan, Italy; Translational Psychiatry Program, Department of Psychiatry and Behavioral Sciences, University of Texas Health Sciences Center at Houston, Houston, TX, United States

Dejan B. Budimirovic
Division of Child and Adolescent Psychiatry, Department of Psychiatry and Behavioral Sciences, Johns Hopkins University School of Medicine; Children's Mental Health Center, Kennedy Krieger Institute, Baltimore, MD, United States

Melissa P. DelBello
Division of Bipolar Disorder Research, Department of Psychiatry and Behavioral Neuroscience, University of Cincinnati College of Medicine, Cincinnati, OH, United States

Giuseppe Delvecchio
Department of Neurosciences and Mental Health, Fondazione IRCCS Ca' Granda Ospedale Maggiore Policlinico, University of Milan, Milan, Italy

Robert L. Findling
Children's Mental Health Center, Kennedy Krieger Institute; Division of Child and Adolescent Psychiatry, Department of Psychiatry and Behavioral Sciences, The Johns Hopkins Hospital, Bloomberg Children's Center, Johns Hopkins University School of Medicine, Baltimore, MD, United States

Gabriel R. Fries
Translational Psychiatry Program, Department of Psychiatry and Behavioral Sciences, University of Texas Health Sciences Center at Houston, Houston, TX, United States

Bradley Grant
Division of Child and Adolescent Psychiatry, Department of Psychiatry and Behavioral Sciences, Johns Hopkins University School of Medicine; Children's Mental Health Center, Kennedy Krieger Institute, Baltimore, MD, United States

Rodrigo Grassi-Oliveira
Developmental Cognitive Neuroscience Lab (DCNL), Pontifical Catholic University of Rio Grande do Sul (PUCRS), Porto Alegre, Brazil

Melissa J. Green
School of Psychiatry, University of New South Wales; Black Dog Institute, Prince of Wales Hospital, Randwick, NSW, Australia

Danella M. Hafeman
Department of Psychiatry, School of Medicine, University of Pittsburgh, Pittsburgh, PA, United States

Dana Baker Kaplin
Division of Child and Adolescent Psychiatry, Department of Psychiatry and Behavioral Sciences, Johns Hopkins University School of Medicine, Baltimore, MD, United States

Iram F. Kazimi
Center of Excellence on Mood Disorders, Department of Psychiatry and Behavioral Sciences, McGovern Medical School, University of Texas Health Science Center at Houston (UTHealth), Houston, TX, United States

Vivian Leung
School of Psychiatry, University of New South Wales; Black Dog Institute, Prince of Wales Hospital, Randwick, NSW, Australia

Mateus L. Levandowski
Developmental Cognitive Neuroscience Lab (DCNL), Pontifical Catholic University of Rio Grande do Sul (PUCRS), Porto Alegre, Brazil

Clare McCormack
Division of Behavioral Medicine, Department of Psychiatry, University Medical Centre, New York, NY, United States

Colm McDonald
Centre for Neuroimaging and Cognitive Genomics (NICOG), Clinical Science Institute, National University of Ireland, Galway, Ireland

Philip B. Mitchell
School of Psychiatry, University of New South Wales; Black Dog Institute, Prince of Wales Hospital, Randwick, NSW, Australia

Fabiano G. Nery
Division of Bipolar Disorder Research, Department of Psychiatry and Behavioral Neuroscience, University of Cincinnati College of Medicine, Cincinnati, OH, United States

Barbara Pavlova
Department of Psychiatry, Dalhousie University; Nova Scotia Health Authority, Halifax, NS, Canada

Alessandro Pigoni
Department of Neurosciences and Mental Health, Fondazione IRCCS Ca' Granda Ospedale Maggiore Policlinico, University of Milan, Milan, Italy

João Quevedo
Neuronal Signaling and Psychopharmacology Laboratory, Graduate Program in Health Sciences, Health Sciences Unit, University of Southern Santa Catarina (UNESC), Criciúma, SC, Brazil; Translational Psychiatry Program, Department of Psychiatry and Behavioral Sciences, University of Texas Health Science Center at Houston (UTHealth); Center of Excellence on Mood Disorders, Department of Psychiatry and Behavioral Sciences, University of Texas Health Science Center at Houston (UTHealth); Neuroscience Graduate Program, University of Texas Graduate School of Biomedical Sciences at Houston, Houston, TX, United States

Gloria Roberts
School of Psychiatry, University of New South Wales; Black Dog Institute, Prince of Wales Hospital, Randwick, NSW, Australia

Carina Sinbandhit
School of Psychiatry, University of New South Wales; Black Dog Institute, Prince of Wales Hospital, Randwick, NSW, Australia

Manpreet K. Singh
Division of Child and Adolescent Psychiatry, Stanford University School of Medicine, Stanford, CA, United States

Jair C. Soares
Center of Excellence on Mood Disorders, Department of Psychiatry and Behavioral Sciences, McGovern Medical School, University of Texas Health Science Center at Houston (UTHealth), Houston, TX, United States

Ekaterina Stepanova
Division of Child and Adolescent Psychiatry, Department of Psychiatry and Behavioral Sciences, Johns Hopkins University School of Medicine; Children's Mental Health Center, Kennedy Krieger Institute, Baltimore, MD, United States

Angela Stuart
School of Psychiatry, University of New South Wales; Black Dog Institute, Prince of Wales Hospital, Randwick, NSW, Australia

Rudolf Uher
Department of Psychiatry, Dalhousie University; Nova Scotia Health Authority, Halifax, NS, Canada

Samira S. Valvassori
Neuronal Signaling and Psychopharmacology Laboratory, Graduate Program in Health Sciences, Health Sciences Unit, University of Southern Santa Catarina (UNESC), Criciúma, SC, Brazil

Consuelo Walss-Bass
Translational Psychiatry Program, Department of Psychiatry and Behavioral Sciences, University of Texas Health Sciences Center at Houston, Houston, TX, United States

Cristian P. Zeni
Center of Excellence on Mood Disorders, Department of Psychiatry and Behavioral Sciences, McGovern Medical School, University of Texas Health Science Center at Houston (UTHealth), Houston, TX, United States

Foreword

The concept of *vulnerability* is capital in mental disorders. At a certain point in history, a bidimensional model called "vulnerability-stress" was developed and many medical conditions, including some traditionally considered as unifactorial, such as infections, were identified as good fits for the model. Currently the model has evolved into a much more complex one, with additional dimensions (epigenetics), modulating, and moderating factors. Psychiatry, arguably the most exciting, challenging, and complex medical specialty, has embraced the vulnerability model as it provides a multifactorial, multifaceted view of the disorders of the brain that feature thought, behavioral, and emotional symptoms.

Another critical concept that has emerged over the past two decades is that of "early intervention." Early intervention implies early detection; again this is challenging in mental disorders, because the phenotypes that are described in our classifications are distal expressions of chronic brain suffering, and the early expressions are blurred and unspecific.

In this amazing book, Jair Soares, Consuelo Walss-Bass, and Paolo Brambilla, who are very well-known experts in the neurobiology of bipolar disorder, have assembled a group of international researchers who are particularly focused on investigating the neurobiological and clinical nuances associated with the concept of vulnerability for development of the syndrome we know as bipolar disorder. The book discusses the prodrome, staging models, genetics and epigenetics, endophenotypes, biomarkers, environmental factors, neuroprogression, and early intervention, including pharmacological and nonpharmacological strategies. By reading the book, one ends up knowing everything that should be known, as of today, about the risk factors that are involved in the development of manic-depressive illness.

I am convinced that what we call now *early intervention* is already *late*, far too late, for effective primary prevention. However, this is not an excuse to stop going in the right direction, which includes the implementation of mental health policies that focus on populations at risk, and interventions that avoid unnecessary medicalization but which may facilitate early action when something goes wrong. For this, the fight against stigma is crucial because that is one of the strongest barriers against voluntary mental health access and treatment. Reading this book, one may understand how much progress we have made in relatively few years. We now know much more about the stress-mediated neurobiological pathways that evolve into bipolar illness in *vulnerable* individuals. Do you want to know who the "vulnerable" are? Probably all of us, but some more than others. For further details, please read the book, and enjoy.

Eduard Vieta
University of Barcelona, Barcelona, Spain
Hospital Clinic, IDIBAPS, CIBERSAM, Barcelona, Spain

The bipolar prodrome

Danella M. Hafeman, Boris Birmaher
Department of Psychiatry, School of Medicine, University of Pittsburgh,
Pittsburgh, PA, United States

CHAPTER OUTLINE

INTRODUCTION

It has long been recognized that bipolar disorder rarely comes "out of the blue," but rather is usually preceded by a period of time, lasting at least a month and up to several years of prodromal symptoms (Correll, Hauser, et al., 2014). A better understanding of prodromal bipolar disorder is essential for several reasons. First, there are often delays in diagnosis and treatment of bipolar disorder lasting an average of 10 years, leading to increased morbidity and poor function (Lish, Dime-Meenan, Whybrow, Price, & Hirschfeld, 1994). Second, even when individuals are identified early, it is often unclear how to treat them, due to concern that some medications (such as antidepressants) might exacerbate their symptoms. A better understanding of the bipolar prodrome might allow us to identify ultra-high-risk individuals, conduct studies to understand the neural correlates of bipolar risk, and test the effects

Bipolar Disorder Vulnerability. **https://doi.org/10.1016/B978-0-12-812347-8.00001-4**

of various classes of medications and/or psychotherapies in this population. In this chapter, we review what is currently known about the bipolar prodrome, from both retrospective and longitudinal studies. We first describe the methods that have been used to assess the bipolar prodrome, including study design and assessment tools. Next, we describe findings from individual studies that shed light on the course and characteristics of this prodrome. Finally, we describe recent directions to integrate these findings (e.g., staged model, risk calculator), and we present a current model for the prodrome based on the extant literature.

METHODS
STUDY DESIGN: CHALLENGES AND STRATEGIES

There are significant challenges with studying prodromal symptoms of a relatively rare disorder like bipolar disorder. It would simply be impractical to well-characterize a sample from the general population, and follow them long enough to have a sufficient number develop bipolar disorder. Thus, there are a variety of strategies that have been used to shed light on this topic, each with its own strengths and limitations. Each provides a different view on this prodromal period, although, encouragingly, the findings from different strategies are quite similar, as we will discuss later. What we know about the bipolar prodrome comes from the following types of studies:

- **Family studies:** There are several longitudinal studies of offspring of parents with bipolar disorder, which have well-characterized these at-risk offspring in childhood and adolescence, and prospectively followed them to assess new-onset disorders, including bipolar disorder (Birmaher, Axelson, Monk, et al., 2009; Duffy, Alda, Crawford, Milin, & Grof, 2007; Egeland et al., 2003; Hillegers et al., 2005). Because the onset of bipolar disorder is much more prevalent in individuals at familial risk, there are sufficient converters to evaluate clinical characteristics that might precede new-onset disorder. In addition, these individuals are at higher risk of other disorders as well, and have high levels of subsyndromal symptoms, so they also represent a clinically at-risk population that is important to characterize and better understand. There are some limitations to this approach, however. First, we don't know that the course of the bipolar prodrome is similar in individuals with vs. without a first-degree relative with bipolar disorder, so it is unclear the degree to which these findings are generalizable to individuals without such family risk. Second, the prevalence of syndromal bipolar disorder (bipolar-I/II) in offspring is still fairly low (e.g., 8.4% in BIOS, though not all participants have passed the risk period), so large samples are required to have sufficient new-onset cases to make inference about predictors (Axelson et al., 2015). To handle this limitation, some groups have instead assessed less stringent outcomes, including bipolar spectrum disorder (which includes bipolar disorder, not otherwise specified) and "mood disorder" (which includes unipolar or bipolar depression). Another approach is to identify high-risk samples within the offspring of parents with bipolar disorder, based on the presence of mood, anxiety, and/or mood lability symptoms. Third, there is the important issue of comorbidity in the bipolar parents, and whether differences observed in offspring are related to the family history of bipolar disorder, per se, or a comorbidity. For example, in the Pittsburgh Bipolar Offspring Study, Attention-Deficit/Hyperactivity Disorder (ADHD) was higher in at-risk offspring than community controls. However, after adjusting for confounders (including nonbipolar psychopathology in both biological parents), this difference was no longer significant (Birmaher, Axelson, Monk, et al., 2009). Studies that recruit healthy controls (as opposed to including parents with nonbipolar psychiatric disorders) cannot necessarily conclude that a particular difference in at-risk vs offspring of healthy parents is due to the bipolar disorder (vs higher rates of ADHD, for example). Fourth, another critical issue when carrying out family risk studies is blinding. If the interviewer knows that a parent has bipolar disorder, ratings might be elevated due to expectations of worse outcomes. Most studies discussed here were blinded, except for the Dutch study, which only included offspring of parents with bipolar disorder (Mesman, Nolen, Reichart, Wals, & Hillegers, 2013). Fifth, when using parent report to assess a child's psychiatric symptoms, it is important to take into account the current mood state of the reporting parent, since this can impact symptom ratings (Maoz, Goldstein, Goldstein, et al., 2014). This is especially crucial in family risk studies where, by definition, at least one parent has bipolar disorder, and thus over-reporting of symptomatology could bias parent-report measures of child psychopathology. Sixth, depending on the age range, participants might not have reached the peak period of conversion to bipolar disorder; thus, there is the possibility that some of the nonconverters might still develop the disorder. These issues have been summarized in previous reviews (DelBello & Geller, 2001; Hauser & Correll, 2013; Hunt, Schwarz, Nye, & Frazier, 2016).

- **Unipolar depression studies:** An episode of unipolar depression, particularly with earlier age of onset and psychotic features, sharply increases the risk of new-onset bipolar disorder (Akiskal, Maser, Zeller, et al., 1995; Kovacs, 1996; Strober & Carlson, 1982). Thus investigators have prospectively assessed depressed individuals, to determine symptom predictors of conversion from unipolar depression to bipolar disorder. The strengths of this approach are the fact that it is prospective, and using a select population with conversion rates that allow for adequate cases of new-onset bipolar disorder, at least over a long follow-up (e.g., 19.6% over a mean follow-up period of 17.5 years) (Fiedorowicz et al., 2011). However, there are also some limitations. First, the rate of conversion is still relatively low over shorter periods of follow-up, thus necessitating a longer follow-up period or larger sample for adequate converters. One way that investigators have handled this is to narrow the selection criteria to participants with depression with psychotic features, thus increasing the base rate for developing bipolar disorder; however, this also makes recruitment more difficult. Second, as with the family studies, there is also the possibility that participants who have not passed the peak age of conversion might be misclassified as nonconverting. Third, selection of a sample based on unipolar depression means that the results might be specific to individuals that debut with a major depressive episode, and not necessarily generalize to those who have a different presentation (e.g., cyclothymia or initial mania/hypomania).

- **Cyclothymia/bipolar disorder-not otherwise specified (BD-NOS) samples:** One of the strongest predictors of new-onset bipolar disorder is subthreshold manic episodes (see below for specific findings), and thus several studies have assessed the clinical variables that predict progression from BD-NOS to BD-I/ II (Akiskal, Djenderedjian, Rosenthal, & Khani, 1977; Alloy, Urošević, et al., 2012; Axelson et al., 2011). Significant strengths of this approach are that conversion is high (30%–50% in 18 months to 5 years; see below for details), meaning that even small studies have enough power to assess prospectively the impact of other clinical variables on progression. One possible limitation, discussed in detail later, is that these rates of conversion call into question whether BD-NOS (or cyclothymia) is a precursor to disorder, or part of a bipolar spectrum disorder; thus it is unclear whether assessing characteristics of individuals with BD-NOS really constitutes studying predictors (vs correlates) of disorder, and/or predictors of disorder progression.

- **Community samples:** Several epidemiologic studies have assessed the effect of subsyndromal symptoms, including symptoms of depression, mania, and psychosis, on the onset of bipolar disorder (Homish, Marshall, Dubovsky, & Leonard, 2013; Regeer et al., 2006; Tijssen et al., 2010b). These studies have large numbers (generally >2000 subjects) over a long follow-up to yield an adequate number of new cases of bipolar disorder. The trade-off in these types of studies is that there is generally infrequent follow-up and diagnoses are usually based on abbreviated assessments (often without review by a psychiatrist). Thus, while there is strength in numbers, and in the generalizability of findings to a community sample, a fine-grained assessment of each individual participant is usually lacking. Instead of a full diagnostic interview, the Composite International Diagnostic Interview (CIDI) is often used; however, this instrument has shown excellent concordance with more comprehensive measures (Kessler et al., 2013).

- **Retrospective:** These studies interview individuals with bipolar disorder, often during the first episode of disorder (though not always), about the nature and timing of symptoms that they were having prior to the onset (Correll, Hauser, et al., 2014; Egeland, Hostetter, Pauls, & Sussex, 2000). One strength of retrospective studies is that they can be carried out with minimal resources, and do not require follow-up of the sample; thus it is feasible to collect multiple samples to confirm findings. Also, unlike prospective studies (see below), the sample is not restricted based on criteria such as family history or an episode of depression. In this way, retrospective studies assess "all comers" and are perhaps more generalizable to the population that will develop bipolar disorder.

However, there is a major limitation, which is that these findings are based on the recall of subjects who have already developed the disorder. Individuals with bipolar disorder might have a more distinct memory of subsyndromal mood symptoms during adolescence, for example, than someone who did not go onto develop bipolar disorder; this could lead to recall bias, and inflate the association between previous symptoms and the development of bipolar disorder. Even in the absence of such bias, subjects in retrospective studies are often asked to recall details about experience more than a decade before; this may lead to incorrect or incomplete reporting of previous symptoms, especially in someone with neurocognitive deficits (as often seen in bipolar disorder). An alternative method is to use records prior to bipolar disorder onset; this removes the problem of recall bias, but clinical records are often incomplete (Egeland et al., 2000). Also, depending on where the bipolar sample is recruited from (e.g., inpatient unit), the reported prodromal symptoms might not be generalizable to individuals with bipolar disorder who do not present to this setting. Thus retrospective studies provide an important source of clinically rich information about what the bipolar prodrome might look like, but these results should be confirmed using additional methods.

In summary, each of these study designs has important strengths, but also significant flaws that might bias findings and/or limit generalizability of results. Thus the strongest conclusions will come from observing similar findings across different study designs.

QUESTIONNAIRES/ASSESSMENTS

Based on a growing knowledge base about the symptoms that predict new-onset bipolar disorder, several scales and instruments have been developed to characterize better this risk. Many of these were reviewed recently (Ratheesh, Berk, Davey, McGorry, & Cotton, 2015); while a few of these questionnaires showed promise in a single study, e.g., the General Behavioral Inventory (GBI) and the Manic Symptom Subscale of the Child Behavioral Checklist (CBCL-MS), none was replicated in high-quality studies. Scales used to screen for bipolar disorder in youth, a closely related though not identical problem, have been evaluated elsewhere (Youngstrom et al., 2004). These authors found that, in general, parent report of manic symptoms better distinguished youth with bipolar disorder from healthy controls than either youth or teacher reports. Here, we discuss scales and assessments that have been used in the attempt to *predict* onset of bipolar disorder. The list below is meant to be not a comprehensive review of questionnaires that could potentially predict bipolar disorder, but rather an overview of strategies that investigators have used to estimate the risk of conversion, and the degree to which these have been validated in prospective studies.

- Child Behavior Checklist (CBCL) subscales (parent-report): One longitudinal study found that a high score on CBCL subscales of attention, aggression, and anxiety/depression predicted new-onset bipolar disorder in youth with ADHD (Biederman et al., 2009); thus they termed this the pediatric bipolar disorder subscale. However, these authors and others found that high scores on this scale predicted not only bipolar disorder, but also other disorders such as depression and conduct disorder (Diler et al., 2009; Meyer et al., 2009); the scale also predicted severity of disorder and poor function, and thus seems to be an indicator of general psychopathology. As such, it is now more often called the dysregulation profile (Althoff, Verhulst, Rettew, Hudziak, & van der Ende, 2010). More recently, Papachristou et al. (2013) developed the CBCL mania scale (CBCL-MS), based on 19 items from the CBCL (Table 1); the scale was found to have high internal consistency, and to discriminate between youth with BD-I and healthy controls (AUC = 0.64) (Papachristou et al., 2013). Youth with BD-I also had higher scores on the CBCL-MS than youth with anxiety ($P = .004$) and major depressive disorder ($P = .002$), but not compared to youth with ODD or ADHD. In a longitudinal community study of Dutch adolescents, the authors found that those in mildly and highly symptomatic classes (based on their CBCL-MS scores at age 11) were at a twofold and fivefold risk, respectively, to develop new-onset bipolar disorder by the age of 19 (Papachristou et al., 2017). After adjustment for confounders, this scale was not predictive of new-onset anxiety or depression, though those in the highly symptomatic class were more likely to have diagnoses of ADHD, oppositional defiant disorder (ODD), and conduct disorder.

- Bipolar Prodrome Symptom Interview and Scale (BPSS clinician-administered): Correll et al. (2007) initially developed a retrospective version of this scale, which included 36 items that assessed subthreshold symptoms of mania, depression, and psychosis. In 52 individuals with child- or adolescent-onset mania, the authors found that all participants had experienced at least one moderately severe manic symptom prior to onset. While approximately half had an "insidious" onset (>1 year of symptoms), most of the remainder had a "subacute" onset (1 month to 1 year); only a small minority (3.8%) had less than a month of symptoms prior to onset. Most common symptoms were subthreshold manic symptoms (irritability, racing thoughts, and increased energy) and depressed mood. The prospective version of this scale (BPSS-P) was developed more recently, and has also been shown to discriminate well between bipolar disorder, other psychopathology, and healthy controls, with expected correlations with other scales of mania and depression (Correll, Olvet, et al., 2014). The prospective utility of this scale has not yet been evaluated.

- General Behavioral Inventory (GBI), Revised (self-report and parent-report): This extensive inventory was developed to screen for bipolar disorder and unipolar depression (Depue, Krauss, Spoont, & Arbisi, 1989); it consists of two subscales, assessing depression and hypomanic/biphasic symptoms. In the Dutch cohort of bipolar offspring, more severe scores on the depression subscale was found to predict future bipolar disorder (vs unipolar depression, nonmood disorders, and no disorder) in the next 5 years; the

Table 1 Studies of bipolar disorder offspring with longitudinal follow-up

Study	Participants	Follow-up	Recruitment source	Assessment tools	At-risk vs healthy controls	Predictors of new-onset BD
Amish (Egeland et al., 2003, 2012; Shaw et al., 2005)	115 at-risk; 106 controls, 8 at-risk developed BD-I	16 years	Community	CARE— Semistructured interview; blinded	Compared to control children, those with BD parent had more problems with attention, energy (high and low), sleep, and talkativeness	Those who later developed BD had more "sensitive" temperament as preschool children, as well as more anxiety/worry; later, lower mood, lower energy, and decreased sleep were evident in those who converted
Dutch (Hillegers et al., 2005; Mesman et al., 2013; Mesman, Nolen, et al., 2017)	140 at-risk; 4 with BD-I/II at intake, 8 developed over follow-up	12 years	Survey + hospitals	KSADS; GBI; YSR; CBCL; not blinded	Did not find increased CBCL scores relative to normative samples, but they did not recruit and follow a control sample	New-onset bipolar disorder was mostly preceded by episode of depression; depressive symptoms (as measured by the General Behavioral Inventory) predicted new-onset bipolar disorder; subthreshold manic symptoms predicted bipolar onset in offspring with mood disorder
Ontario Cohort (Duffy et al., 2007, 2014; Duffy, Alda, Hajek, Sherry, & Grof, 2010)	229 at-risk (133 LiR, 96 LiNR) and 86 controls; 8 with BD at intake, 31 lifetime BD	16 years	Outpatient clinics	KSADS; blinded	Higher risk for anxiety and mood disorders	Anxiety disorders predicts mood disorders in general; bipolar disorder generally debuts with a depressive episode

Continued

Table 1 Studies of bipolar disorder offspring with longitudinal follow-up—cont'd

Study	Participants	Follow-up	Recruitment source	Assessment tools	At-risk vs healthy controls	Predictors of new-onset BD
Pittsburgh Bipolar Offspring Study (Axelson et al., 2015; Birmaher et al., 2013; Birmaher, Axelson, Monk, et al., 2009; Diler et al., 2011; Hafeman et al., 2016; Levenson et al., 2015, 2017; Sparks et al., 2014)	359 at-risk and 220 controls; 33 had BD at baseline, 44 with new-onset	12 years	Advertisements, research studies, outpatient clinics	KSADS; CALS; MFQ; SCARED; CBCL; SSHS; blinded	More mood lability, internalizing and externalizing symptoms in at-risk vs controls. Also higher rates of anxiety, disruptive behavioral disorders, ADHD, major depressive episodes (but not MDD)	Diagnostic predictors of BD: subthreshold manic episode, major depression, disruptive behavior disorder. Dimensional predictors: subthreshold manic symptoms, anxiety/depression, mood lability. Sleep difficulties also predicted new-onset BP
Pittsburgh Bipolar Offspring Study—Preschool Study (2–5 years) (Birmaher et al., 2010; Maoz, Goldstein, Axelson, et al., 2014)	122 at-risk, 102 controls	Ongoing; only baseline reported	Advertisements, research studies, outpatient clinics	CBCL; ECI-4; EAS	More aggression, mood dysregulation, sleep disturbances, and somatic complaints; higher rates of ADHD	n/a
Australia Cohort (Perich et al., 2015)	118 at-risk, 110 controls	Ongoing; only baseline reported	Advertisements, research studies, outpatient clinics; consumer organizations	K-SADS-PL; DIGS v4 (for adults)	Higher rates of depressive, anxiety, and behavioral disorders in at-risk	n/a
Multisite NIMH-Funded Cohort (Nurnberger Jr et al., 2011)	141 at-risk, 91 controls	Ongoing; only baseline reported	Genetic studies	K-SADS-PL	Higher rates of affective disorders in general, and bipolar disorder	Retrospective assessment: anxiety and externalizing disorders predicted major affective disorder, but only in at-risk offspring

(Radke-Yarrow et al., 1992)	44 at-risk, 72 controls (+82 offspring of mother with MDD)	3 years	Daycare centers, advertisements, outpatient clinics	CBCL; CAS	Higher rates of depression, anxiety, and disruptive behavioral disorders in at-risk (also in offspring of MDD)	n/a
(Hammen et al., 1987, 1990)	18 at-risk, 38 controls + offspring of mothers with MDD (n = 19) and medical illness (n = 18)	3 years	Inpatient units, outpatient clinics, private referrals	K-SADS	Higher rates of mood d/o (67%) and MDD (22%) than controls (17% and 11%, respectively)	n/a
(LaRoche et al., 1985, 1987)	39 at-risk	3 years	Outpatient clinics	Children's Psychiatric Rating Scale	18% mood disorder; 13% "cyclothymic traits"	n/a

BD, Bipolar Disorder; CARE, Child and Adolescent Research Evaluation; CAS, Child Assessment Scale; CBCL, Child Behavioral Checklist; DIGS, Diagnostic Interview for Genetic Studies; EAS, Emotionality Activity Sociability Survey; ECI-4, Early Childhood Inventory-4; K-SADS-PL, Kiddie-Schedule for Affective Disorders and Schizophrenia for School-Aged Children; LINR, Offspring of lithium-nonresponsive parent; LIR, Offspring of lithium-responsive parent; MD, Major Depressive Disorder.

hypomanic/biphasic scale was not predictive (Reichart et al., 2005). The authors provide several explanations for this, including the low power (only nine conversions to bipolar disorder). Several scales have been developed based on the GBI, including shortened scales (Mesman, Youngstrom, Juliana, Nolen, & Hillegers, 2017), a shortened parent-report of the GBI (Youngstrom, Frazier, Demeter, Calabrese, & Findling, 2008), and self-report of the GBI in adolescents (Danielson, Youngstrom, Findling, & Calabrese, 2003). While many of these adaptations have shown promise regarding screening and discrimination between disorders, they have not yet been evaluated in a longitudinal study to determine if high scores are predictive of new-onset bipolar disorder.

- Hypomanic Personality Scale (self-report): This is a measure of an "overactive, gregarious style" of interacting with others (Eckblad & Chapman, 1986) that was found to predict bipolar disorder at 13-year follow-up: 25% (9/36) with the highest scores developed bipolar disorder (Kwapil et al., 2000); however, it was not predictive of bipolar disorder in another larger sample (Klein, Lewinsohn, & Seeley, 1996)
- Bipolar At-Risk Criteria (clinician-administered): Building on previous findings about risk for bipolar disorder, Bechdolf and colleagues defined the Bipolar At-Risk (BAR) Criteria to be: (1) subthreshold manic symptoms (≥ 2 days, <4 days); (2) depression ($\times 1$ week) + cyclothymia; or (3) depression ($\times 1$ week) + family history of bipolar. They prospectively evaluated 35 adults who met this criteria, along with 35 controls, for 12 months; while 5 in the BAR group converted, none in the non-BAR group developed bipolar disorder (Bechdolf et al., 2014).
- BIS/BAS (self-report): The Behavioral Approach System (BAS) is central to motivation to obtain and response to reward, and is thought to be hypersensitive in individuals with and at-risk for bipolar disorder. To test whether BAS hypersensitivity would predict new-onset bipolar disorder, Alloy and colleagues assessed community adolescents for reward sensitivity, using the BAS subscales of the BIS/BAS and the reward sensitivity subscale of the Sensitivity to Reward/ Punishment Questionnaire). In a year of follow-up, 12.3% (21/171) of the adolescents in the High BAS group (highest 15th percentile on both scales) developed bipolar spectrum disorder vs 4.2% (5/119) in the Medium BAS group (Alloy, Bender, et al., 2012). Thus the BIS/BAS, in combination with another measure of reward sensitivity, predicted bipolar disorder onset in adolescents.
- CALS (self-report and parent-report): The Child Affective Lability Scale (CALS) was developed by Gerson et al. (1996); it includes both self- and parent-report versions that assess for sudden and intense changes in mood. Based on an analysis from the Pittsburgh Bipolar Offspring Study, the CALS has three factors (anxiety/depression, irritability, and mania), all of which were highest in offspring with bipolar disorder, intermediate in nonbipolar offspring of parents with bipolar disorder, and lowest in control offspring; the irritability factor was most important (Birmaher et al., 2013). The CALS score was also found to predict new-onset bipolar disorder over follow-up (Hafeman et al., 2016).

In summary, many scales have been developed to assess part or all of the bipolar prodrome. While none has yet been replicated in high-quality prospective trials, many of these scales have been shown to predict new-onset bipolar disorder in at least one longitudinal study. A further consideration for the value of these scales is whether they are transportable to a community setting. Several of these scales are self- or parent-reports that are free and available for use, and can easily be integrated into a community psychiatry evaluation. The clinician-administered assessments are more laborious, and require training, but also represent potential tools for the clinician (and/or researcher) to assess bipolar disorder risk.

RESULTS
FAMILY STUDIES

Offspring of parents with bipolar disorder are at sharply elevated risk for developing bipolar disorder, with 10%–20% developing bipolar spectrum, and the majority developing a psychiatric disorder (Axelson et al., 2015; Duffy et al., 2014; Mesman et al., 2016). Thus family risk studies provide two important pieces of information. First, how do offspring of parents with bipolar disorder compare to offspring of community controls (who do not have bipolar disorder)? Such comparisons begin to point to possible precursors of the disorder, though many of these offspring will never actually develop bipolar disorder. In particular, it is also important to assess which differences are specific to bipolar disorder (and not due to other parental psychopathology), thus highlighting the importance of including "controls" with nonbipolar psychopathology. Second, which symptoms or diagnoses in these at-risk youth predict the onset of bipolar disorder? This provides a more targeted set of symptoms that may predict onset of bipolar disorder; however, most studies have limited power to conduct such an analysis, which requires large numbers of at-risk youth with quite a lengthy follow-up.

Four major prospective studies of bipolar offspring have shed light on the bipolar prodrome (Birmaher, Axelson, Monk, et al., 2009; Duffy et al., 2007; Egeland et al., 2003; Hafeman et al., 2016; Hillegers et al., 2005); there are two additional ongoing studies, but only baseline findings have been reported at the time of writing (Nurnberger Jr et al., 2011; Perich et al., 2015). In addition to these larger studies, there were some earlier studies with small sample sizes and short follow-up; these studies nonetheless paved the way for these larger and more well-powered cohorts (Hammen, Burge, Burney, & Adrian, 1990; LaRoche et al., 1987; Radke-Yarrow, Nottelmann, Martinez, Fox, & Belmont, 1992), and have also been previously reviewed (DelBello & Geller, 2001; Lapalme, Hodgins, & LaRoche, 1997). Methodology and findings from longitudinal family studies are summarized in Table 1.

One major study has followed an Amish sample of 115 bipolar offspring for over 16 years; eight converted to BD-I. They found that at-risk offspring (compared to controls) had difficulties with attention, energy, and sleep, all of which were present in the individuals who converted to BD-I (Egeland et al., 2012; Shaw, Egeland, Endicott, Allen, & Hostetter, 2005). The Ontario cohort has followed 229 at-risk offspring, 31 who have bipolar disorder, and 86 healthy controls for over 16 years; these offspring were also characterized according to whether the parent responded to lithium. Investigators assessed diagnoses in at-risk vs healthy controls, and predicted onset of bipolar disorder. They found that at-risk individuals had higher incidence of anxiety and mood disorders, and that anxiety disorders appeared to precede and predict mood disorders (Duffy et al., 2014). The Pittsburgh Bipolar Offspring Study has followed 359 offspring for over 12 years (44 had Bipolar Spectrum Disorder (BPSD) at baseline, 31 converted over follow-up) and 220 community controls (Axelson et al., 2015; Birmaher, Axelson, Monk, et al., 2009; Hafeman et al., 2016). Data on dimensional symptoms and diagnosis have been collected every 2 years in this sample, with follow-up ongoing. Findings of this study include: (1) increased mood lability, anxiety, and depression in at-risk vs controls (Birmaher et al., 2013; Diler et al., 2011); (2) increased rates of psychiatric disorders, including anxiety, bipolar disorder, and mood disorders in general (Birmaher, Axelson, Monk, et al., 2009); (3) the best *dimensional* predictors of new-onset BPSD were symptoms of mood lability, anxiety/depression, and, closer to onset, subthreshold manic symptoms (Hafeman et al., 2016); (4) the best *diagnostic or categorical* predictors of BPSD were depression and subthreshold hypomania (Axelson et al., 2015); and (5) sleep problems (particularly night time awakenings) were also predictive of new-onset bipolar disorder in the at-risk offspring (Levenson et al., 2015). These differences were found to be highly specific to bipolar disorder, since adjustment for parental nonbipolar psychopathology and other confounders did not alter findings. There was also a separate sample of participants recruited during preschool years (2–5 years old). Preschool-aged offspring of parents with bipolar disorder were found to have higher rates of ADHD, aggression, mood dysregulation, and sleep problems, as compared to community controls (Birmaher et al., 2010; Maoz, Goldstein, Axelson, et al., 2014). The **Dutch Study** has followed 140 offspring of parents with bipolar disorder for over 12 years, assessing mainly categorical disorders in these offspring (though with some dimensional measures at baseline); this study did not include a control group. Of the 14 individuals with new-onset bipolar disorder, 13 developed a depressive disorder first, indicating that bipolar disorder tends to debut with depression (Mesman et al., 2013). A more recent analysis found that, of 29 offspring with mood disorders at baseline, 10 developed new-onset bipolar disorder over follow-up; the most important predictor of progression was subsyndromal manic symptoms (Mesman, Nolen, Keijsers, & Hillegers, 2017).

In summary, offspring of parents with bipolar disorder have difficulties with a variety of dimensional domains, including mood lability, anxiety, attention, and sleep (Birmaher et al., 2013; Diler et al., 2011; Egeland et al., 2003; Shaw et al., 2005). In terms of categorical disorders, they also have a higher rate of psychiatric diagnoses, including anxiety and mood disorders (Birmaher, Axelson, Monk, et al., 2009; Duffy et al., 2014; Nurnberger Jr et al., 2011; Perich et al., 2015). These seem to precede the onset of bipolar disorder, except for substance abuse, which appears to occur primarily after bipolar disorder (Axelson et al., 2015).

However, which of these symptoms and disorders actually predict the new onset of bipolar disorder in these at-risk samples? While many of the studies in Table 1 are powered to assess differences between offspring and controls, the number of converters is generally small, thus limiting conclusions. The most consistent predictor of new-onset bipolar disorder is subthreshold manic symptoms (Axelson et al., 2015; Hafeman et al., 2016; Mesman, Nolen, et al., 2017). Subthreshold manic symptoms appear to be predictive closer (within 5 years) of bipolar onset, while symptoms of anxiety and depression appear to be earlier predictors; mood lability increases the risk of new-onset bipolar disorder in both the short and long term (Hafeman et al., 2016). Persistent and severe symptoms are the most predictive of bipolar onset (Hafeman et al., 2016). The majority of bipolar disorder cases seem to debut with a depressive episode (Axelson et al., 2015; Duffy et al., 2014; Mesman et al., 2013). While anxiety disorders precede mood disorders in these offspring (Duffy et al., 2014), it is unclear whether anxiety predicts bipolar disorder specifically, or simply the onset of depression. Sleep disturbances also predict new-onset bipolar spectrum disorder in the offspring (Levenson et al., 2015). These prodromal symptoms might be especially potent if the parent had an earlier age of mood disorder onset and/or two parents with bipolar disorder, two factors which substantially increase the risk of conversion (Goldstein et al., 2010; Hafeman et al., 2016).

DEPRESSED SAMPLES

Adults and adolescents who present with major depression are also at increased risk of developing bipolar disorder, and a number of studies have assessed clinical features that might predict onset in these individuals, and thus might be considered prodromal. For a full description of these studies, see the review by Faedda et al. (2015). In one of the first studies to assess prospectively children and adolescents with depression, Strober and Carlson followed 60 adolescents, hospitalized for depression, for 3–4 years. They found that 20% of these individuals developed bipolar disorder; the most important predictors were rapid onset of symptoms, psychotic features, family history of affective disorders (especially bipolar disorder), and psychopharmacologic-induced hypomania (Strober & Carlson, 1982). Kochman et al. conducted a 2-year prospective study in 80 children and adolescents with major depression, evaluating dimensional symptoms and cyclothymic-hypersensitive temperament (CHT) using a questionnaire derived from the Cyclothymic Subscale of the Temperament Evaluation of the Memphis, Pisa, Paris, and San Diego Autoquestionnaire (TEMPS-A); 43% of these youth converted to bipolar disorder during follow-up, and this conversion was more likely in those individuals who had high CHT scores (64% converted) (Kochman et al., 2005). Studies with adults have produced similar results. Tohen et al. followed a sample of adults with psychotic depression for 2 years ($n = 49$), and found that 33% converted to bipolar disorder; the most important predictors of conversion were mood lability and subthreshold hypomanic symptoms, though family history was not assessed (Tohen et al., 2012). Using data from the NIMH Collaborative Depression Study, Akiskal et al. assessed predictors of conversion to bipolar disorder-I and bipolar disorder-II in 559 adults over 11 years of follow-up, during which 8.6% converted to bipolar disorder-II and 3.9% developed bipolar disorder-I. The authors found that there were no significant predictors of conversion to bipolar disorder-I; however, earlier age of depression onset predicted bipolar disorder-II, as well as certain personality factors, specifically mood lability and high energy-activity (Akiskal et al., 1995). Several years later, Fiedorowicz et al. also utilized data from the NIMH Collaborative Depression Study to assess the rate and predictors of conversion to bipolar disorder in 550 adults with unipolar depression, over a mean follow-up period of 17.5 years. They found that 19.6% of the sample converted to bipolar disorder-I or bipolar disorder-II, and predictors were psychosis, subthreshold hypomanic symptoms, and early age of onset (Fiedorowicz et al., 2011).

In summary, as with the family at-risk studies, subsyndromal manic symptoms and mood lability are among the most consistent factors to predict the onset of bipolar disorder. In depressed adolescents, cyclothymic temperament (characterized by mood lability and emotion reactivity) is an important predictor of bipolar disorder (Kochman et al., 2005). In individuals with psychotic depression, both mood lability and subsyndromal hypomanic symptoms predict 2-year conversion to hypomania (Tohen et al., 2012). To our knowledge, the impact of family history of bipolar disorder in those with psychotic depression has not yet been assessed, a limitation of this literature. In depressed adults, subsyndromal hypomanic symptoms (especially decreased need for sleep, increased energy, and increased goal-related activity) predict new-onset bipolar disorder (Fiedorowicz et al., 2011). Other clinically predictive features in depressed individuals are younger age of onset, family history of bipolar disorder, and psychotic symptoms (Akiskal et al., 1995; Fiedorowicz et al., 2011; Kovacs, 1996; Strober & Carlson, 1982).

CYCLOTHYMIA/BD-NOS SAMPLES

While subsyndromal manic symptoms increase risk for conversion to bipolar disorder, episodes of manic symptoms, even if they do not meet full duration or symptom criteria for a hypomanic episode, are highly predictive of future bipolar-I and -II disorder. One of the first studies to show this was by Akiskal and colleagues in the late 1970s. They assessed 46 adults with cyclothymia (defined as "recurrent and short cycles of depression and hypomania that clearly fell short of the full syndromal criteria"), recruited from outpatient treatment, and followed them for 2–3 years. The authors found that 35% of these individuals developed full-threshold bipolar disorder over this period (Akiskal et al., 1977). Similarly, studies prospectively assessing adolescents and young adults have found very high rates of conversion from subsyndromal bipolar disorder (BD-NOS) to bipolar disorder-I or -II. In the Course and Outcome of Bipolar Youth (COBY) study, children and adolescents with BD-I, BD-II, and BD-NOS were followed prospectively; follow-up is ongoing. BD-NOS was operationalized to include participants who did not meet the criteria for BD-I or BD-II but had distinct periods of abnormally elevated or irritable mood plus: (1) at least two associated DSM-IV manic symptoms (three if the mood is irritability only); (2) a clear change in functioning; and (3) symptom duration of at least 4 h for a minimum of 4 days. Using these criteria, 45% of youth with BD-NOS (63/140) converted to BD-I or -II within 5 years; the strongest predictor of conversion was a family history of mania/hypomania ($P = .006$) (Axelson et al., 2011). Using a similar definition of subsyndromal bipolar disorder, the Longitudinal Investigation of Bipolar Spectrum project assessed college-age participants who met criteria for BD-NOS ($n = 57$) and followed them prospectively for 4 years; 52% of these young adults converted to BD-I (10.5%) or BD-II (41.1%) within 4.5 years (Alloy, Urošević, et al., 2012). Finally, Martinez and Fristad followed 27 children (8–11 years old) with BD-NOS for 18 months, as part of a study of Multifamily Psychoeducational Psychotherapy. Over this time period, one-third of these children converted to BD-I or -II; these values were even higher in those with a positive family history (Martinez & Fristad, 2013).

In summary, these studies indicate a high rate of conversion (between 30% and 50%) from BD-NOS to syndromal bipolar disorder over a period of 1–5 years; studies across multiple age ranges, from middle childhood to adulthood, showed similarly high rates. Risk of conversion was even higher in individuals with a family history of bipolar disorder. These rates of conversion do call into question whether subsyndromal bipolar disorder is part of the prodrome for bipolar disorder, or whether it is in fact a manifestation of the disorder itself. Indeed, youth with BD-NOS have comparable family history of bipolar disorder, risk for suicidality and substance abuse, and psychosocial impairment to those with bipolar disorder-I/II (Axelson et al., 2006; Goldstein et al., 2010, 2011; Hafeman et al., 2013). At the very least, BD-NOS represents a very late prodrome, which causes significant morbidity regardless of further progression.

COMMUNITY SAMPLES

Subsyndromal manic symptoms have also been found to be predictive of new-onset bipolar disorder in epidemiologic samples. Regeer et al. analyzed data from the Netherlands Mental Health Survey and Incidence Study (NEMESIS), which included a 3-year follow-up of 7076 adults. Of the 4628 individuals with no previous major depression or bipolar disorder, who had data from at least one follow-up visit, 14 developed new-onset bipolar disorder; subsyndromal manic symptoms and depressive symptoms predicted this outcome (Regeer et al., 2006). A further analysis in this dataset assessed the joint predictive power of both subsyndromal manic and psychotic symptoms; these authors found that while 3.0% of participants with subthreshold manic symptoms (and no psychosis) developed bipolar disorder during follow-up, 9.5% with both subthreshold manic and psychotic symptoms developed this diagnosis; thus subsyndromal manic symptoms were especially predictive of bipolar disorder in this sample in the presence of subthreshold psychotic symptoms (Kaymaz et al., 2007). Homish et al. analyzed data from the National Epidemiologic Survey on Alcohol and Related Conditions (NESARC), to assess whether subthreshold elation or irritability at Wave 1 predicted bipolar disorder at Wave 2 (3 years later). Using data from 40,512 adult participants who did not yet meet criteria for bipolar disorder at Wave 1, the authors found that both elation and irritability (found in 6.8% of the sample) predicted new-onset bipolar disorder; 13% of those with both symptoms developed bipolar disorder, compared to 8% of those with only one symptom (elation or irritability), and 2.6% without either symptom. These results were significant after adjusting for demographic variables and comorbidity (Homish et al., 2013). Tijssen et al. used data from the Early Developmental Stages of Psychopathology (EDSP), which included 3021 adolescents and young adults (14–24 years old), to assess whether fleeting vs. persistent subthreshold mood symptoms predicted bipolar disorder at a mean 8-year follow-up. They found that there were increased odds of later developing (hypo)manic episodes with fleeting hypomanic symptoms, which was even more pronounced when these symptoms were persistent; there was also an increased risk of (hypo)manic episodes in those who had persistent depression (Tijssen et al., 2010b). From the Tracking Adolescents' Individual Lives Survey (TRAILS), a prospective community-based survey, authors assessed the relationship between subthreshold manic symptoms at 11 years old (as measured by the CBCL-MS) and bipolar disorder onset by the age of 19 in 1429 adolescents. They found that those who were highly symptomatic were five times as likely to develop new-onset bipolar disorder compared to the normative group; those with mild symptoms were twice as likely as the normative youth to develop bipolar disorder by 19 years old (Papachristou et al., 2017). Lewinsohn et al. followed 1507 adolescents, 893 into adulthood (24 years old), as part of the Oregon Adolescent Depression Project (OADP); 48 had subthreshold bipolar disorder during adolescence. These authors found that subthreshold bipolar disorder during adolescence was predictive of psychopathology in general, major depression, anxiety, and personality disorders; however, only 2.1% of adolescents with subsyndromal bipolar disorder developed full-threshold bipolar disorder by early adulthood (Lewinsohn, Klein, & Seeley, 2000).

Thus, the majority of epidemiologic studies find that hypomanic symptoms predict new-onset bipolar disorder in adults (Homish et al., 2013; Regeer et al., 2006) and adolescents (Papachristou et al., 2017; Tijssen et al., 2010b). Of note, while differences are significant, the percentage of individuals who go on to convert from subthreshold to full-threshold bipolar disorder is generally less in the community setting (vs clinical studies). Based on other population studies, subclinical symptoms of mania (and perhaps even episodes of mania and hypomania) are fairly common in the general population, and often do not cause impairment (Paaren et al., 2013; Stringaris, Santosh, Leibenluft, & Goodman, 2010; Tijssen et al., 2010a). Thus it is possible that subsyndromal hypomanic symptoms found in a clinical sample (with some functional impairment) might have a different meaning, and a more severe course, than those found in the general population (a form of Berkson's bias). One study (Lewinsohn et al., 2000) did not show a relationship between subthreshold bipolar disorder in adolescence and later bipolar disorder (at 24 years old), though those with subthreshold bipolar disorder were more impaired in several other ways (with higher rates of major depressive disorder (MDD), etc.). These authors used a relatively liberal definition for subthreshold bipolar disorder (irritable or elated mood for a distinct period of time, plus one additional symptom), which could explain the absence of progression; they also had a fairly small sample size.

RETROSPECTIVE STUDIES: METAANALYSES AND REVIEWS

Due to the large number of studies, we will only review a subset; previous reviews have described these studies in greater detail (Van Meter, Burke, Youngstrom, Faedda, & Correll, 2016). One of the first studies to assess retrospectively prodromal symptoms in adults with bipolar disorder was conducted by Lish et al., using survey data from 500 adults in the National Depressive and Manic-Depressive Association. These authors found that 59% of individuals reported their symptoms starting in childhood or adolescence; the most frequent initial symptoms were manic symptoms, depressive symptoms, sleep difficulties, and mood lability; and 39% of individuals reported a period of 10 years or more between seeking treatment and being diagnosed with bipolar disorder (Lish et al., 1994). Because of the design of this study, it was unclear the degree to which this time period (and associated symptoms) represented a prodromal period vs. bipolar disorder without accurate diagnosis. One of the best-known and seminal retrospective studies of prodromal symptoms in bipolar disorder was conducted by Egeland et al. (2000). The authors assessed records of 58 hospitalized adults with bipolar disorder-I, to determine prodromal symptoms present prior to first onset. They found that mood and energy changes, as well as a difficulty controlling anger, preceded onset of bipolar disorder-I by 9–12 years. Assessing the bipolar prodrome in youth, Correll et al. used the Bipolar Prodrome Symptom Scale—Retrospective (as described above) to characterize symptoms prior to bipolar disorder onset in 52 adolescents. They found that a prodromal period was common in these youth, lasting about 10 months, with subsyndromal manic symptoms, depressive symptoms, and a decrease in psychosocial functioning prior to onset (Correll, Hauser, et al., 2014). A recent metaanalysis including adult and youth with bipolar disorder looked at prodromal symptoms that preceded the onset of bipolar disorder across 11 studies (predominantly retrospective), including the studies discussed above (Van Meter et al., 2016). While the literature is very heterogeneous, the authors found the most prevalent symptoms prior to bipolar disorder onset were manic symptoms (e.g., increased energy, elated mood, talkativeness), inability to think, academic/work difficulties, depressed mood, and sleep disturbances; these symptoms were found in >50% of individuals who eventually developed bipolar disorder. The prodrome was found to range from 4.6 to 130 months.

In summary, retrospective studies using different methods (records and interviews) point to the presence of a prodromal period, consistent with prospective studies. This prodrome includes subthreshold symptoms of mania (such as elation, elevated energy), as well as symptoms of depression and decreased function that last, in most cases, for a period of at least several months.

STAGING MODELS

In other areas of medicine, the presence or absence of diagnosis is only part of the story; as important is the stage of the disorder, a characteristic that informs both prognosis and treatment decisions. From this perspective, it might not be sufficient to define one bipolar prodrome, but rather it might be more accurate and useful to describe multiple stages of this prodrome. These stages were defined as 1a and 1b within the staging model first proposed by McGorry in 2006, as a framework for the development of interventions for severe mood and psychotic disorders. Stage 1a is a more general, nonspecific stage with little clinical impairment, whereas Stage 1b becomes much more specific and is also associated with functional deterioration (Fig. 1) (McGorry, Hickie, Yung, Pantelis, & Jackson, 2006). This framework was specifically applied to bipolar disorder by Berk and colleagues, who explored how this staging might apply to adults and adolescents with bipolar disorder (Berk et al., 2014).

Converging evidence indicates that such a framework is useful to describe the bipolar prodrome. Stage 1a appears to be defined mostly by symptoms of anxiety, depression, and mood lability; in contrast, Stage 1b seems to be characterized by manic symptomatology (Duffy et al., 2014; Hafeman et al., 2016). In the BIOS longitudinal study, we found that early symptoms of anxiety/depression (as measured by the CBCL) were most predictive, while closer to onset, manic symptoms were most important. Of note, we found that mood lability was important throughout follow-up, both at baseline and the visit proximal to conversion (Fig. 1) (Hafeman et al., 2016). Most comorbidities appear to precede the development of bipolar disorder; substance abuse is an exception to this rule, and occurs following bipolar disorder onset (Axelson et al., 2015).

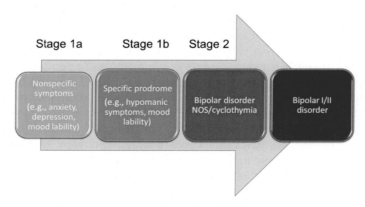

FIG. 1

A staged model for the development of bipolar disorder.

Modified from McGorry, P. D., Hickie, I. B., Yung, A. R., Pantelis, C. & Jackson, H. J. (2006).
"Clinical staging of psychiatric disorders: a heuristic framework for choosing earlier, safer and more
effective interventions." Australian and New Zealand Journal of Psychiatry, 40(8), 616–622; Berk, M., Berk,
L., Dodd, S., Cotton, S., Macneil, C., Daglas, R., et al. (2014). "Stage managing bipolar disorder." Bipolar
Disorders, 16(5), 471–477.

One interesting question is whether BD-NOS or cyclothymia is considered part of Stage 1b—i.e., a specific prodrome for bipolar disorder-I/II—or part of Stage 2—i.e., the onset of disorder. Youth with BD-NOS have a high degree of functional impairment, and rates of negative outcomes such as substance abuse and suicide attempts that are comparable to what is observed in bipolar disorder-I/II (Birmaher, Axelson, Goldstein, et al., 2009; Goldstein et al., 2009, 2012). Similarly, cyclothymia is associated with a high level of morbidity and impairment, as reviewed in Van Meter, Youngstrom, and Findling (2012). Thus BD-NOS can present as clinically severe as bipolar disorder-I/II, and might be viewed as Stage 2 (Fig. 1).

RISK CALCULATOR

While an understanding of which factors contribute to the bipolar prodrome is a critical first step, ultimately the clinical question is how all of these risk factors combine to impact the prognosis of a particular patient. To this end, risk calculators have been used in other areas of medicine to estimate the risk of a particular outcome within a specified time period. In psychiatry, recent work has focused on developing risk calculators for predicting psychosis in (1) clinically defined high-risk populations (Cannon et al., 2016; Carrión et al., 2016) and (2) general outpatient clinics (Fusar-Poli, Rutigliano, Stahl, et al., 2017). We recently extended this work by developing a risk calculator to predict the 5-year new-onset bipolar spectrum disorder in 412 youth at familial risk for the disorder. Using predictors identified from a metaanalysis of prodromal symptoms (Van Meter, Moreira, & Youngstrom, 2011), we found that risk calculator including subsyndromal manic symptoms, depressive symptoms, mood lability, anxiety, global psychosocial function, and parental age of mood disorder onset discriminated well between individuals who would develop new-onset bipolar spectrum disorder in the next 5 years versus those who would not; we used internal validation (1000 bootstrap resamples) to decrease bias due to overfitting. The area under the curve (AUC) for this model was 0.76, indicating good discrimination, and comparable to risk calculators used in other areas of medicine. This risk calculator is the first step toward developing tools that will help clinicians and researchers estimate individual risk, and use this information to inform management and treatment options. This tool is available on our website for use by clinicians (http://www.pediatricbipolar.pitt.edu). In the future, we will empirically test whether this risk calculator improves with the addition of biomarkers and other objective measures, such as genetic findings and neural circuitry.

SUMMARY AND FUTURE DIRECTIONS

While each design reviewed in this chapter has limitations, taken together, results from disparate study designs point to the presence of a bipolar prodrome, often at least a year long (and sometimes upwards of a decade), that might represent a target for early intervention. This prodrome is likely nonspecific initially, including persistent and clinically significant symptoms of anxiety, depression, sleep disturbances, and mood lability; however, over time, subthreshold manic symptoms become more prominent, finally consolidating in episodes that meet criteria for BD-NOS/cyclothymia, and eventually, hypomania or mania (Fig. 1). These symptoms may be especially predictive when the individual has a parent with bipolar disorder, even more so if the parent had early onset of mood disorder, and/or if both parents have bipolar disorder. It is important to note that the progression of disorder is likely more heterogeneous than indicated in Fig. 1, with individuals spending different amounts of time in each stage, and possibly skipping stages as well. However, this model provides an illustration of the basic progression of the disorder, and might later be used to guide early intervention and treatment (Berk et al., 2014). Several scales and questionnaires have been designed to evaluate the risk of conversion to bipolar disorder, and as discussed previously, none has been replicated in high-quality longitudinal studies. This in part reflects the difficulty of doing such research, given the low baserate of bipolar disorder and a difficult-to-recruit sample. However, many of these instruments show promise, and the overlapping symptoms that they measure reflect a relative consensus about the phenomenology of the bipolar prodrome. In addition, the new development of a risk calculator for predicting bipolar disorder provides a potentially helpful strategy for integrating previous research, and understanding what these predictors mean for the individual patient.

What are the next steps to better understand and identify the bipolar prodrome, and use this knowledge to help our patients? First, it is important to be able to quantify risk of conversion for an individual, to inform decisions about monitoring and treatment. The risk calculator that we recently developed for youth at familial risk for bipolar disorder is an initial step toward this; however, our results need to be replicated and externally validated. Also, additional risk calculators should be developed for other potentially at-risk populations (e.g., adolescents with major depressive disorder). The risk calculator should eventually incorporate other possible predictors, including biomarkers (e.g., objective measures of sleep, neuroimaging, etc.), to perhaps further improve our ability to discriminate between those who will develop bipolar disorder versus those who will not, in the specified time frame. Second, as the bipolar prodrome becomes better defined, it is imperative that we begin to carry out well-designed studies to assess potential early intervention, including pharmacologic (e.g., antidepressants vs mood stabilizers) and psychotherapy (e.g., family-focused therapy). Defining an ultra-high-risk population will also be helpful from a research perspective, to test potential early interventions in a sample with a high rate of new-onset bipolar disorder. Third, it will be important to relate these individual clinical observations (e.g., mood lability, reward sensitivity, subthreshold manic symptoms) to each other, and to other potential biomarkers.

REFERENCES

Akiskal, H. S., Djenderedjian, A. M., Rosenthal, R. H., & Khani, M. K. (1977). Cyclothymic disorder: validating criteria for inclusion in the bipolar affective group. *The American Journal of Psychiatry*, *134*(11), 1227–1233.

Akiskal, H. S., Maser, J. D., Zeller, P. J., et al. (1995). Switching from 'unipolar' to bipolar II: an 11-year prospective study of clinical and temperamental predictors in 559 patients. *Archives of General Psychiatry, 52*(2), 114–123.

Alloy, L. B., Bender, R. E., Whitehouse, W. G., Wagner, C. A., Liu, R. T., Grant, D. A., et al. (2012). High behavioral approach system (BAS) sensitivity, reward responsiveness, and goal-striving predict first onset of bipolar spectrum disorders: a prospective behavioral high-risk design. *Journal of Abnormal Psychology, 121*(2), 339–351.

Alloy, L. B., Urošević, S., Abramson, L. Y., Jager-Hyman, S., Nusslock, R., Whitehouse, W. G., et al. (2012). Progression along the bipolar spectrum: a longitudinal study of predictors of conversion from bipolar spectrum conditions to bipolar I and II disorders. *Journal of Abnormal Psychology, 121*(1), 16–27.

Althoff, R. R., Verhulst, F., Rettew, D. C., Hudziak, J. J., & van der Ende, J. (2010). Adult outcomes of childhood dysregulation: a 14-year follow-up study. *Journal of the American Academy of Child and Adolescent Psychiatry, 49*(11), 1105–1116.e1101.

Axelson, D., Birmaher, B., Strober, M., Gill, M. K., Valeri, S., Chiappetta, L., et al. (2006). Phenomenology of children and adolescents with bipolar spectrum disorders. *Archives of General Psychiatry, 63*(10), 1139–1148.

Axelson, D. A., Birmaher, B., Strober, M. A., Goldstein, B. I., Ha, W., Gill, M. K., et al. (2011). Course of subthreshold bipolar disorder in youth: diagnostic progression from bipolar disorder not otherwise specified. *Journal of the American Academy of Child and Adolescent Psychiatry, 50*(10), 1001–1016. [e1003].

Axelson, D., Goldstein, B., Goldstein, T., Monk, K., Yu, H., Hickey, M. B., et al. (2015). Diagnostic precursors to bipolar disorder in offspring of parents with bipolar disorder: a longitudinal study. *The American Journal of Psychiatry, 172*(7), 638–646.

Bechdolf, A., Ratheesh, A., Cotton, S. M., Nelson, B., Chanen, A. M., Betts, J., et al. (2014). The predictive validity of bipolar at-risk (prodromal) criteria in help-seeking adolescents and young adults: a prospective study. *Bipolar Disorders, 16*(5), 493–504.

Berk, M., Berk, L., Dodd, S., Cotton, S., Macneil, C., Daglas, R., et al. (2014). Stage managing bipolar disorder. *Bipolar Disorders, 16*(5), 471–477.

Biederman, J., Petty, C. R., Monuteaux, M. C., Evans, M., Parcell, T., Faraone, S. V., et al. (2009). The child behavior checklist-pediatric bipolar disorder profile predicts a subsequent diagnosis of bipolar disorder and associated impairments in ADHD youth growing up: a longitudinal analysis. *The Journal of Clinical Psychiatry, 70*(5), 732–740.

Birmaher, B., Axelson, D., Goldstein, B., Monk, K., Kalas, C., Obreja, M., et al. (2010). Psychiatric disorders in preschool offspring of parents with bipolar disorder: the Pittsburgh bipolar offspring study (BIOS). *The American Journal of Psychiatry, 167*(3), 321–330.

Birmaher, B., Axelson, D., Goldstein, B., Strober, M., Gill, M. K., Hunt, J., et al. (2009). Four-year longitudinal course of children and adolescents with bipolar spectrum disorders: the course and outcome of bipolar youth (COBY) study. *The American Journal of Psychiatry, 166*(7), 795–804.

Birmaher, B., Axelson, D., Monk, K., Kalas, C., Goldstein, B., Hickey, M. B., et al. (2009). Lifetime psychiatric disorders in school-aged offspring of parents with bipolar disorder: the Pittsburgh bipolar offspring study. *Archives of General Psychiatry, 66*(3), 287–296.

Birmaher, B., Goldstein, B. I., Axelson, D. A., Monk, K., Hickey, M. B., Fan, J., et al. (2013). Mood lability among offspring of parents with bipolar disorder and community controls. *Bipolar Disorders, 15*(3), 253–263.

Cannon, T. D., Changhong, Y., Addington, J., Bearden, C. E., Cadenhead, K. S., Cornblatt, B. A., et al. (2016). An individualized risk calculator for research in prodromal psychosis. *American Journal of Psychiatry, 173*(10), 980–988.

Carrión, R. E., Cornblatt, B. A., Burton, C. Z., Tso, I. F., Auther, A. M., Adelsheim, S., et al. (2016). Personalized prediction of psychosis: external validation of the NAPLS-2 psychosis risk calculator with the EDIPPP project. *American Journal of Psychiatry, 173*(10), 989–996.

Correll, C. U., Hauser, M., Penzner, J. B., Auther, A. M., Kafantaris, V., Saito, E., et al. (2014). Type and duration of subsyndromal symptoms in youth with bipolar I disorder prior to their first manic episode. *Bipolar Disorders, 16*(5), 478–492.

Correll, C. U., Olvet, D. M., Auther, A. M., Hauser, M., Kishimoto, T., Carrion, R. E., et al. (2014). The bipolar prodrome symptom interview and scale-prospective (BPSS-P): description and validation in a psychiatric sample and healthy controls. *Bipolar Disorders, 16*(5), 505–522.

Correll, C. U., Penzner, J. B., Frederickson, A. M., Richter, J. J., Auther, A. M., Smith, C. W., et al. (2007). Differentiation in the preonset phases of schizophrenia and mood disorders: evidence in support of a bipolar mania prodrome. *Schizophrenia Bulletin, 33*(3), 703–714.

Danielson, C. K., Youngstrom, E. A., Findling, R. L., & Calabrese, J. R. (2003). Discriminative validity of the general behavior inventory using youth report. *Journal of Abnormal Child Psychology, 31*(1), 29–39.

DelBello, M. P., & Geller, B. (2001). Review of studies of child and adolescent offspring of bipolar parents. *Bipolar Disorders, 3*(6), 325–334.

Depue, R. A., Krauss, S., Spoont, M. R., & Arbisi, P. (1989). General behavior inventory identification of unipolar and bipolar affective conditions in a nonclinical university population. *Journal of Abnormal Psychology, 98*(2), 117.

Diler, R. S., Birmaher, B., Axelson, D., Goldstein, B., Gill, M., Strober, M., et al. (2009). The child behavior checklist (CBCL) and the CBCL-bipolar phenotype are not useful in diagnosing pediatric bipolar disorder. *Journal of Child and Adolescent Psychopharmacology, 19*(1), 23–30.

Diler, R. S., Birmaher, B., Axelson, D., Obreja, M., Monk, K., Hickey, M. B., et al. (2011). Dimensional psychopathology in offspring of parents with bipolar disorder. *Bipolar Disorders, 13*(7–8), 670–678.

Duffy, A., Alda, M., Crawford, L., Milin, R., & Grof, P. (2007). The early manifestations of bipolar disorder: a longitudinal prospective study of the offspring of bipolar parents. *Bipolar Disorders, 9*(8), 828–838.

Duffy, A., Alda, M., Hajek, T., Sherry, S. B., & Grof, P. (2010). Early stages in the development of bipolar disorder. *Journal of Affective Disorders, 121*(1–2), 127–135.

Duffy, A., Horrocks, J., Doucette, S., Keown-Stoneman, C., McCloskey, S., & Grof, P. (2014). The developmental trajectory of bipolar disorder. *The British Journal of Psychiatry, 204*(2), 122–128.

Eckblad, M., & Chapman, L. J. (1986). Development and validation of a scale for hypomanic personality. *Journal of Abnormal Psychology, 95*(3), 214–222.

Egeland, J. A., Endicott, J., Hostetter, A. M., Allen, C. R., Pauls, D. L., & Shaw, J. A. (2012). A 16-year prospective study of prodromal features prior to BPI onset in well Amish children. *Journal of Affective Disorders, 142*(1–3), 186–192.

Egeland, J. A., Hostetter, A. M., Pauls, D. L., & Sussex, J. N. (2000). Prodromal symptoms before onset of manic-depressive disorder suggested by first hospital admission histories. *Journal of the American Academy of Child and Adolescent Psychiatry, 39*(10), 1245–1252.

Egeland, J. A., Shaw, J. A., Endicott, J., Pauls, D. L., Allen, C. R., Hostetter, A. M., et al. (2003). Prospective study of prodromal features for bipolarity in well Amish children. *Journal of the American Academy of Child and Adolescent Psychiatry, 42*(7), 786–796.

Faedda, G. L., Marangoni, C., Serra, G., Salvatore, P., Sani, G., Vázquez, G. H., et al. (2015). Precursors of bipolar disorders: a systematic literature review of prospective studies. *Journal of Clinical Psychiatry, 76*(5), 614–624.

Fiedorowicz, J. G., Endicott, J., Leon, A. C., Solomon, D. A., Keller, M. B., & Coryell, W. H. (2011). Subthreshold hypomanic symptoms in progression from unipolar major depression to bipolar disorder. *The American Journal of Psychiatry, 168*(1), 40–48.

Fusar-Poli, P., Rutigliano, G., Stahl, D., et al. (2017). Development and validation of a clinically based risk calculator for the transdiagnostic prediction of psychosis. *JAMA Psychiatry, 74*(5), 493–500.

Gerson, A. C., Gerring, J. P., Freund, L., Joshi, P. T., Capozzoli, J., Brady, K., et al. (1996). The children's affective lability scale: a psychometric evaluation of reliability. *Psychiatry Research, 65*(3), 189–198.

Goldstein, T. R., Birmaher, B., Axelson, D., Goldstein, B. I., Gill, M. K., Esposito-Smythers, C., et al. (2009). Psychosocial functioning among bipolar youth. *Journal of Affective Disorders, 114*(1–3), 174–183.

Goldstein, T. R., Ha, W., Axelson, D. A., Goldstein, B. I., Liao, F., Gill, M. K., et al. (2012). Predictors of prospectively examined suicide attempts among youth with bipolar disorder. *Archives of General Psychiatry, 69*(11), 1113–1122.

Goldstein, T. R., Obreja, M., Shamseddeen, W., Iyengar, S., Axelson, D. A., Goldstein, B. I., et al. (2011). Risk for suicidal ideation among the offspring of bipolar parents: results from the bipolar offspring study (BIOS). *Archives of Suicide Research, 15*(3), 207–222.

Goldstein, B. I., Shamseddeen, W., Axelson, D. A., Kalas, C., Monk, K., Brent, D. A., et al. (2010). Clinical, demographic, and familial correlates of bipolar spectrum disorders among offspring of parents with bipolar disorder. *Journal of the American Academy of Child and Adolescent Psychiatry, 49*(4), 388–396.

Hafeman, D., Axelson, D., Demeter, C., Findling, R. L., Fristad, M. A., Kowatch, R. A., et al. (2013). Phenomenology of bipolar disorder not otherwise specified in youth: a comparison of clinical characteristics across the spectrum of manic symptoms. *Bipolar Disorders, 15*(3), 240–252.

Hafeman, D. M., Merranko, J., Axelson, D., Goldstein, B. I., Goldstein, T., Monk, K., et al. (2016). Toward the definition of a bipolar prodrome: dimensional predictors of bipolar spectrum disorders in at-risk youths. *American Journal of Psychiatry, 173*(7), 695–704.

Hammen, C., Burge, D., Burney, E., & Adrian, C. (1990). Longitudinal study of diagnoses in children of women with unipolar and bipolar affective disorder. *Archives of General Psychiatry, 47*(12), 1112–1117.

Hammen, C., Gordon, D., Burge, D., Adrian, C., Jaenicke, C., & Hiroto, D. (1987). Maternal affective disorders, illness, and stress: risk for children's psychopathology. *The American Journal of Psychiatry, 144*(6), 736–741.

Hauser, M., & Correll, C. U. (2013). The significance of at-risk or prodromal symptoms for bipolar I disorder in children and adolescents. *Canadian Journal of Psychiatry, 58*(1), 22–31.

Hillegers, M. H., Reichart, C. G., Wals, M., Verhulst, F. C., Ormel, J., & Nolen, W. A. (2005). Five-year prospective outcome of psychopathology in the adolescent offspring of bipolar parents. *Bipolar Disorders, 7*(4), 344–350.

Homish, G. G., Marshall, D., Dubovsky, S. L., & Leonard, K. (2013). Predictors of later bipolar disorder in patients with subthreshold symptoms. *Journal of Affective Disorders*, *144*(1–2), 129–133.

Hunt, J., Schwarz, C. M., Nye, P., & Frazier, E. (2016). Is there a bipolar prodrome among children and adolescents? *Current Psychiatry Reports*, *18*(4), 35.

Kaymaz, N., van Os, J., de Graaf, R., ten Have, M., Nolen, W., & Krabbendam, L. (2007). The impact of subclinical psychosis on the transition from subclinicial mania to bipolar disorder. *Journal of Affective Disorders*, *98*(1–2), 55–64.

Kessler, R. C., Calabrese, J. R., Farley, P. A., Gruber, M. J., Jewell, M. A., Katon, W., et al. (2013). Composite international diagnostic interview screening scales for DSM-IV anxiety and mood disorders. *Psychological Medicine*, *43*(8), 1625–1637.

Klein, D. N., Lewinsohn, P. M., & Seeley, J. R. (1996). Hypomanic personality traits in a community sample of adolescents. *Journal of Affective Disorders*, *38*(2), 135–143.

Kochman, F. J., Hantouche, E. G., Ferrari, P., Lancrenon, S., Bayart, D., & Akiskal, H. S. (2005). Cyclothymic temperament as a prospective predictor of bipolarity and suicidality in children and adolescents with major depressive disorder. *Journal of Affective Disorders*, *85*(1–2), 181–189.

Kovacs, M. (1996). Presentation and course of major depressive disorder during childhood and later years of the life span. *Journal of the American Academy of Child and Adolescent Psychiatry*, *35*(6), 705–715.

Kwapil, T. R., Miller, M. B., Zinser, M. C., Chapman, L. J., Chapman, J., & Eckblad, M. (2000). A longitudinal study of high scorers on the hypomanic personality scale. *Journal of Abnormal Psychology*, *109*(2), 222–226.

Lapalme, M., Hodgins, S., & LaRoche, C. (1997). Children of parents with bipolar disorder: a metaanalysis of risk for mental disorders. *Canadian Journal of Psychiatry*, *42*(6), 623–631.

LaRoche, C., Cheifetz, P., Lester, E., Schibuk, L., DiTommaso, E., & Engelsmann, F. (1985). Psychopathology in the offspring of parents with bipolar affective disorders. *The Canadian Journal of Psychiatry*, *30*(5), 337–343.

LaRoche, C., Sheiner, R., Lester, E., Benierakis, C., Marrache, M., Engelsmann, F., et al. (1987). Children of parents with manic-depressive illness: a follow-up study. *The Canadian Journal of Psychiatry*, *32*(7), 563–569.

Levenson, J. C., Axelson, D. A., Merranko, J., Angulo, M., Goldstein, T. R., Mullin, B. C., et al. (2015). Differences in sleep disturbances among offspring of parents with and without bipolar disorder: association with conversion to bipolar disorder. *Bipolar Disorders*, *17*(8), 836–848.

Levenson, J. C., Soehner, A., Rooks, B., Goldstein, T. R., Diler, R., Merranko, J., et al. (2017). Longitudinal sleep phenotypes among offspring of bipolar parents and community controls. *Journal of Affective Disorders*, *215*, 30–36.

Lewinsohn, P. M., Klein, D. N., & Seeley, J. R. (2000). Bipolar disorder during adolescence and young adulthood in a community sample. *Bipolar Disorders*, *2*(3 Pt 2), 281–293.

Lish, J. D., Dime-Meenan, S., Whybrow, P. C., Price, R. A., & Hirschfeld, R. M. (1994). The National Depressive and Manic-depressive Association (DMDA) survey of bipolar members. *Journal of Affective Disorders*, *31*(4), 281–294.

Maoz, H., Goldstein, T., Axelson, D. A., Goldstein, B. I., Fan, J., Hickey, M. B., et al. (2014). Dimensional psychopathology in preschool offspring of parents with bipolar disorder. *Journal of Child Psychology and Psychiatry*, *55*(2), 144–153.

Maoz, H., Goldstein, T., Goldstein, B. I., Axelson, D. A., Fan, J., Hickey, M. B., et al. (2014). The effects of parental mood on reports of their children's psychopathology. *Journal of the American Academy of Child and Adolescent Psychiatry*, *53*(10), 1111–1122.e1115.

Martinez, M. S., & Fristad, M. A. (2013). Conversion from bipolar disorder not otherwise specified (BP-NOS) to bipolar I or II in youth with family history as a predictor of conversion. *Journal of Affective Disorders, 148*(2–3), 431–434.

McGorry, P. D., Hickie, I. B., Yung, A. R., Pantelis, C., & Jackson, H. J. (2006). Clinical staging of psychiatric disorders: a heuristic framework for choosing earlier, safer and more effective interventions. *Australian and New Zealand Journal of Psychiatry, 40*(8), 616–622.

Mesman, E., Birmaher, B. B., Goldstein, B. I., Goldstein, T., Derks, E. M., Vleeschouwer, M., et al. (2016). Categorical and dimensional psychopathology in Dutch and US offspring of parents with bipolar disorder: a preliminary cross-national comparison. *Journal of Affective Disorders, 205*, 95–102.

Mesman, E., Nolen, W. A., Keijsers, L., & Hillegers, M. H. J. (2017). Baseline dimensional psychopathology and future mood disorder onset: findings from the Dutch Bipolar Offspring Study. *Acta Psychiatrica Scandinavica, 136*(2), 201–209.

Mesman, E., Nolen, W. A., Reichart, C. G., Wals, M., & Hillegers, M. H. (2013). The Dutch bipolar offspring study: 12-year follow-up. *The American Journal of Psychiatry, 170*(5), 542–549.

Mesman, E., Youngstrom, E. A., Juliana, N. K., Nolen, W. A., & Hillegers, M. H. (2017). Validation of the seven up seven down inventory in bipolar offspring: screening and prediction of mood disorders. Findings from the Dutch bipolar offspring study. *Journal of Affective Disorders, 207*, 95–101.

Meyer, S. E., Carlson, G. A., Youngstrom, E., Ronsaville, D. S., Martinez, P. E., Gold, P. W., et al. (2009). Long-term outcomes of youth who manifested the CBCL-pediatric bipolar disorder phenotype during childhood and/or adolescence. *Journal of Affective Disorders, 113*(3), 227–235.

Nurnberger, J. I., Jr., McInnis, M., Reich, W., Kastelic, E., Wilcox, H. C., Glowinski, A., et al. (2011). A high-risk study of bipolar disorder. Childhood clinical phenotypes as precursors of major mood disorders. *Archives of General Psychiatry, 68*(10), 1012–1020.

Paaren, A., von Knorring, A. L., Olsson, G., von Knorring, L., Bohman, H., & Jonsson, U. (2013). Hypomania spectrum disorders from adolescence to adulthood: a 15-year follow-up of a community sample. *Journal of Affective Disorders, 145*(2), 190–199.

Papachristou, E., Oldehinkel, A. J., Ormel, J., Raven, D., Hartman, C. A., Frangou, S., et al. (2017). The predictive value of childhood subthreshold manic symptoms for adolescent and adult psychiatric outcomes. *Journal of Affective Disorders, 212*, 86–92.

Papachristou, E., Ormel, J., Oldehinkel, A. J., Kyriakopoulos, M., Reinares, M., Reichenberg, A., et al. (2013). Child behavior checklist—Mania scale (CBCL-MS): development and evaluation of a population-based screening scale for bipolar disorder. *PLoS One, 8*(8), e69459.

Perich, T., Lau, P., Hadzi-Pavlovic, D., Roberts, G., Frankland, A., Wright, A., et al. (2015). What clinical features precede the onset of bipolar disorder? *Journal of Psychiatric Research, 62*, 71–77.

Radke-Yarrow, M., Nottelmann, E., Martinez, P., Fox, M. B., & Belmont, B. (1992). Young children of affectively ill parents: a longitudinal study of psychosocial development. *Journal of the American Academy of Child and Adolescent Psychiatry, 31*(1), 68–77.

Ratheesh, A., Berk, M., Davey, C. G., McGorry, P. D., & Cotton, S. M. (2015). Instruments that prospectively predict bipolar disorder—a systematic review. *Journal of Affective Disorders, 179*, 65–73.

Regeer, E. J., Krabbendam, L., de Graaf, R., ten Have, M., Nolen, W. A., & van Os, J. (2006). A prospective study of the transition rates of subthreshold (hypo)mania and depression in the general population. *Psychological Medicine, 36*(5), 619–627.

Reichart, C. G., van der Ende, J., Wals, M., Hillegers, M. H., Nolen, W. A., Ormel, J., et al. (2005). The use of the GBI as predictor of bipolar disorder in a population of adolescent offspring of parents with a bipolar disorder. *Journal of Affective Disorders, 89*(1–3), 147–155.

Shaw, J. A., Egeland, J. A., Endicott, J., Allen, C. R., & Hostetter, A. M. (2005). A 10-year prospective study of prodromal patterns for bipolar disorder among Amish youth. *Journal of the American Academy of Child and Adolescent Psychiatry, 44*(11), 1104–1111.

Sparks, G. M., Axelson, D. A., Yu, H., Ha, W., Ballester, J., Diler, R. S., et al. (2014). Disruptive mood dysregulation disorder and chronic irritability in youth at familial risk for bipolar disorder. *Journal of the American Academy of Child and Adolescent Psychiatry, 53*(4), 408–416.

Stringaris, A., Santosh, P., Leibenluft, E., & Goodman, R. (2010). Youth meeting symptom and impairment criteria for mania-like episodes lasting less than four days: an epidemiological enquiry. *Journal of Child Psychology and Psychiatry, and Allied Disciplines, 51*(1), 31–38.

Strober, M., & Carlson, G. (1982). Bipolar illness in adolescents with major depression: clinical, genetic, and psychopharmacologic predictors in a three- to four-year prospective follow-up investigation. *Archives of General Psychiatry, 39*(5), 549–555.

Tijssen, M. J., van Os, J., Wittchen, H. U., Lieb, R., Beesdo, K., Mengelers, R., et al. (2010a). Evidence that bipolar disorder is the poor outcome fraction of a common developmental phenotype: an 8-year cohort study in young people. *Psychological Medicine, 40*(2), 289–299.

Tijssen, M. J., van Os, J., Wittchen, H. U., Lieb, R., Beesdo, K., Mengelers, R., et al. (2010b). Prediction of transition from common adolescent bipolar experiences to bipolar disorder: 10-year study. *The British Journal of Psychiatry, 196*(2), 102–108.

Tohen, M., Khalsa, H. M., Salvatore, P., Vieta, E., Ravichandran, C., & Baldessarini, R. J. (2012). Two-year outcomes in first-episode psychotic depression the McLean-Harvard first-episode project. *Journal of Affective Disorders, 136*(1–2), 1–8.

Van Meter, A. R., Burke, C., Youngstrom, E. A., Faedda, G. L., & Correll, C. U. (2016). The bipolar Prodrome: meta-analysis of symptom prevalence prior to initial or recurrent mood episodes. *Journal of the American Academy of Child and Adolescent Psychiatry, 55*(7), 543–555.

Van Meter, A. R., Moreira, A. L., & Youngstrom, E. A. (2011). Meta-analysis of epidemiologic studies of pediatric bipolar disorder. *The Journal of Clinical Psychiatry, 72*(9), 1250–1256.

Van Meter, A. R., Youngstrom, E. A., & Findling, R. L. (2012). Cyclothymic disorder: a critical review. *Clinical Psychology Review, 32*(4), 229–243.

Youngstrom, E. A., Findling, R. L., Calabrese, J. R., Gracious, B. L., Demeter, C., Bedoya, D. D., et al. (2004). Comparing the diagnostic accuracy of six potential screening instruments for bipolar disorder in youths aged 5 to 17 years. *Journal of the American Academy of Child and Adolescent Psychiatry, 43*(7), 847–858.

Youngstrom, E. A., Frazier, T. W., Demeter, C., Calabrese, J. R., & Findling, R. L. (2008). Developing a 10-item mania scale from the parent general behavior inventory for children and adolescents. *The Journal of Clinical Psychiatry, 69*(5), 831–839.

Staging models and neuroprogression in bipolar disorder

2

Daniela V. Bavaresco*,†, João Quevedo*,‡,§,¶,
Jair C. Soares§, Samira S. Valvassori*

Neuronal Signaling and Psychopharmacology Laboratory, Graduate Program in Health Sciences, Health Sciences Unit, University of Southern Santa Catarina (UNESC), Criciúma, SC, Brazil* Neurosciences Laboratory, Graduate Program in Health Sciences, Health Sciences Unit, University of Southern Santa Catarina (UNESC), Criciúma, SC, Brazil† Translational Psychiatry Program, Department of Psychiatry and Behavioral Sciences, University of Texas Health Science Center at Houston (UTHealth), Houston, TX, United States‡ Center of Excellence on Mood Disorders, Department of Psychiatry and Behavioral Sciences, University of Texas Health Science Center at Houston (UTHealth), Houston, TX, United States§ Neuroscience Graduate Program, University of Texas Graduate School of Biomedical Sciences at Houston, Houston, TX, United States¶

CHAPTER OUTLINE

INTRODUCTION

Bipolar disorder (BD) is a common mental disorder, and is among the most severe and debilitating psychiatric disorders. BD is characterized by recurrent episodes of depression and mania or hypomania. The pharmacological treatments available for BD are often insufficient for many patients. Despite continuing treatment, many BD patients may show recurrent mood episodes, residual symptoms, functional impairment, psychosocial disability, and significant medical and psychiatric comorbidity. Several drugs, such as lithium, valproate, anticonvulsants, and antipsychotics, have shown efficacy in controlling acute manic symptoms; however, these drugs have only moderate effects on depressive episodes.

It is known that multiple factors may be involved in the pathophysiology of BD, such as genetic, biochemical, psychodynamic, social, and environmental factors. Studies have described a progressive deterioration course in many patients. Some

Bipolar Disorder Vulnerability. https://doi.org/10.1016/B978-0-12-812347-8.00002-6

authors have suggested that sensitization to stress, mood episodes, and stimulants can lead to implications for BD progression. Despite the importance of BD, the precise neurobiology underlying this disorder is currently unknown.

NEUROPROGRESSION AND BIPOLAR DISORDER

BD has been described as a cyclic (periodic) disease; however, recently, several studies have shown progressive characteristics in the manifestation of symptoms in the disorder. The term "neuroprogression" has been used to define the disease process acceleration and its underlying factors, such as changes in peripheral biomarkers, cognitive functions, neuroimaging, and functionality, which appear to varying degrees depending on evolution stage. Neuroprogression can be described as a progressive and gradual neuron loss, either by brain structure damage, or by neuronal death, leading to nervous system dysfunction. In BD, the alternation between depression and mania periods, and their recurrence, leads to progression and increased severity and frequency of episodes.

There is gradual cellular damage from the disorder's first episode. As the crisis progresses, the brain has difficulties in maintaining homeostasis; this reduces its ability to recover, resulting in damage. The long-term course of BD patients commonly presents as higher degrees of impairment, progressive decrease in duration of euthymic periods and inversely increased mood episodes, as well as poor response to psychosocial and pharmacological treatments.

Clinical staging is widely used in many medical specialties, for determining the development and progress of disease. Although Fava and Kellner (1993) first proposed clinical staging in psychiatry, much earlier Kraepelin (1921) had suggested a longitudinal course for BD development and evolution, describing signs of progression and increased latency between mood episodes. It is important to take into account conditions that may influence the prognosis for BD, such as potential comorbidities, substance abuse, tobacco and alcohol, stressors, and social problems, among other recurrences and complications.

Several divisions of BD phases arise from staging models. McGorry, Hickie, Yung, Pantelis, and Jackson (2006) postulated some BD clinical stages, widely accepted in several studies:

- **0 clinical stage**—the at-risk phase, no psychiatric disorders or symptoms present, but rather increased risk of mood disorder (such as family history and exposure to environmental stress);
- **1a clinical stage**—mild or nonspecific symptoms of mood disorder, patients show mild functional change or decline; this could be considered the prodromal phase, as discussed in Chapter 1;
- **1b clinical stage**—the ultra-high-risk phase, presents moderate but subthreshold symptoms;
- **2 clinical stage**—defined as the first episode threshold mood disorder, with moderate-severe symptoms;

- **3a clinical stage**—recurrence of subthreshold mood symptoms;
- **3b clinical stage**—first threshold relapse, recurrence of mood disorder, residual symptoms, or neurocognition below the best level achieved following the first episode;
- **3c clinical stage**—multiple relapses, provided worsening in clinical extent and impact of illness is objectively present;
- **4 clinical stage**—persistent unremitting illness, reflecting persisting unremitting symptoms that are potentially nonresponsive to treatment.

Studies have proposed other divisions and subdivisions, but similarly based. Kapczinski et al. (2009) suggested the following stages:

- latent clinical stage (risk for BD development);
- clinical stage I (without psychiatric symptoms, with well-defined periods of euthymia); clinical stage II (early stage, patient presents cycling of humor episodes);
- clinical stage III (patient presents altered biomarkers and a clinically relevant pattern of cognitive and functioning deterioration); and
- clinical stage IV (loss of patient autonomy, significant presence of altered brain biomarkers).

Post (2010) expanded these BD clinical stages by proposing a structure composed of seven clinical stages:

- clinical stage 1 (vulnerability);
- clinical stage 2 (presymptomatic interval);
- clinical stage 3 (prodrome);
- clinical stage 4 (full syndrome);
- clinical stage 5 (recurrence);
- clinical stage 6 (progression); and
- clinical stage 7 (treatment resistance).

Specific neurobiological markers may have different roles in early compared to late stages. Pathophysiological changes are observed at molecular, cellular, and structural levels. These changes, in turn, promote deleterious effects contributing to overstated inflammation, oxidative stress, mitochondrial and endoplasmic reticulum dysfunction, decreased neurotrophin levels, telomere shortening, epigenetic alterations, neuroendocrine dysfunction, and activation of neuronal death pathways.

Oxidative stress can be identified in the early stages of BD and tends to be more pronounced over time, suggesting that oxidative stress may be a mechanism of neuroprogression and a plasma biomarker for monitoring the disorder course (Berk et al., 2011; Rizzo et al., 2014). Similarly, several studies have suggested that some mediators of inflammation are related to disease progression in BD (Brietzke & Kapczinski, 2008; Brietzke, Kauer-Sant'Anna, Teixeira, & Kapczinski, 2009; Brietzke, Stertz, et al., 2009; Kapczinski et al., 2009). This is based on the findings that inflammatory molecules are activated during acute mania but not during euthymia. Notably, TNFα, while elevated throughout the course, is known to be higher later in the disorder, suggesting the inflammatory state is more perturbed at the late stage of illness.

Neurotrophic factors such as brain derived neurotrophic factor (BDNF), nerve growth factor (NGF), and glial cell-derived neurotrophic factor (GDNF) have been highly studied in BD and may be key molecules to explain BD pathophysiology, as well as its neuroprogression. Clinical studies have demonstrated that BD patients show decreased BDNF levels during manic and depressive episodes compared to euthymic patients and control subjects. Barbosa et al. (2011) demonstrated that during manic episodes, BD patients have lower NGF levels than euthymic and control subjects; decreased levels are related to manic episode severity. GDNF levels have been related to different mood episodes in BD, supporting the neurotrophic factor role in BD pathophysiology and disease progression.

The mechanisms underlying pathophysiological alterations in neuroprogression are still unknown. The Allostatic Load (AL) theory is one of the few models to explain the BD progression course. Allostasis is conceptualized as a condition of constant modification of the internal environment of living organisms. AL refers to chronic stress-induced hyperactivation of the homeostatic mechanism, which can lead to clinical and neurobiological progression in BD.

The "Kindling" hypothesis, according to the heuristic model of Post and Weiss (1996), suggests that BD episodes are increasingly frequent due to permanent sensitization and frequent stress. In this way, after several repeated stimuli the frequent autonomic response occurs. The Kindling and Allostatic Load models reinforce the concept that delay of treatment in BD exposes the patient to more episodes of humor and this is related to greater number of hospitalizations, increased suicide risk, development of comorbidities, social damage, complications, and decreased quality of life.

Another theory is the concept of Accelerated Aging (AA) in BD, suggesting that BD has premature senescence (Fries et al., 2017). Natural aging can be understood as a multidimensional process involving physical, psychological, and social changes, including neurobiological, structural, functional, and chemical alterations. Aging leads to increase in several biological markers and cellular alterations, which are also seen in BD patients, regardless of age, but directly related to BD neuroprogression.

EARLY STAGES OF BIPOLAR DISORDER

Clinical staging evaluates where the patient is within the course of a given disorder. Thus staging can be an efficient tool in the identification and differentiation of the initial and milder phenomena from those that accompany the extent, progression, and chronicity of the disease. Consequently, it allows for selection of the most relevant interventions in the earliest stages, and those instituted in advanced stages. In recent years, research that focused on the prodromal and early stages of BD has shown that early detection and intervention strategies have the potential to delay, reduce severity, or prevent complete episodes of BD (see Chapter 1).

Recognizing the existence of a vulnerability is to believe that BD does not start abruptly, but rather that there are different dimensions of symptoms prior to the first episode of mood. Genetic polymorphisms, positive family history, and mild symptoms below the diagnostic threshold are described as precursors and risk factors for BD. Vulnerability is a particularity that indicates a state of weakness, delicacy, and instability. In the case of individuals who demonstrate vulnerability, an indicated prevention strategy is to reduce exposure to risk factors such as environmental stressors.

Early detection can be considered a recognition of predisease stages, from prodromal characteristics to the first point in which the clinical manifestation meets the diagnostic criteria. The prodromal phase can be clinically noticeable and the search for professional attention should begin at this stage.

Prodrome has been recognized as a valid and clinically useful construct in disorders with progressive onset and progressive evolution. Conceptually, prodrome is a sign or group of symptoms that may indicate the onset of a disease before specific symptoms arise. Identifying the prodrome signs and symptoms in BD is an extremely complex task. However, an early prodrome in BD can be defined as a time interval between the first signs and symptoms until the disorder becomes diagnosed. As discussed in Chapter 1, these symptoms include irritability, impulsiveness, anxious symptoms, sleep disturbances, aggressiveness, hyperactivity, and lability of humor. Retrospective studies have shown that patients with BD often exhibit earlier periods of these signs and symptoms before the diagnosis of a mood episode is complete. It is important to emphasize that in general the studies that point out prodromal symptoms are retrospective studies, thus, they may have recall bias. Another important point is the possible overlap and similarity of BD symptoms with other psychiatric disorders, leading to difficulties in defining the specific prodrome symptoms for BD.

Incorrect initial diagnoses are common and this leads to inadequate treatment, potentially inducing changes in mood and acceleration of the humor cycle. This highlights the importance and benefits of early detection, leading to possible prevention and adoption of social and functional assistance strategies for the patients. In conclusion, in the face of the increased acceptance of neuroprogression in BD, it is important to emphasize the need for early detection and primary prevention, searching for symptoms and signs indicative of early stages of BD.

REFERENCES

Barbosa, I. G., Huguet, R. B., Neves, F. S., Reis, H. J., Bauer, M. E., Janka, Z., et al. (2011). Impaired nerve growth factor homeostasis in patients with bipolar disorder. *The World Journal of Biological Psychiatry, 12,* 228–232.

Berk, M., Kapczinski, F., Andreazza, A. C., Dean, O. M., Giorlando, F., Maes, M., et al. (2011). Pathways underlying neuroprogression in bipolar disorder: focus on inflammation, oxidative stress and neurotrophic factors. *Neuroscience and Biobehavioral Reviews, 35,* 804–817.

Brietzke, E., & Kapczinski, F. (2008). TNF-alpha as a molecular target in bipolar disorder. *Progress in Neuro-Psychopharmacology & Biological Psychiatry, 32,* 1355–1361.

Brietzke, E., Kauer-Sant'Anna, M., Teixeira, A. L., & Kapczinski, F. (2009). Abnormalities in serum chemokine levels in euthymic patients with bipolar disorder. *Brain, Behavior, and Immunity, 23,* 1079–1082.

Brietzke, E., Stertz, L., Fernandes, B. S., Kauer-Sant'Anna, M., Mascarenhas, M., Escosteguy Vargas, A., et al. (2009). Comparison of cytokine levels in depressed, manic and euthymic patients with bipolar disorder. *Journal of Affective Disorders, 116*, 214–217.

Fava, G. A., & Kellner, R. (1993). Staging: a neglected dimension in psychiatric classification. *Acta Psychiatrica Scandinavica, 87*, 225–230.

Fries, G. R., Bauer, I. E., Scaini, G., Wu, M.-J., Kazimi, I. F., Valvassori, S. S., et al. Accelerated epigenetic aging and mitochondrial DNA copy number in bipolar disorder. *Translational Psychiatry, 7*(12), (2017). 1283. https://doi.org/10.1038/s41398-017-0048-8.

Kapczinski, F., Dias, V. V., Kauer-Sant'Anna, M., Frey, B. N., Grassi-Oliveira, R., Colom, F., et al. (2009). Clinical implications of a staging model for bipolar disorders. *Expert Review of Neurotherapeutics, 9*, 957–966.

Kraepelin, E. (1921). *Manic-depressive insanity and paranoia*. Edinburgh: Livingston.

McGorry, P. D., Hickie, I. B., Yung, A. R., Pantelis, C., & Jackson, H. J. (2006). Clinical staging of psychiatric disorders: a heuristic framework for choosing earlier, safer and more effective interventions. *The Australian and New Zealand Journal of Psychiatry, 40*, 616–622.

Post, R. M. (2010). Mechanisms of illness progression in the recurrent affective disorders. *Neurotoxicity Research, 18*, 256–271.

Post, R. M., & Weiss, S. R. (1996). A speculative model of affective illness cyclicity based on patterns of drug tolerance observed in amygdala-kindled seizures. *Molecular Neurobiology, 13*, 33–60.

Rizzo, L. B., Costa, L. G., Mansur, R. B., Swardfager, W., Belangero, S. I., Grassi-Oliveira, R., et al. (2014). The theory of bipolar disorder as an illness of accelerated aging: implications for clinical care and research. *Neuroscience and Biobehavioral Reviews, 42*, 157–169.

Influence of early childhood trauma on risk for bipolar disorder

3

Mateus L. Levandowski, Rodrigo Grassi-Oliveira,
Developmental Cognitive Neuroscience Lab (DCNL), Pontifical Catholic University of Rio Grande do Sul (PUCRS), Porto Alegre, Brazil

CHAPTER OUTLINE

BIPOLAR DISORDER AS A STRESS SENSITIVE ILLNESS

Bipolar disorder (BD) is a complex phenomenon with multifactorial factors that could account for development of the disorder (Grande, Berk, Birmaher, & Vieta, 2016; Phillips & Kupfer, 2013). Stressful experiences are a well-studied factor in BD, are highly related to the early stages of the disorder (e.g., prodromal symptoms and first mood episode), and are a strong predictor of relapse (Kemner et al., 2015; Kim, Miklowitz, Biuckians, & Mullen, 2007; Lex, Bazner, & Meyer, 2017). Therefore, the interest in how stress is related to BD has given rise to theoretical models for this relationship (for a review about these models, see Brietzke, Mansur, Soczynska, Powell, & McIntyre, 2012).

One studied model is the stress-diathesis model, which indicates that biological or genetic traits (diatheses) interact with environmental influences (stress) predisposing individuals towards developing a certain disorder (Monroe & Simons, 1991). According to this theory, each person has a threshold for vulnerability, and when the critical amount of stress has been reached, the person will develop a particular disorder. Thus, this model formulates a linear relationship between stress exposure and mental disorders, but does not consider other mechanisms, such as adaptation processes towards stress.

Another widely known paradigm is the allostasis model, offering a more comprehensive understanding of disease processes (McEwen & Wingfield, 2010).

Bipolar Disorder Vulnerability. https://doi.org/10.1016/B978-0-12-812347-8.00003-8

This model focused on the allostatic process, which is an adaptive mechanism in the stress response, setting a biological homeostasis according to environmental needs in order for an individual to survive. In order to maximize survival, biological processes have been molded by evolution to maintain stability (homoeostasis) when facing changes/adversities in the environment. In this sense, our central nervous system (CNS) is responsible for regulating and anticipating behavioral and biological changes in order to maintain a functional balance (McEwen, 2007).

However, the chronic challenge in this adaptive process could result in allostatic load (e.g., detrimental physiological consequences due to chronic stress). The allostatic load induces alterations in our neuronal, endocrine, and immunological systems, which could lead to health problems. Thus, in this model, the onset of a mental disorder starts when the allostatic process fails due to intense and repeated exposure to stress, causing an overactivity or inactivity in the physiological adaptation (Brietzke et al., 2012; Kapczinski et al., 2008).

In this sense, BD, as other stress-sensitive disorders, is assumed to be the result of a complex interaction in which the environment has a central role (Aldinger & Schulze, 2017; Etain, Henry, Bellivier, Mathieu, & Leboyer, 2008). From an environmental perspective, childhood maltreatment attracts attention due its relation with early onset and worse lifetime course of BD (Agnew-Blais & Danese, 2016; Daruy-Filho, Brietzke, Lafer, & Grassi-Oliveira, 2011). Childhood maltreatment (see Table 1 for a big-picture view of the problem), often operationalized as childhood abuse and/or neglect, is recognized to reprogram molecular and neurobiological systems; in turn, this could precipitate the allostatic load in a still-developing organism (Grassi-Oliveira, Ashy, & Stein, 2008). Thus, when the person who should protect is, in fact, the person who is creating stressors or threatening the child, this generates a stressor with far-reaching impact, overloading the allostatic process (Danese & McEwen, 2012; Widom, Horan, & Brzustowicz, 2015). In short, studies assessing stress in childhood have provided compelling reasons to believe that this early exposure increases risk for psychopathology, such as BD.

In this chapter we will introduce how childhood maltreatment disrupts neurodevelopment and is related to BD outcomes. Moreover, we will discuss the influence of childhood maltreatment on the early and different phenotypic expression of prodromal manifestations and consequently on early BD onset.

CHILDHOOD MALTREATMENT AS A RISK FOR DEVELOPING BIPOLAR DISORDER

Bipolar disorder is considered to be among the most heritable mental disorders; however, the science behind the genes-disorder relationship is complicated by a list of environmental risk factors that should be taken into account (Aldinger & Schulze, 2017; Etain et al., 2008; Grande et al., 2016). In fact, several cross-sectional studies have documented that childhood trauma is more common among individuals with BD, as compared to the general population, with rates as high as 50%

Table 1 Childhood maltreatment landscape: extension and description of the problem

Scope of the problem	Risk factors	Consequences of maltreatment
– Child maltreatment is an important global problem with serious life-long consequences – One in four US children experience some form of child maltreatment in their lifetime – The economic burden of child maltreatment in the United States is approximately $124 billion – The rate of death as a consequence of maltreatment is 2.2 per 100,000 children	– Children younger than 4 years of age – Lack of understanding of children's needs, child development, and parenting skills – Family history of child maltreatment – Family substance abuse and/or mental health issues – Family disorganization, dissolution, and violence, including intimate partner violence – Concentrated neighborhood disadvantage	– The suffering can have long-lasting consequences – Disruption in early brain development – Disruption in development of the endocrine and immune systems – Risk for behavioral, physical, and mental health problems

Maltreatment type	Definition	Estimated prevalence[a]
Acts of Commission (Child Abuse)		
Physical abuse	A parent, stepparent, or adult living in your home pushed, grabbed, slapped, threw something at you, or hit you so hard that you had marks or were injured	28%
Sexual abuse	An adult, relative, family friend, or stranger who was at least 5 years older than you ever touched or fondled your body in a sexual way, made you touch his/her body in a sexual way, or attempted to have any type of sexual intercourse with you	10%
Psychological abuse	A parent, stepparent, or adult living in your home swore at you, insulted you, put you down, or acted in a way that made you afraid that you might be physically hurt	9%
Acts of Omission (Child Neglect)		
Physical neglect	There was not someone to take care of you, protect you, and take you to the doctor if you needed it, you did not have enough to eat, your parents were too drunk or too high to take care of you, and you had to wear dirty clothes	78%
Emotional neglect	Someone in your family did not help you feel important or special, you did not feel loved, people in your family did not look out for each other and did not feel close to each other, and your family was not a source of strength and support	

[a]A child that has suffered from one type of childhood maltreatment is more likely to suffer other types of maltreatment during childhood (40% suffer from two or more types of maltreatment).

(Daruy-Filho et al., 2011; Sala, Goldstein, Wang, & Blanco, 2014), and reaching up to 63% when multiple trauma events are experienced by an individual (Etain et al., 2010).

To explore a degree of causality or dose-response interaction between childhood maltreatment and BD, some studies investigated the odds ratio (OR) for this relation. The National Epidemiologic Survey on Alcohol and Related Conditions (NESARC) has investigated the impact of exposure to childhood physical abuse on risk for several psychiatric conditions. They found that early stress exposure increases the risk for BD with an OR of 1.48 [1.26–1.74] (adjusted for sociodemographic characteristics and psychiatric comorbidities) (Sugaya et al., 2012). Importantly, this study only evaluated the impact of child physical abuse, and this type of exposure usually happens together with other types of maltreatment.

In this sense, some authors has found that between all types of abuse and neglect, the most prevalent in BD patients is emotional abuse exposure (Alvarez et al., 2011), with a larger odds ratio (1.88 [1.23–2.86]) for BD risk compared to other types of abuse and neglect (Etain et al., 2010). Furthermore, one study found a dose-response relation between number of maltreatment exposures and BD course, showing the existence of differences in the course of BD based on the type of traumatic exposure and the number of exposures (Sala et al., 2014); this will be further highlighted in the next section. The prevalence of maltreatment in BD and the influence that the exposure has on the disorder are even greater than are found in unipolar patients (Hyun, Friedman, & Dunner, 2000), in whom early life stress exposure is linked with the pathophysiological model of the disorder (Saleh et al., 2017).

Considering these studies, childhood maltreatment, especially cases of emotional abuse, are an important risk factor for BD. There is considerable data showing that emotional abuse early in life may impair emotional features (Teicher, Samson, Anderson, & Ohashi, 2016), something that is a central problem in BD patients. Therefore, growing up with traumatic events may cause deficits in emotion regulation and expression, social cognition, and social competence, and creates a risk for development of poor coping strategies, such as engaging in fights and substance abuse (Alvarez et al., 2011; Kapczinski et al., 2008). Living with a family without emotional contact, or explosive and inadequate emotional reactions, can result in learning deficient emotion regulation and coping strategies in the face of stress (DiLillo, Tremblay, & Peterson, 2000).

A proper development of emotional, social, biological, and behavioral systems early in life is important in order to cope well with adverse situations (Grassi-Oliveira et al., 2008; Teicher et al., 2016). Disruption during development, with a constant challenge in these systems, could result in overload (allostatic load) and lead to long-term physical and mental health outcomes through neurobiological pathways. Allostatic load may cause brain rewiring, impairing executive functions due to neuroanatomical and neurofunctional changes; these changes contribute to poor coping strategies and affect deregulation during stressful situations (Grassi-Oliveira, Daruy-Filho, & Brietzke, 2010; Kapczinski et al., 2008). The coping and affect impairment in turn increase the susceptibility to be involved in further stressful situations.

Indeed, stressful life events appear to precede the occurrence of mood symptoms, playing a role in the onset and course of BD (Kemner et al., 2015; Kim et al., 2007; Koenders et al., 2014). Thus, it has been hypothesized that childhood trauma disrupts the allostatic mechanisms, rendering the individual vulnerable to future stressors in life, which are strong predictors of increases in subsequent mood symptoms (Aas et al., 2016). Indeed, stressful life events predict the first mood episode, while subsequent episodes appear to require less stress to activate the disorder.

It is worth mentioning that over the years, some hypotheses have been made to explain the relation between early life trauma and BD (Etain et al., 2008). Fig. 1 illustrates and explains some of these hypotheses, and shows an integrative view of them. They range from explaining that maltreatment can occur as a consequence of child disruptive behavior during the early prodromal phases of BD, to explaining that genetic substrates can affect the offspring directly (heritability) and indirectly (parent symptoms lead to inadequate parenting bond). Genetic features are important to understand BD risk, and they are the focus of following chapters. Much data have highlighted that genetic factors interact with environmental vulnerability for BD (McGuffin et al., 2003), and more recently, studies have shown the role of epigenetic mechanisms mediating these risk (Fries, Walss-Bass, Soares, & Quevedo, 2016).

We have discussed in this section the relationship between childhood trauma and risk for developing BD. There are considerable data showing that early life trauma not only plays a role in the risk for bipolar disorder, but also may influence differences in the course or expression of the illness.

BIPOLAR DISORDER WITH CHILDHOOD MALTREATMENT IS CLINICALLY DIFFERENT

Bipolar disorder is ranked as one of the most disabling illnesses, with a reduction in function, cognition, and quality of life (Grande et al., 2016). The fluctuation in mood state is highly harmful for the individual; however, what seems to be bad can become worse. There are important studies suggesting that childhood maltreatment plays a role in modulating the clinical phenotype of the disorder (Aas et al., 2016; Daruy-Filho et al., 2011; Sala et al., 2014), showing that bipolar disorder has many facets when followed by child trauma experiences. These facets include more suicidal rates, higher number of hospitalizations, lower treatment adherence, and a rapid cycling course. Moreover, childhood maltreatment also modulates cognitive, functional, and biological features as presented in detail in Table 2. The results presented in this table appear to be consistently replicated across studies, providing support that bipolar disorder patients exposed to childhood trauma have more adverse illness characteristics and outcomes compared to nonexposed patients.

In this sense, there are consistent indications that childhood maltreatment not only is a risk for developing a mental disorder, but also is critical to the different clinical manifestations (Teicher & Samson, 2013). This is not something exclusive to bipolar disorder patients, but happens across many other psychiatric disorders,

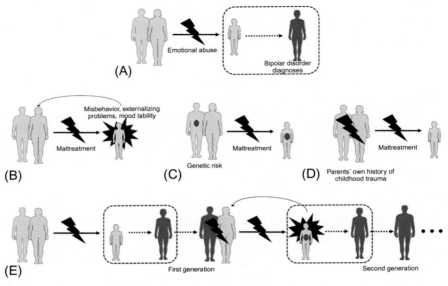

FIG. 1

Illustration of hypothetical models explaining the relationship between childhood maltreatment and bipolar disorder. (A) Studies have documented that childhood maltreatment is present in 51% of bipolar disorder cases and that emotional abuse is the most frequent type of maltreatment in bipolar patients. Some studies show that individuals who looked after the patients when they were a child were more likely to be critical, hostile, or intrusive, and have alcohol misuse, which may constitute emotional abuse. The emotional and affective instability during early life may enhance the long-lasting chances of this child becoming an adult with emotional and affective problems (core problems in bipolar disorder). (B) Prodromal, early disease manifestations, and/or comorbidities associated with bipolar disorder may be an antecedent risk factor for childhood maltreatment. As discussed in Chapter 1, it is common that children present irritability and mood lability prior to being diagnosed with bipolar disorder. Also, early onset bipolar disorder has been associated with externalizing problems, such as disruptive behavior disorders, attention deficit/hyperactivity, and conduct disorder. These characteristics may increase the risk of the child being punished (e.g., physical and emotional abuse) by parents with poor coping strategies. (C) Bipolar disorder is known to be highly heritable, with studies of families and twins showing the importance of genetic factors affecting susceptibility to the disorder. Moreover, bipolar disorder offspring reported higher levels of childhood trauma. Thus, being a child of a parent with bipolar disorder may put the individual at risk to be exposed to family violence, less organized homes, and unstable parents, which can also result in childhood maltreatment. In conclusion, this model suggests that the genetic substrate of the parents leads both to the abuse and to the illness in the children. (D) Intergenerational transmission of childhood maltreatment. The cycle of violence learned through modeling and observational learning. The parents' own history of childhood trauma may drive inadequate parenting attitudes towards their offspring. (E) The last model integrates all of them. Some of the characteristics presented in the other models are likely to occur together and contribute to a vicious cycle, moving from generation to generation. Early disturbances in emotional and affective interactions may predispose deficits in the emotional and affective system later in life. The constant environmental challenge would alter normal brain development through the allostatic load process. Moreover, in these models, there is a possible involvement of learned behaviors from their parents (e.g., irritability, poor coping strategy).

Table 2 Selected studies of the impact of childhood maltreatment on bipolar disorder phenotypes

Association of CM and:	(1) Disorder course and symptom features	↓ Age of illness onset (Etain et al., 2013; Garno, Goldberg, Ramirez, & Ritzler, 2005; Leverich et al., 2002; Post et al., 2016; Romero et al., 2009); ↑ number of hospitalizations (Alvarez et al., 2011; Brown, McBride, Bauer, Williford, & Cooperative Studies Program 430 Study, 2005; Marchand, Wirth, & Simon, 2005); ↓ final response to BD treatment (Marchand et al., 2005); ↑ duration of treatment (Marchand et al., 2005); ↑ cycling frequencies (Brown et al., 2005; Etain et al., 2013; Leverich et al., 2002); ↑ suicide attempts (Etain et al., 2013; Garno et al., 2005; Leverich et al., 2002); ↑ delay on diagnosis (Leverich et al., 2002; Romero et al., 2009); ↑ depressive episodes (Garno et al., 2005); ↑ manic episodes (Etain et al., 2013; Garno et al., 2005); ↑ PTSD comorbidity (Brown et al., 2005; Garno et al., 2005; Romero et al., 2009); ↑ psychotic symptoms (Shevlin, Dorahy, & Adamson, 2007); ↑ SUD comorbidity (Etain et al., 2013; Garno et al., 2005; Sala et al., 2014); ↓ treatment adherence (Conus et al., 2010); ↑ medical co-morbidities (Post, Altshuler, Leverich, Frye, et al., 2013)
	(2) Cognitive features	↓ Emotional processing (Russo et al., 2015); ↓ IQ (Bucker et al., 2013; Poletti et al., 2016); ↓ attention (Bucker et al., 2013; Poletti et al., 2016; Stevanin et al., 2017); ↓ verbal memory (Bucker et al., 2013; Poletti et al., 2016; Savitz, van der Merwe, Stein, Solms, & Ramesar, 2008; Stevanin et al., 2017); ↓ working memory (Bucker et al., 2013; Poletti et al., 2016); ↓ executive function (Poletti et al., 2016; Savitz et al., 2008); ↑ cognitive distortions (Poletti, Colombo, & Benedetti, 2014)
	(3) Function features	↑ Psychosocial stressors (Conus et al., 2010; Leverich et al., 2002; Sala et al., 2014); ↓ quality of life (Brown et al., 2005; Romero et al., 2009); ↑ disability pension (Brown et al., 2005); ↑ forensic problems (Conus et al., 2010); ↑ unemployment (Conus et al., 2010)
	(4) Biological features	↓ BNDF levels (Kauer-Sant'Anna et al., 2007); ↓ BDNF mRNA levels (Aas et al., 2014); ↑ NR3C1 methylation (Perroud et al., 2014); ↓ corpus callosum volume (Bucker et al., 2014); disruption of white matter integrity (Benedetti et al., 2014); ↓ dorsolateral prefrontal cortex (Duarte et al., 2016); ↓ right thalamus (Duarte et al., 2016)

Notes: BD, bipolar disorder; BNDF, brain-derived neurotrophic factor; CM, childhood maltreatment; IQ, intelligence quotient; PTSD, posttraumatic stress disorder; SUD, substance use disorder.

↑, increased feature in bipolar disorder patients with CM exposure comparing with patients without exposure.
↓, decreased feature in bipolar disorder patients with CM exposure comparing with patients without exposure.

such as depression (Nemeroff et al., 2003), posttraumatic stress disorder (Schoedl et al., 2010), and drug addiction (Levandowski et al., 2016). These disorders have in common distinct clinical and biological phenotypes when preceded by a history of childhood maltreatment.

This observation led Teicher and Samson to develop a term that indicates this phenomenon: ecophenotype. This term represents a phenotypic expression influenced by a maltreatment exposure (Teicher & Samson, 2013). Thus, they argue that maltreated and nonmaltreated individuals with the same diagnosis have different disorder phenotypes and this should be taken into account for future research and treatments. Moreover, a recent meta-analysis with 30 studies has shown that childhood maltreatment predicts unfavorable bipolar disorder course, suggesting that childhood maltreatment may be used as an indicator of disease progression (Agnew-Blais & Danese, 2016).

Interestingly, Post et al. (2011) have found that bipolar disorder outpatients from the United States have more severe symptoms than patients in some countries in Europe (Netherlands and Germany). One possible explanation postulated by the authors is the early age of onset observed in the US patients, since early onset can result in more time with the disorder without treatment, resulting in worsening of symptoms. Both early onset and time without treatment are related to adversity early in life. The average age of bipolar disorder onset differed by around 20 years in those exposed or not to adversity early in life (5.9 vs 25.8 years of age) (Post et al., 2016). Moreover, bipolar disorder patients who reported child abuse also reported a longer duration of time ill without treatment in comparison to patients without such history (13 vs 8 years) (Leverich et al., 2002). These results corroborate the findings of the comparisons between the United States and Europe, showing that in the United States, there are more cases of early adversity than in the Netherlands and Germany (Post, Altshuler, Leverich, Nolen, et al., 2013).

Therefore, early adversity may increase bipolar disorder severity by lowering the disorder age of onset, which can influence a different and worse clinical manifestation. Early onset of bipolar disorder has been associated with treatment delay, a known factor for an adverse course of illness. Finally, childhood maltreatment also influences early prodromal manifestations, which contribute to symptoms worsening and difficulty of illness identification, since these prodromal manifestations occur when children are very young.

CHILDHOOD MALTREATMENT INFLUENCE ON THE EARLY PHASE OF BIPOLAR DISORDER: PRODROMAL MANIFESTATIONS

As discussed in Chapter 1, several studies have identified that bipolar disorder patients start to experience mild symptoms, but progressively, months or years before the illness onset (see Figs. 2A and 3) (Bechdolf et al., 2012; Rios et al., 2015). These symptoms could start in adolescence or even during childhood, and gradually

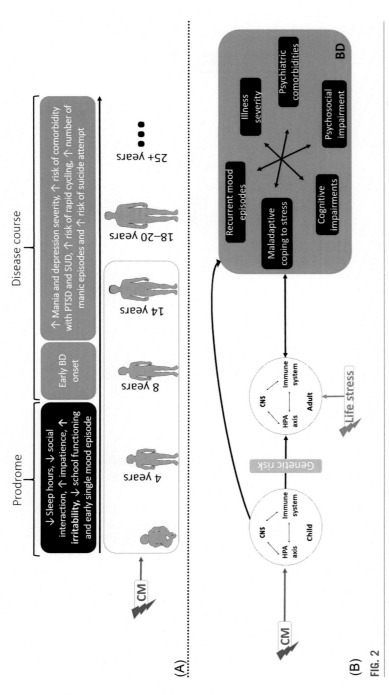

FIG. 2

(A) Illustrative scheme reflecting the impact of childhood maltreatment on early prodromal symptoms, early onset and different behavior manifestations of BD. Data suggest that early trauma modulates the clinical expression and course of the disorder. (B) This model illustrates how childhood maltreatment that occurs within the sensitive neurodevelopmental period can shape the organism to suit the environmental needs (allostatic load). Here, we show the dynamic interplay between nervous, endocrine, and immunological systems after early life stress exposure predisposing the expression of bipolar disorder later in life. Chronic early life stress may impact the endocrine system, producing a permanent reprogramming of the HPA axis. This HPA axis reprogramming may result in prolonged inflammatory response and neurotoxicity, inducing structural and functional changes in the brain still under development. Further, genetic factors in interaction with early life stress may mediate the same system changes (nervous, endocrine, and immune), leading to overproduction of inflammatory mediators and contributing to maintaining the HPA hyperactivity and neurotoxicity in the brain. These factors may moderate impairment of specific brain regions, including regions of the prefrontal cortex, lowering the coping strategies of the individual. In turn, the poor coping would allow the increase of life stress experiences during early adulthood, which appear to increase vulnerability to BD. Epigenetic mechanisms could help explain how early stressful experiences may go forward to influence later vulnerability to psychopathology (discussed in Chapter 4).

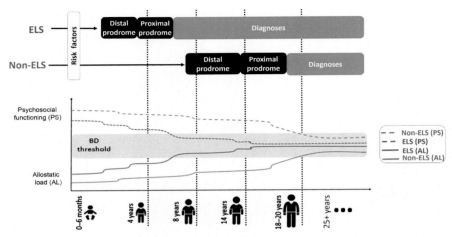

FIG. 3

Childhood maltreatment is strongly linked with BD diagnosis. Moreover, early life stress (ELS) may modulate the clinical expression of the disorder, here represented by early distal and proximal prodrome and, consequently, early BD onset. Independent of presence or not of childhood trauma in early life, disease expression is followed by a decrease of psychosocial function and increase of allostatic load. This phenomenon starts in the prodromal phase until it reaches the threshold for bipolar disorder. The difference on this trajectory between psychosocial function and allostatic load is that patients with a history of childhood maltreatment have early and swift impairment on this system. Exposure to ELS has cascading effects on stress physiology and psychological and social functioning that continue to accumulate allostatic load even when there is no more stress, thus making the model of "childhood maltreatment + bipolar disorder" more severe than the "bipolar disorder without childhood stress" model.

increase until the first mood episode. This set of symptoms prediagnosis is called distal and proximal prodrome, progressing when the symptoms became more specific to full bipolar disorder diagnosis (Skjelstad, Malt, & Holte, 2010).

The earlier the identification of the prodromal manifestation, the better the chances are of prevention and/or early treatment. However, assessment of these symptoms is complicated, due to their heterogeneity of presentation. Most of the prodromal symptoms could start during adolescence, when the individual has many changes going on, and it becomes hard for some parents and professionals to distinguish between what is normal or not for this age (Leverich & Post, 2006; Malhi, Bargh, Coulston, Das, & Berk, 2014). This becomes even harder when these prodromal symptoms start in childhood, mainly in cases related to child abuse and neglect, since the prodrome of the disorder is earlier than for other bipolar disorder patients (Noto et al., 2015).

The prodrome symptoms in bipolar disorder patients exposed to childhood maltreatment may be different (Fig. 2A) and manifest earlier (Fig. 3) than in patients not exposed to this stress. These symptoms could be in the form of fewer sleep hours, more impatience, and mood lability (Malhi et al., 2014; Noto et al., 2015). If present during the early phase (distal prodrome) these symptoms may further lead to constant stress for the child (e.g., being punished by parents for their behavior) and so predispose to proximal prodrome and later to bipolar disorder diagnosis (Duffy, Jones, Goodday, & Bentall, 2015).

The chronic stress activation may shape the nervous, endocrine, and immunological systems to future psychopathology (Fig. 2B illustrates this model). Children that have suffered from child abuse or neglect may have sleep and conduct problems years before the first mood episode and this results in constant stress reactivity (at behavioral and biological levels). The early impact on the endocrine system could permanently reprogram the hypothalamus pituitary adrenal axis (HPA), inducing consequences in CNS areas sensitive to glucocorticoids, such as the prefrontal cortex, amygdala, and hippocampus (Lupien, McEwen, Gunnar, & Heim, 2009; Teicher et al., 2016). These areas are crucial for inhibitory activity, emotional response, learning, and memory: all problems that are central for bipolar disorder. The early reprogramming may interact with genetic risk and predispose the individual to other life stress situations due inability to cope with situations, predisposing in turn to bipolar disorder onset.

CONCLUSION

Health caregivers should know that bipolar disorder could occur before adolescence and that there is a high probability that the disorder is not only associated with genetic risk, but also with environmental stressors. Childhood maltreatment history is often found in bipolar disorder patients, with several lines of evidence showing the harmful effect of early stress on the course of disease, such as early age of onset and higher symptomatology levels in comparison with patients without such history. Fig. 3 depicts the working model of this chapter, showing that bipolar diagnosis is followed by a decrease of psychosocial functioning and increase of allostatic load. However, this imbalance happens faster in bipolar disorder patients with childhood maltreatment in their history. This may lead to early and more severe distal and proximal prodrome symptoms, modulating later the clinical expression of the disorder.

Preventing childhood maltreatment is a key component to reduce childhood suffering and mental disorders later in life. Bipolar disorder patients with childhood maltreatment have a poor response to usual treatments. Further research is required to move forward on alternative treatment options for this specific population. Addressing childhood maltreatment as a clinical routine could be the focus of early intervention strategies based on reducing stress sensitivity and reactivity (important bipolar disorder features increased by trauma history). Finally, prodromal manifestations of the disorder may occur during early childhood, and psychiatrists and pediatricians should be aware of this possibility for a more opportune intervention.

REFERENCES

Aas, M., Haukvik, U. K., Djurovic, S., Tesli, M., Athanasiu, L., Bjella, T., et al. (2014). Interplay between childhood trauma and BDNF val66met variants on blood BDNF mRNA levels and on hippocampus subfields volumes in schizophrenia spectrum and bipolar disorders. *Journal of Psychiatric Research*, *59*, 14–21. https://doi.org/10.1016/j.jpsychires.2014.08.011.

Aas, M., Henry, C., Andreassen, O. A., Bellivier, F., Melle, I., & Etain, B. (2016). The role of childhood trauma in bipolar disorders. *International Journal of Bipolar Disorders*, *4*(1), 2. https://doi.org/10.1186/s40345-015-0042-0.

Agnew-Blais, J., & Danese, A. (2016). Childhood maltreatment and unfavourable clinical outcomes in bipolar disorder: a systematic review and meta-analysis. *Lancet Psychiatry*, *3*(4), 342–349. https://doi.org/10.1016/S2215-0366(15)00544-1.

Aldinger, F., & Schulze, T. G. (2017). Environmental factors, life events, and trauma in the course of bipolar disorder. *Psychiatry and Clinical Neurosciences*, *71*(1), 6–17. https://doi.org/10.1111/pcn.12433.

Alvarez, M. J., Roura, P., Oses, A., Foguet, Q., Sola, J., & Arrufat, F. X. (2011). Prevalence and clinical impact of childhood trauma in patients with severe mental disorders. *The Journal of Nervous and Mental Disease*, *199*(3), 156–161. https://doi.org/10.1097/NMD.0b013e31820c751c.

Bechdolf, A., Ratheesh, A., Wood, S. J., Tecic, T., Conus, P., Nelson, B., et al. (2012). Rationale and first results of developing at-risk (prodromal) criteria for bipolar disorder. *Current Pharmaceutical Design*, *18*(4), 358–375.

Benedetti, F., Bollettini, I., Radaelli, D., Poletti, S., Locatelli, C., Falini, A., et al. (2014). Adverse childhood experiences influence white matter microstructure in patients with bipolar disorder. *Psychological Medicine*, *44*(14), 3069–3082. https://doi.org/10.1017/S0033291714000506.

Brietzke, E., Mansur, R. B., Soczynska, J., Powell, A. M., & McIntyre, R. S. (2012). A theoretical framework informing research about the role of stress in the pathophysiology of bipolar disorder. *Progress in Neuro-Psychopharmacology & Biological Psychiatry*, *39*(1), 1–8. https://doi.org/10.1016/j.pnpbp.2012.05.004.

Brown, G. R., McBride, L., Bauer, M. S., Williford, W. O., & Cooperative Studies Program 430 Study Team. (2005). Impact of childhood abuse on the course of bipolar disorder: a replication study in U.S. veterans. *Journal of Affective Disorders*, *89*(1–3), 57–67. https://doi.org/10.1016/j.jad.2005.06.012.

Bucker, J., Kozicky, J., Torres, I. J., Kauer-Sant'anna, M., Silveira, L. E., Bond, D. J., et al. (2013). The impact of childhood trauma on cognitive functioning in patients recently recovered from a first manic episode: data from the systematic treatment optimization program for early mania (STOP-EM). *Journal of Affective Disorders*, *148*(2–3), 424–430. https://doi.org/10.1016/j.jad.2012.11.022.

Bucker, J., Muralidharan, K., Torres, I. J., Su, W., Kozicky, J., Silveira, L. E., et al. (2014). Childhood maltreatment and corpus callosum volume in recently diagnosed patients with bipolar I disorder: data from the systematic treatment optimization program for early mania (STOP-EM). *Journal of Psychiatric Research*, *48*(1), 65–72. https://doi.org/10.1016/j.jpsychires.2013.10.012.

Conus, P., Cotton, S., Schimmelmann, B. G., Berk, M., Daglas, R., McGorry, P. D., et al. (2010). Pretreatment and outcome correlates of past sexual and physical trauma in 118 bipolar I disorder patients with a first episode of psychotic mania. *Bipolar Disorders*, *12*(3), 244–252. https://doi.org/10.1111/j.1399-5618.2010.00813.x.

Danese, A., & McEwen, B. S. (2012). Adverse childhood experiences, allostasis, allostatic load, and age-related disease. *Physiology & Behavior, 106*(1), 29–39. https://doi.org/10.1016/j.physbeh.2011.08.019.

Daruy-Filho, L., Brietzke, E., Lafer, B., & Grassi-Oliveira, R. (2011). Childhood maltreatment and clinical outcomes of bipolar disorder. *Acta Psychiatrica Scandinavica, 124*(6), 427–434. https://doi.org/10.1111/j.1600-0447.2011.01756.x.

DiLillo, D., Tremblay, G. C., & Peterson, L. (2000). Linking childhood sexual abuse and abusive parenting: the mediating role of maternal anger. *Child Abuse & Neglect, 24*(6), 767–779.

Duarte, D. G., Neves Mde, C., Albuquerque, M. R., de Souza-Duran, F. L., Busatto, G., & Correa, H. (2016). Gray matter brain volumes in childhood-maltreated patients with bipolar disorder type I: a voxel-based morphometric study. *Journal of Affective Disorders, 197,* 74–80. https://doi.org/10.1016/j.jad.2016.02.068.

Duffy, A., Jones, S., Goodday, S., & Bentall, R. (2015). Candidate risks indicators for bipolar disorder: early intervention opportunities in high-risk youth. *The International Journal of Neuropsychopharmacology, 19*(1), https://doi.org/10.1093/ijnp/pyv071.

Etain, B., Aas, M., Andreassen, O. A., Lorentzen, S., Dieset, I., Gard, S., et al. (2013). Childhood trauma is associated with severe clinical characteristics of bipolar disorders. *The Journal of Clinical Psychiatry, 74*(10), 991–998. https://doi.org/10.4088/JCP.13m08353.

Etain, B., Henry, C., Bellivier, F., Mathieu, F., & Leboyer, M. (2008). Beyond genetics: childhood affective trauma in bipolar disorder. *Bipolar Disorders, 10*(8), 867–876. https://doi.org/10.1111/j.1399-5618.2008.00635.x.

Etain, B., Mathieu, F., Henry, C., Raust, A., Roy, I., Germain, A., et al. (2010). Preferential association between childhood emotional abuse and bipolar disorder. *Journal of Traumatic Stress, 23*(3), 376–383. https://doi.org/10.1002/jts.20532.

Fries, G. R., Walss-Bass, C., Soares, J. C., & Quevedo, J. (2016). Non-genetic transgenerational transmission of bipolar disorder: targeting DNA methyltransferases. *Molecular Psychiatry, 21*(12), 1653–1654. https://doi.org/10.1038/mp.2016.172.

Garno, J. L., Goldberg, J. F., Ramirez, P. M., & Ritzler, B. A. (2005). Impact of childhood abuse on the clinical course of bipolar disorder. *The British Journal of Psychiatry, 186,* 121–125. https://doi.org/10.1192/bjp.186.2.121.

Grande, I., Berk, M., Birmaher, B., & Vieta, E. (2016). Bipolar disorder. *Lancet, 387*(10027), 1561–1572. https://doi.org/10.1016/S0140-6736(15)00241-X.

Grassi-Oliveira, R., Ashy, M., & Stein, L. M. (2008). Psychobiology of childhood maltreatment: effects of allostatic load? *Revista Brasileira de Psiquiatria, 30*(1), 60–68.

Grassi-Oliveira, R., Daruy-Filho, L., & Brietzke, E. (2010). New perspectives on coping in bipolar disorder. *Psychology & Neuroscience, 3*(2), 161–165.

Hyun, M., Friedman, S. D., & Dunner, D. L. (2000). Relationship of childhood physical and sexual abuse to adult bipolar disorder. *Bipolar Disorders, 2*(2), 131–135.

Kapczinski, F., Vieta, E., Andreazza, A. C., Frey, B. N., Gomes, F. A., Tramontina, J., et al. (2008). Allostatic load in bipolar disorder: implications for pathophysiology and treatment. *Neuroscience and Biobehavioral Reviews, 32*(4), 675–692. https://doi.org/10.1016/j.neubiorev.2007.10.005.

Kauer-Sant'Anna, M., Tramontina, J., Andreazza, A. C., Cereser, K., da Costa, S., Santin, A., et al. (2007). Traumatic life events in bipolar disorder: impact on BDNF levels and psychopathology. *Bipolar Disorders, 9*(Suppl 1), 128–135. https://doi.org/10.1111/j.1399-5618.2007.00478.x.

Kemner, S. M., van Haren, N. E., Bootsman, F., Eijkemans, M. J., Vonk, R., van der Schot, A. C., et al. (2015). The influence of life events on first and recurrent admissions in bipolar disorder. *International Journal of Bipolar Disorder, 3,* 6. https://doi.org/10.1186/s40345-015-0022-4.

Kim, E. Y., Miklowitz, D. J., Biuckians, A., & Mullen, K. (2007). Life stress and the course of early-onset bipolar disorder. *Journal of Affective Disorders, 99*(1–3), 37–44. https://doi.org/10.1016/j.jad.2006.08.022.

Koenders, M. A., Giltay, E. J., Spijker, A. T., Hoencamp, E., Spinhoven, P., & Elzinga, B. M. (2014). Stressful life events in bipolar I and II disorder: cause or consequence of mood symptoms? *Journal of Affective Disorders, 161*, 55–64. https://doi.org/10.1016/j.jad.2014.02.036.

Levandowski, M. L., Viola, T. W., Prado, C. H., Wieck, A., Bauer, M. E., Brietzke, E., et al. (2016). Distinct behavioral and immunoendocrine parameters during crack cocaine abstinence in women reporting childhood abuse and neglect. *Drug and Alcohol Dependence, 167*, 140–148. https://doi.org/10.1016/j.drugalcdep.2016.08.010.

Leverich, G. S., McElroy, S. L., Suppes, T., Keck, P. E., Jr., Denicoff, K. D., Nolen, W. A., et al. (2002). Early physical and sexual abuse associated with an adverse course of bipolar illness. *Biological Psychiatry, 51*(4), 288–297.

Leverich, G. S., & Post, R. M. (2006). Course of bipolar illness after history of childhood trauma. *Lancet, 367*(9516), 1040–1042. https://doi.org/10.1016/S0140-6736(06)68450-X.

Lex, C., Bazner, E., & Meyer, T. D. (2017). Does stress play a significant role in bipolar disorder? A meta-analysis. *Journal of Affective Disorders, 208*, 298–308. https://doi.org/10.1016/j.jad.2016.08.057.

Lupien, S. J., McEwen, B. S., Gunnar, M. R., & Heim, C. (2009). Effects of stress throughout the lifespan on the brain, behaviour and cognition. *Nature Reviews. Neuroscience, 10*(6), 434–445. https://doi.org/10.1038/nrn2639.

Malhi, G. S., Bargh, D. M., Coulston, C. M., Das, P., & Berk, M. (2014). Predicting bipolar disorder on the basis of phenomenology: implications for prevention and early intervention. *Bipolar Disorders, 16*(5), 455–470. https://doi.org/10.1111/bdi.12133.

Marchand, W. R., Wirth, L., & Simon, C. (2005). Adverse life events and pediatric bipolar disorder in a community mental health setting. *Community Mental Health Journal, 41*(1), 67–75.

McEwen, B. S. (2007). Physiology and neurobiology of stress and adaptation: central role of the brain. *Physiological Reviews, 87*(3), 873–904. https://doi.org/10.1152/physrev.00041.2006.

McEwen, B. S., & Wingfield, J. C. (2010). What is in a name? Integrating homeostasis, allostasis and stress. *Hormones and Behavior, 57*(2), 105–111. https://doi.org/10.1016/j.yhbeh.2009.09.011.

McGuffin, P., Rijsdijk, F., Andrew, M., Sham, P., Katz, R., & Cardno, A. (2003). The heritability of bipolar affective disorder and the genetic relationship to unipolar depression. *Archives of General Psychiatry, 60*(5), 497–502. https://doi.org/10.1001/archpsyc.60.5.497.

Monroe, S. M., & Simons, A. D. (1991). Diathesis-stress theories in the context of life stress research: implications for the depressive disorders. *Psychological Bulletin, 110*(3), 406–425.

Nemeroff, C. B., Heim, C. M., Thase, M. E., Klein, D. N., Rush, A. J., Schatzberg, A. F., et al. (2003). Differential responses to psychotherapy versus pharmacotherapy in patients with chronic forms of major depression and childhood trauma. *Proceedings of the National Academy of Sciences of the United States of America, 100*(24), 14293–14296. https://doi.org/10.1073/pnas.2336126100.

Noto, M. N., Noto, C., Caribe, A. C., Miranda-Scippa, A., Nunes, S. O., Chaves, A. C., et al. (2015). Clinical characteristics and influence of childhood trauma on the prodrome of bipolar disorder. *Revista Brasileira de Psiquiatria, 37*(4), 280–288. https://doi.org/10.1590/1516-4446-2014-1641.

Perroud, N., Dayer, A., Piguet, C., Nallet, A., Favre, S., Malafosse, A., et al. (2014). Childhood maltreatment and methylation of the glucocorticoid receptor gene NR3C1 in bipolar disorder. *The British Journal of Psychiatry, 204*(1), 30–35. https://doi.org/10.1192/bjp.bp.112.120055.

Phillips, M. L., & Kupfer, D. J. (2013). Bipolar disorder diagnosis: challenges and future directions. *Lancet, 381*(9878), 1663–1671. https://doi.org/10.1016/S0140-6736(13)60989-7.

Poletti, S., Aggio, V., Brioschi, S., Dallaspezia, S., Colombo, C., & Benedetti, F. (2016). Multidimensional cognitive impairment in unipolar and bipolar depression and the moderator effect of adverse childhood experiences. *Psychiatry and Clinical Neurosciences*, https://doi.org/10.1111/pcn.12497.

Poletti, S., Colombo, C., & Benedetti, F. (2014). Adverse childhood experiences worsen cognitive distortion during adult bipolar depression. *Comprehensive Psychiatry, 55*(8), 1803–1808. https://doi.org/10.1016/j.comppsych.2014.07.013.

Post, R. M., Altshuler, L. L., Kupka, R., McElroy, S. L., Frye, M. A., Rowe, M., et al. (2016). Age of onset of bipolar disorder: combined effect of childhood adversity and familial loading of psychiatric disorders. *Journal of Psychiatric Research, 81*, 63–70. https://doi.org/10.1016/j.jpsychires.2016.06.008.

Post, R. M., Altshuler, L. L., Leverich, G. S., Frye, M. A., Suppes, T., McElroy, S. L., et al. (2013). Role of childhood adversity in the development of medical co-morbidities associated with bipolar disorder. *Journal of Affective Disorders, 147*(1–3), 288–294. https://doi.org/10.1016/j.jad.2012.11.020.

Post, R. M., Altshuler, L., Leverich, G., Nolen, W., Kupka, R., Grunze, H., et al. (2013). More stressors prior to and during the course of bipolar illness in patients from the United States compared with the Netherlands and Germany. *Psychiatry Research, 210*(3), 880–886. https://doi.org/10.1016/j.psychres.2013.08.007.

Post, R. M., Leverich, G. S., Altshuler, L. L., Frye, M. A., Suppes, T., Keck, P. E., et al. (2011). Differential clinical characteristics, medication usage, and treatment response of bipolar disorder in the US versus The Netherlands and Germany. *International Clinical Psychopharmacology, 26*(2), 96–106. https://doi.org/10.1097/YIC.0b013e3283409419.

Rios, A. C., Noto, M. N., Rizzo, L. B., Mansur, R., Martins, F. E., Jr., Grassi-Oliveira, R., et al. (2015). Early stages of bipolar disorder: characterization and strategies for early intervention. *Revista Brasileira de Psiquiatria, 37*(4), 343–349. https://doi.org/10.1590/1516-4446-2014-1620.

Romero, S., Birmaher, B., Axelson, D., Goldstein, T., Goldstein, B. I., Gill, M. K., et al. (2009). Prevalence and correlates of physical and sexual abuse in children and adolescents with bipolar disorder. *Journal of Affective Disorders, 112*(1–3), 144–150. https://doi.org/10.1016/j.jad.2008.04.005.

Russo, M., Mahon, K., Shanahan, M., Solon, C., Ramjas, E., Turpin, J., et al. (2015). The association between childhood trauma and facial emotion recognition in adults with bipolar disorder. *Psychiatry Research, 229*(3), 771–776. https://doi.org/10.1016/j.psychres.2015.08.004.

Sala, R., Goldstein, B. I., Wang, S., & Blanco, C. (2014). Childhood maltreatment and the course of bipolar disorders among adults: epidemiologic evidence of dose-response effects. *Journal of Affective Disorders, 165*, 74–80. https://doi.org/10.1016/j.jad.2014.04.035.

Saleh, A., Potter, G. G., McQuoid, D. R., Boyd, B., Turner, R., MacFall, J. R., et al. (2017). Effects of early life stress on depression, cognitive performance and brain morphology. *Psychological Medicine, 47*(1), 171–181. https://doi.org/10.1017/S0033291716002403.

Savitz, J. B., van der Merwe, L., Stein, D. J., Solms, M., & Ramesar, R. S. (2008). Neuropsychological task performance in bipolar spectrum illness: genetics, alcohol abuse, medication and childhood trauma. *Bipolar Disorders, 10*(4), 479–494. https://doi.org/10.1111/j.1399-5618.2008.00591.x.

Schoedl, A. F., Costa, M. C., Mari, J. J., Mello, M. F., Tyrka, A. R., Carpenter, L. L., et al. (2010). The clinical correlates of reported childhood sexual abuse: an association between age at trauma onset and severity of depression and PTSD in adults. *Journal of Child Sexual Abuse*, *19*(2), 156–170. https://doi.org/10.1080/10538711003615038.

Shevlin, M., Dorahy, M. J., & Adamson, G. (2007). Trauma and psychosis: an analysis of the National Comorbidity Survey. *The American Journal of Psychiatry*, *164*(1), 166–169. https://doi.org/10.1176/ajp.2007.164.1.166.

Skjelstad, D. V., Malt, U. F., & Holte, A. (2010). Symptoms and signs of the initial prodrome of bipolar disorder: a systematic review. *Journal of Affective Disorders*, *126*(1–2), 1–13. https://doi.org/10.1016/j.jad.2009.10.003.

Stevanin, S., Bressan, V., Vehvilainen-Julkunen, K., Pagani, L., Poletti, P., & Kvist, T. (2017). The multidimensional nursing generations questionnaire: development, reliability, and validity assessments. *Journal of Nursing Management*. https://doi.org/10.1111/jonm.12465.

Sugaya, L., Hasin, D. S., Olfson, M., Lin, K. H., Grant, B. F., & Blanco, C. (2012). Child physical abuse and adult mental health: a national study. *Journal of Traumatic Stress*, *25*(4), 384–392. https://doi.org/10.1002/jts.21719.

Teicher, M. H., & Samson, J. A. (2013). Childhood maltreatment and psychopathology: a case for ecophenotypic variants as clinically and neurobiologically distinct subtypes. *The American Journal of Psychiatry*, *170*(10), 1114–1133. https://doi.org/10.1176/appi.ajp.2013.12070957.

Teicher, M. H., Samson, J. A., Anderson, C. M., & Ohashi, K. (2016). The effects of childhood maltreatment on brain structure, function and connectivity. *Nature Reviews. Neuroscience*, *17*(10), 652–666. https://doi.org/10.1038/nrn.2016.111.

Widom, C. S., Horan, J., & Brzustowicz, L. (2015). Childhood maltreatment predicts allostatic load in adulthood. *Child Abuse & Neglect*, *47*, 59–69. https://doi.org/10.1016/j.chiabu.2015.01.016.

Gene-environment interactions in high-risk populations

4

Gabriel R. Fries, Consuelo Walss-Bass

Translational Psychiatry Program, Department of Psychiatry and Behavioral Sciences, University of Texas Health Sciences Center at Houston, Houston, TX, United States

CHAPTER OUTLINE

INTRODUCTION

Bipolar disorder (BD) is a very complex disorder from both the clinical and neurobiological perspectives. Studies aimed at identification of the biological basis of BD have suggested multiple pathways to be involved in its pathophysiology, but so far have failed to convey a unique and/or a specific mechanism that can solely explain the onset, presentation, and progression of the illness. The one consistent finding shown by multiple independent studies is that genetics plays a very important role in BD. Specifically, it is known that the "heritability" of BD, i.e., the degree to which BD can be attributed to inherited genetic factors (typically assessed in twin studies), is very high, estimated to be 70%–80% (Uher, 2014). Accordingly, first-degree relatives of patients with BD (who share genetic markers with the probands) are known to have a higher risk of developing BD and other psychiatric disorders than healthy controls (Lichtenstein et al., 2009; Rasic, Hajek, Alda, & Uher, 2014).

Bipolar Disorder Vulnerability. https://doi.org/10.1016/B978-0-12-812347-8.00004-X

Based on this, the advent of sophisticated techniques for the analysis of molecular genetic data has led to a series of studies aimed at identifying the genetic basis of BD. Although informative, studies employing genome-wide measurements have identified only a few common variants with small effects that, alone, do not account for the high heritability of BD. In addition to rare genetics variants (which are now being investigated and may account for some of BD's heritability), it has been proposed that the interaction between genes and environmental stimuli might help explain a significant part of this "heritability gap" (Halldorsdottir & Binder, 2017; Uher, 2014). A "gene×environment" interaction relates to the ability of one's genotype to modulate the way he/she responds to specific environmental stimuli (Dalle Molle et al., 2017). As described by Uher (2014), this interaction relates to the ability of the genotype to modulate one's *sensitivity* to an environmental factor (rather than the probability of being exposed to it, which would constitute a gene×environment *correlation*) (Uher, 2014). In fact, as we will discuss later in this chapter, several environmental stressors have been associated with an increased risk of BD, particularly early in life, but a great variability is seen in the outcome to these exposures.

Several different models have been proposed for the understanding of gene×environment interactions. The oldest and most common one is known as the "diathesis-stress model," in which a subject's genotype is believed to confer a higher susceptibility to a negative life event. According to this model, genetic variants are considered to be either "protective" or "risk-conferring"; individuals presenting the "risk" variant will develop the disorder when exposed to adversity, whereas those presenting the "protective" variant will not develop the disorder when exposed to the same negative environment (Halldorsdottir & Binder, 2017).

An alternative to this model is known as the "differential-susceptibility perspective," which was proposed to overcome the diathesis-stress model limitation of only focusing on negative environmental exposures. According to this model, there are no "good" or "bad" genetic variants; rather, they are seen as *plasticity* variants that influence one's vulnerability to the disorder depending on the environment (be it negative or positive) (Boyce, 2016). In other words, the same variant that confers a higher risk in a negative environment can also be protective in a positive setting. This model is more tangible from an evolutionary perspective, considering that natural selection would have already eliminated certain genetic variants if they were purely negative (Dalle Molle et al., 2017).

So far, the vast majority of studies investigating gene×environment interactions in BD have focused on a diathesis-stress model, which is understandable given the large body of evidence suggesting association of BD with mostly negative life experiences. Nevertheless, one should keep in mind that the mechanisms underlying the reported interactions are not yet understood, and we are still taking our first steps towards their clarification. In this chapter we will initially discuss findings of environmental and genetic factors that have been shown to independently contribute to the risk of BD, followed by a summary of gene×environment interaction findings reported for BD. Finally, potential mechanisms responsible for these interactions and the future outcomes and perspectives of the field will be discussed.

ENVIRONMENTAL RISK FACTORS FOR BIPOLAR DISORDER

Several environmental factors have been suggested to contribute to severe mental illnesses, including mood disorders. Their effects to increase the risk for these disorders are dependent on the period of time in which an individual is exposed (prenatal, perinatal, or postnatal period), the duration and frequency of the environmental exposure, and the degree of susceptibility presented by the individual (this degree, as discussed earlier, is likely mediated by genetic factors). Specific to BD, relatively limited evidence of direct influence of environmental exposures are available. A series of risk factors have been identified and some even show strong associations with disease onset, but in general they present very low specificity for BD and seem to increase the risk of psychopathology in several diagnostic categories (Marangoni, Hernandez, & Faedda, 2016).

In this section we will briefly discuss the most relevant environmental factors linked to BD risk in early life, i.e., in prenatal and perinatal periods, as well as during childhood and adolescence. It is known that environmental factors continue to play key roles in modulating BD during adulthood, but in this case they are associated primarily with modulation of the course of illness (symptomatology, severity, progression, recurrence, suicide attempts, and comorbidities, among others) (Aldinger & Schulze, 2017). In this chapter we will focus primarily on the factors that have been shown to modulate the risk of BD in high-risk youth.

PRENATAL ENVIRONMENTAL RISK FACTORS

Some of the prenatal environmental risk factors associated with BD include maternal infection, smoking, and stress exposure during pregnancy. All of these are thought to interfere directly or indirectly with fetal and postnatal neurodevelopment, ultimately increasing the offspring's vulnerability to the disorder later in life.

Specifically, gestational influenza infection has been shown to increase significantly the risk of BD (fourfold), being even higher (sixfold) for exposures occurring during the third trimester (Parboosing, Bao, Shen, Schaefer, & Brown, 2013). A later study also showed that prenatal flu exposure significantly increases the risk of BD with psychotic symptoms but not BD in general (Canetta et al., 2014; Marangoni et al., 2016). By activating inflammatory pathways, the infection is thought to induce the release of several pro-inflammatory molecules that can cross the placental barrier and cause damages to cells in the central nervous system of the developing fetus. Moreover, the in utero inflammation caused by the influenza infection can program individuals for a pro-inflammatory immune phenotype later in life (Simanek & Meier, 2015), which can ultimately increase an individual's susceptibility to BD and other mental disorders. Finally, gestational infection might also lead to the fetal programming of offspring stress reactivity by modulating the placental enzyme 11-β-hydroxysteroid dehydrogenase type 2, which is responsible for regulating fetal exposure to maternal glucocorticoids (Simanek & Meier, 2015). Of note, although of extremely high significance, these finding have not been consistently replicated across populations, and some studies suggest no association between prenatal influenza exposure and BD (Anderson et al., 2016; Machon, Mednick, & Huttunen, 1997).

Other infectious agents have been interrogated, as well, and do not seem to be significantly linked to BD risk, such as prenatal *Toxoplasma gondii* exposure (which has been consistently linked to an increased risk of schizophrenia) (Del Grande, Galli, Schiavi, Dell'Osso, & Bruschi, 2017), herpes simplex virus (HSV) type 1, HSV type 2, or cytomegalovirus (Mortensen et al., 2011). Importantly, even though the prenatal exposure to these agents has not been shown to increase the risk for BD in the offspring, some of them can have important effects on adult BD patients and modify the course and presentation of the illness (Barichello et al., 2016).

In addition to influenza infection, maternal smoking during pregnancy, which can interfere with fetal growth and development, has also been shown to cause a 1.41- to 2-fold increase of BD risk in the offspring (Chudal, Brown, Gissler, Suominen, & Sourander, 2015; Talati et al., 2013), even though some of it might be confounded by other familial background factors (Chudal et al., 2015). Gestational smoking can also lead to more severe forms of psychopathology in patients with BD, such as the presentation of psychotic symptoms (Mackay, Anderson, Pell, Zammit, & Smith, 2017). Finally, prenatal exposure to war during the first trimester was shown to increase the risk of BD in the offspring, suggesting that exposure to stressful/traumatic events might also play a role in increasing BD risk (Kleinhaus et al., 2013).

PERINATAL ENVIRONMENTAL RISK FACTORS

The perinatal period refers to the short period of time, usually weeks, immediately before and after birth. Environmental factors at this time have been significantly understudied in regards to BD risk (at least compared to pre- and postnatal environmental risk factors), and can include obstetrical complications, type of delivery, and preterm birth, among others.

Preterm birth has been shown to increase significantly the risk of BD in adulthood (Nosarti et al., 2012; Ogendahl et al., 2006). Accordingly, these findings are based on the hypothesis that preterm birth is associated with impaired neurodevelopment, including evidence that shows that young adults who were born very preterm show neuroanatomical alterations that resemble those of patients with psychiatric disorders (Nosarti et al., 2012).

Interestingly, the type of delivery can also have long-lasting effects that can ultimately influence one's vulnerability to psychiatric disorder. For instance, elective, but not emergency caesarean section has been associated with an increased BD risk (Chudal et al., 2014), as well as with the development of psychosis (O'Neill et al., 2016). Because these results might be confounded by other environmental and familial factors, a causal relationship between caesarean section and these conditions is still weakly supported (O'Neill et al., 2016).

Finally, perinatal complications have been reported to be greater in at-risk offspring than in children of healthy controls (as assessed by the Rochester Research

Obstetrical Scale, or ROS, which assesses information about risk factors and complications during labor, delivery, and conditions of the newborn immediately after birth) (Singh et al., 2007). Moreover, overall obstetric complication scores have been shown to be significantly higher in BD than their unaffected siblings (Kinney et al., 1993; Kinney, Yurgelun-Todd, Tohen, & Tramer, 1998), although some studies found no association between labor and delivery complications and BD (Browne et al., 2000; Buka, Tsuang, & Lipsitt, 1993).

CHILDHOOD/ADOLESCENCE ENVIRONMENTAL RISK FACTORS

Childhood is known as a period of enhanced vulnerability to environmental factors due to the ongoing maturation of the central nervous system (Etain, Henry, Bellivier, Mathieu, & Leboyer, 2008). Not surprisingly, a series of environmental factors during early life have been shown to increase the risk of BD in adults. These include early parental loss (Agid et al., 1999; Mortensen, Pedersen, Melbye, Mors, & Ewald, 2003), physical and emotional trauma, such as childhood abuse and neglect, substance use, dysfunctions in family environment and functioning, and even the degree of light exposure early in life (Bauer et al., 2014, 2015).

Substance use

There is evidence to suggest that, with many patients with BD, substance abuse precedes the onset of BD by means of inducing affective deregulation (Leite et al., 2015). Accordingly, individuals with adolescent and young-adult substance use disorder have been shown to present an increased risk of developing a secondary mood disorder (Kenneson, Funderburk, & Maisto, 2013; Wilens, Biederman, Abrantes, & Spencer, 1997). Cannabis use has been associated with a younger age of onset of BD (Gonzalez-Pinto et al., 2008; Lagerberg et al., 2011), even showing a dose-response relationship (Lagerberg et al., 2011). In this same vein, an increased risk for BD has also been associated with opioid dependence from nonmedical use (Kenneson et al., 2013; Martins, Keyes, Storr, Zhu, & Chilcoat, 2009). Altogether, these studies might suggest a "causative" role for substance use in the onset of BD, but an interesting hypothesis has also suggested that the substance use might be a result of self-medication of prodromal mood disorder symptoms in these susceptible individuals (Kenneson et al., 2013).

Childhood trauma and family adversity

The role of childhood maltreatment in risk for developing BD was discussed in detail in Chapter 3. Here we give a brief summary. Childhood trauma is often reported by patients with BD (Etain et al., 2008), with the most common type of traumatic experience being emotional abuse (Brietzke et al., 2012). In fact, childhood trauma and abuse have been associated with a younger age of onset (Garno, Goldberg, Ramirez, & Ritzler, 2005; Leverich et al., 2002), recurrent depressive symptoms in adulthood, premorbid functional levels, occurrence of suicidal behavior, substance abuse, and poorer adherence to treatment (Brietzke et al., 2012; Jaworska-Andryszewska & Rybakowski, 2016). Accordingly, it has been hypothesized that childhood trauma and abuse might occur before the development of BD and trigger the first episode in some cases (Brietzke et al., 2012). Importantly, an alternative hypothesis discusses the possibility that the genetic risk for BD can lead to a higher likelihood of experiencing trauma during childhood (Etain et al., 2008). Regardless of the answer to this "chicken and egg" question, the association between genetic/family risk × traumatic/abusive events is a clear representation of the role that gene × environment can play in the onset of BD.

The evidence related to childhood trauma/abuse strongly emphasizes the important role of the family and the environment it creates in the risk of BD in vulnerable youth. Early disturbances in the interactions between parents and their children might make the latter more susceptible to affective disturbances in adulthood (Etain et al., 2008). Offspring of patients with BD, for instance, might present an increased risk of psychiatric disorders not only based purely on their genetic risk but also by experiencing a more challenging family environment associated with the parental diagnosis. Accordingly, BD offspring are more likely to have experienced moderate to severe interpersonal stress compared to controls (Ostiguy et al., 2009), which is supported by reports of disruptive home environments driven by personality features, greater emotion dysregulation, and high levels of family tension with low emotional support in BD families (Benti, Manicavasagar, Proudfoot, & Parker, 2014). BD patients also report physical, verbal, or sexual abuse within their families (Benti et al., 2014).

With the use of the Family Environment Scale (FES), several studies have reported that families with a parent with BD may experience an environment with higher levels of conflict and lower levels of cohesion, expressiveness, organization, intellectual-cultural orientation, and active-recreational orientation (Barron et al., 2014; Belardinelli et al., 2008; Chang, Blasey, Ketter, & Steiner, 2001; Ferreira et al., 2013; Romero, Delbello, Soutullo, Stanford, & Strakowski, 2005). In addition, BD parents have been shown to endorse more negative communication styles and to be less expressive than control parents (Vance, Huntley Jones, Espie, Bentall, & Tai, 2008). As a consequence, a longer duration of exposure to parental BD has been associated with increased risks of psychopathology and substance use disorders, especially exposure during the first 2 years of life (Goodday et al., 2015).

Importantly, despite this growing body of evidence, it does not appear that the environment alone of families with a BD parent determines the psychiatric outcome in the offspring (Chang et al., 2001), supporting the hypothesis of gene × environment interactions. Nonetheless, these findings warrant the investigation of family-based interventions focusing on psychoeducation and improved communication between family members, with the ultimate goal of addressing issues of conflict, organization, and expressiveness (Barron et al., 2014).

EVIDENCE OF GENE-ENVIRONMENT INTERACTIONS IN BD

As suggested by the previous chapter, the genetic basis of BD has been shown to be extremely complex and hard to tackle. Several genome-wide association studies (GWAS) have been published with growing number of patients and controls

(Baum et al., 2008; Chen et al., 2013; Cichon et al., 2011; Consortium, 2007; Green et al., 2013; Muhleisen et al., 2014; PGC, 2011; Scott et al., 2009; Sklar et al., 2008; Smith et al., 2009), but most of the results are still poorly replicated in independent populations. Besides the still relatively small sample size for the GWAS analysis of BD—the last meta-analysis by the Psychiatric Genetic Consortium included "only" 11,974 cases and 51,792 controls (PGC, 2011)—this lack of replication might also be explained by the extreme clinical complexity of BD, differential (and still unexplored) patterns of BD inheritance, and the possibility of genetic heterogeneity to BD (Alsabban, Rivera, & McGuffin, 2011). Overall, GWAS studies suggest that BD is a polygenic disease associated with several common genetic alterations with very small individual effect sizes (Goes, 2016). Of note, even when considering the addition of rare genetic variants to this puzzle (Ament et al., 2015), such studies are still not able to explain completely the high heritability seen in BD, as discussed in the previous sections. In this context, gene×environment studies, in which environmental factors are included as a variable, represent a relevant extension of traditional GWAS.

Evidence of interactions between genes and environment in BD comes from several types of study, including twin studies. These studies, by comparing biochemical and neuroanatomical parameters between monozygotic and dizygotic twins, have found that gene×environment interactions can influence stress response (Vinberg et al., 2009), cortical gray matter (van der Schot et al., 2009), inflammation (Padmos et al., 2009), and subcortical volumes (Bootsman et al., 2015) in patients with BD. Moreover, interactions are also suggested in regards to suicidal behavior among patients (Brezo et al., 2010; Gonzalez-Castro et al., 2015). A summary of gene×environment interaction findings in BD is presented in Table 1. Other interesting gene×environment associations have also been suggested to take place between season of birth, childhood trauma, and socioeconomic status with specific genotypes in determining psychiatric symptoms and disorders (Bradley et al., 2008; Chotai, Serretti, Lattuada, Lorenzi, & Lilli, 2003; Debnath et al., 2013; Swartz, Hariri, & Williamson, 2017), as well.

MECHANISMS OF GENE-ENVIRONMENT INTERACTIONS

Gene-environment interactions, as summarized in the previous section, can influence an individual's risk of disease by different mechanisms (Fig. 1). Some types of environmental exposure, for example, can cause a direct effect on DNA and induce mutations (mutagens), such as radiation and cigarette smoking. Although this is possible, it is likely that these environmental effects are not the most relevant for the pathogenesis of psychiatric disorders like BD. Conversely, environment-induced gene-gene interactions, activation of specific transcription factors, or epigenetic mechanisms may be more relevant in the context of BD. The latter mechanism may be particularly relevant based on its dependency on both genotype and environment and the growing body of evidence suggesting epigenetic alterations in BD (Ludwig & Dwivedi, 2016).

Table 1 Gene-environment interaction studies in bipolar disorder

Gene/loci	Environmental factor	Sample	Outcome	Reference
5-HTT	Childhood sexual abuse and cannabis abuse	BD patients with or without lifetime psychotic symptoms (n = 137)	Short allele of the 5-HTTLPR polymorphism and cannabis use were more frequent among patients with psychotic symptoms. Complex interactions between presence of the short allele, cannabis use or dependence and childhood sexual abuse	De Pradier, Gorwood, Beaufils, Ades, and Dubertret (2010)
BDNF	Childhood trauma	Patients with a broad DSM-IV schizophrenia spectrum disorder or BD (n = 323)	Interaction between history of childhood trauma and BDNF met allele determines lower BDNF mRNA levels and reduced hippocampal subfield volumes CA2/3 and CA4 dentate gyrus	Aas et al. (2014)
BDNF	Family cohesion	Pediatric BD patients (n = 29) and controls (n = 22)	Interaction between low family cohesion and BDNF met allele determines left hippocampal volume in patients	Zeni et al. (2016)
BDNF	Stressful life events	BD patients (n = 487) and controls (n = 598)	Interaction between stressful life events and BDNF val6met polymorphism on determining worse depressive episodes	Hosang et al. (2010)
BDNF	Family dysfunction	BD offspring (n = 64) and controls (n = 51)	BD offspring with the Val/Val genotype showed higher levels of anxiety than offspring with other genotypes. Family dysfunction moderated the association between the BDNF genotype and anxiety symptoms	Park et al. (2015)
BDNF	History of childhood trauma	BD patients (n = 80)	Interaction between BDNF Met allele and childhood sexual abuse on determining BD illness severity and chronicity	Miller et al. (2013)
COMT	Stressful life events	BD patients (n = 482) and controls (n = 205)	The impact of stressful life events was moderated by the COMT genotype for the worst depressive episode. For the worst manic episodes no significant interaction was found	Hosang, Fisher, Cohen-Woods, McGuffin, and Farmer (2017)

Gene	Environment	Sample	Finding	Reference
COMT	Herpes simplex virus type 1 infection	BD patients ($n=107$) and controls ($n=95$)	Interaction between COMT Val158Met SNP and serological evidence of HSV-1 infection on determining worse cognitive functioning in patients	Dickerson et al. (2006)
CRHR1	Early childhood abuse	BD suicide attempters ($n=631$) and BD nonattempters ($n=657$)	Nominal interaction ($P=1.22 \times 10^{-2}$) between CRHR1 rs2664008 and early childhood abuse on determining suicide behavior in BD	Breen, Seifuddin, Zandi, Potash, and Willour (2015)
Genome-wide	Childhood trauma	Patients with BD ($n=1119$)	Interaction between SNPs in or near genes coding for calcium channel activity-related proteins and childhood trauma determines age of illness onset	Anand, Koller, Lawson, Gershon, and Nurnberger (2015)
SLC1A2	Lithium treatment	BD patients ($n=110$)	Interaction between lithium treatment and SLC1A2 -181A>C genotype to influence the history of the illness	Dallaspezia et al. (2012)
TLR2	Toxoplasma gondii infection	BD patients ($n=138$) and controls ($n=167$)	Trend for an interaction between the TLR2 rs3804099 SNP and T. gondii seropositivity in conferring BD risk	Oliveira et al. (2016)
TLR2	History of childhood trauma	BD patients ($n=531$)	Interaction between TLR2 rs3804099 TT genotype and reported sexual abuse on determining an earlier age at onset of BD	Oliveira et al. (2015)
TPH2 and ADARB1	History of child abuse	Patients with MDD, BD and schizophrenia (suicide attempters, $n=165$; and suicide nonattempters, $n=188$)	Interaction between TPH2 rs4290270 and general traumas on determining an increased risk for suicide attempt. Interaction of general traumas, TPH2 rs4290270 and ADARB1 rs4819035 and highest predisposition to suicide attempts	Karanovic et al. (2017)

5-HTT: serotonin transporter; 5-HTTLPR: serotonin transporter length polymorphic region; ADARB1: adenosine deaminase, RNA specific B1; BD: bipolar disorder; BDNF: brain-derived neurotrophic factor; COMT: catechol-O-methyltransferase; CRHR1: corticotropin-releasing hormone receptor 1; SLC1A2: solute carrier family 1 member 2; TLR2: toll-like receptor 2; TPH2: tryptophan hydroxylase 2.

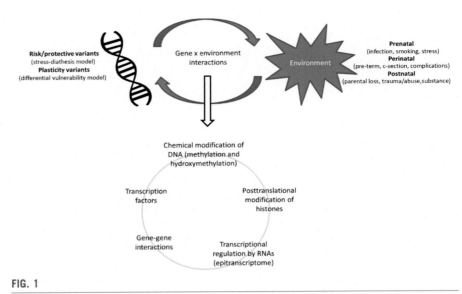

FIG. 1

Molecular mechanisms underlying gene×environment interactions.

The term "epigenetics" refers to the mechanisms by which the expression and function of the DNA is altered without any alteration in its primary sequence. In fact, these mechanisms can have long-lasting effects on gene expression that are similar to those induced by actual DNA mutations (therefore being known as "epimutations"). They include covalent chemical alterations to the DNA (methylation and hydroxylation, for instance) and histones (the proteins that help compact the DNA in the nucleus), as well as posttranscriptional alterations involving numerous noncoding RNAs.

Among known epigenetic alterations, DNA methylation—the covalent addition of a methyl (CH_3) group to the DNA (normally to cytosine residues) by DNA methyltransferases—is among the most studied epigenetic modifications in psychiatry due to its relative stability compared to others. This modification has been shown to induce both activation and repression of gene expression (depending on the location within the gene where it occurs), but it typically leads to repression if present at promoter regions. In fact, a plethora of studies in preclinical models and in humans have suggested that DNA methylation alterations can be induced early in life and sometimes maintained into adulthood (Jawahar, Murgatroyd, Harrison, & Baune, 2015; Mitchell, Schneper, & Notterman, 2016), meeting the criteria for a neurodevelopmental marker influenced by environmental exposures. In particular, a landmark study on the association between childhood trauma and genetic variants of the gene coding for the FK506-binding protein 51 (FKBP51, which is tightly linked to the regulation of the stress response) has shown that childhood trauma can lead to a genotype-specific DNA methylation change of this gene, ultimately determining changes in the response to stress in adults (Klengel et al., 2013). This study empirically demonstrated the role of DNA methylation in mediating the interaction between gene and environment, which laid the foundation for several other investigations. Not surprisingly, several studies have reported alterations in DNA methylation in patients with BD and in the mechanisms of action of psychotropic medications (Fries, Li, et al., 2016), potentially playing a role in the BD intergenerational transmission as well (Fries, Walss-Bass, Soares, & Quevedo, 2016).

Of note, several of the environmental factors associated with BD risk (discussed above) have been shown to influence epigenetic markers and downstream gene expression. Specifically, family functioning parameters among pediatric patients with BD or unaffected offspring of BD parents have recently been shown to correlate with gene expression markers of risk of BD in children (Fig. 2) (Fries et al., 2017).

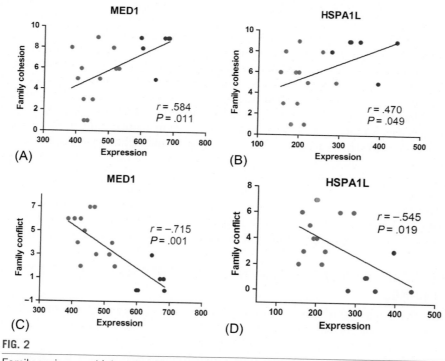

FIG. 2

Family environment interacts with expression of risk markers in blood from youth at high risk for bipolar disorder. (A and B) Family cohesion assessed by the Family Environment Scale (FES) positively correlates with the expression of *MED1* and *HSPA1L*. (C and D) Family conflict assessed by the FES negatively correlates with the expression of *MED1* and *HSPA1L*. *Red*: pediatric patients with bipolar disorder; *green*: unaffected offspring of parents with bipolar disorder (high-risk youth); *blue*: healthy controls. *MED1*: Mediator Complex Subunit 1; *HSPA1L*: Heat Shock Protein Family A (Hsp70) Member 1 Like. Modified from Fries, G.R., Quevedo, J., Zeni, C. P., Kazimi, I. F., Zunta-Soares, G., Spiker, D. E., et al. (2017). Integrated transcriptome and methylome analysis in youth at high risk for bipolar disorder: a preliminary analysis. Translational Psychiatry, 7(3), e1059.

PERSPECTIVES AND FUTURE DIRECTIONS
METHODOLOGICAL LIMITATIONS OF GENE × ENVIRONMENT INTERACTION STUDIES

Evidence gathered so far consistently suggests an important role of gene × environment interactions in the pathophysiology of BD and in its risk in vulnerable populations. However, due to several current methodological limitations, the exact mediators of these interactions in BD are far from being fully understood. For instance, as seen in Table 1, most gene × environment interaction studies so far have focused on candidate genes and single nucleotide polymorphisms (SNPs). Considering the polygenic nature of BD and the high complexity of its genetic basis, it is likely that significantly robust findings will only be discovered when using larger genome-wide approaches. In addition, the majority of the studies being published in the field were not originally designed to investigate gene × environment interactions directly; rather, most of them involve posthoc analyses of previously collected data. Although this is not a limitation per se, it has resulted in limited and poorly described environmental measures for the specific type of analysis required for interaction studies.

Moreover, most of the current databases containing GWAS data from patients lack detailed information on environmental measures, which also hinders the use of such data for this purpose. Particularly, this leads to another limitation of current studies: lack of statistical power. Gene × environment interaction analyses, by including one more dependent variable (i.e., environment), require even larger samples sizes than GWAS studies do. These will likely only be achievable in collaborative settings such as the Psychiatric Genomics Consortium, which combines data from several research groups that are willing to share their results with the community. Of note, the integration of data for such an analysis will also require standardized protocols for assessing the environmental exposure ("exposomes") (Blumberg, 2016).

CLINICAL TRANSLATION OF GENE × ENVIRONMENT INTERACTION FINDINGS

An obvious pay-off of gene × environment interaction studies is the potential ability to identify subjects that are more likely to develop the illness in the face of known environmental stimuli, providing an opportunity for the delivery of preventive therapies. Moreover, a better understanding of gene × environment interactions is expected to translate eventually into more targeted treatments (considering the possibility of using an individual's genotype and environmental measures to determine which medications will be more efficacious in the future).

These studies also allow the identification of "toxic" environments and which subjects would be at higher risk when exposed to them. That opens a plethora of opportunities for treatments focused on modifying such environments to prevent the dangerous gene × environment interaction that may occur. This also applies to the targeting of family environment, especially in families with a parent diagnosed with BD. For instance, targeting diseased family environments can counteract the expression of risk genes that would confer a risk of BD (Fries et al., 2017). In this same vein, according to the differential susceptibility hypothesis, we can propose that individuals with genetic variants related to BD could have a chance of not developing the disease if they could engage in an enriched/healthier environment.

CONCLUSION

In summary, the risk and onset of BD is likely determined by a combination of genetic make-up and environmental exposure, which we traditionally call gene × environment interaction. This interaction is thought to underlie at least part of the heritability gap seen between the heritability rates shown in twin studies and the effect of genetic variants identified by GWAS. The systematic longitudinal analysis of the genome, epigenome, and "exposome" of a large number of individuals will eventually lead to the identification of reliable genetic markers of risk of BD, preventive treatments, and modification of "toxic" environments with the aim of preventing the onset of illness and of identifying groups of patients that can benefit from a more targeted and successful treatment.

REFERENCES

Aas, M., Haukvik, U. K., Djurovic, S., Tesli, M., Athanasiu, L., Bjella, T., et al. (2014). Interplay between childhood trauma and BDNF val66met variants on blood BDNF mRNA levels and on hippocampus subfields volumes in schizophrenia spectrum and bipolar disorders. *Journal of Psychiatric Research*, *59*, 14–21. https://doi.org/10.1016/j.jpsychires.2014.08.011.

Agid, O., Shapira, B., Zislin, J., Ritsner, M., Hanin, B., Murad, H., et al. (1999). Environment and vulnerability to major psychiatric illness: a case control study of early parental loss in major depression, bipolar disorder and schizophrenia. *Molecular Psychiatry*, *4*(2), 163–172.

Aldinger, F., & Schulze, T. G. (2017). Environmental factors, life events, and trauma in the course of bipolar disorder. *Psychiatry and Clinical Neurosciences*, *71*(1), 6–17. https://doi.org/10.1111/pcn.12433.

Alsabban, S., Rivera, M., & McGuffin, P. (2011). Genome-wide searches for bipolar disorder genes. *Current Psychiatry Reports*, *13*(6), 522–527. https://doi.org/10.1007/s11920-011-0226-y.

Ament, S. A., Szelinger, S., Glusman, G., Ashworth, J., Hou, L., Akula, N., et al. (2015). Rare variants in neuronal excitability genes influence risk for bipolar disorder. *Proceedings of the National Academy of Sciences of the United States of America*, *112*(11), 3576–3581. https://doi.org/10.1073/pnas.1424958112.

Anand, A., Koller, D. L., Lawson, W. B., Gershon, E. S., & Nurnberger, J. I. (2015). Genetic and childhood trauma interaction effect on age of onset in bipolar disorder: an exploratory analysis. *Journal of Affective Disorders*, *179*, 1–5. https://doi.org/10.1016/j.jad.2015.02.029.

Anderson, J. J., Hoath, S., Zammit, S., Meyer, T. D., Pell, J. P., Mackay, D., et al. (2016). Gestational influenza and risk of hypomania in young adulthood: prospective birth cohort study. *Journal of Affective Disorders*, *200*, 182–188. https://doi.org/10.1016/j.jad.2016.04.048.

Barichello, T., Badawy, M., Pitcher, M. R., Saigal, P., Generoso, J. S., Goularte, J. A., et al. (2016). Exposure to perinatal infections and bipolar disorder: a systematic review. *Current Molecular Medicine*, *16*(2), 106–118.

Barron, E., Sharma, A., Le Couteur, J., Rushton, S., Close, A., Kelly, T., et al. (2014). Family environment of bipolar families: a UK study. *Journal of Affective Disorders*, *152–154*, 522–525. https://doi.org/10.1016/j.jad.2013.08.016.

Bauer, M., Glenn, T., Alda, M., Andreassen, O. A., Angelopoulos, E., Ardau, R., et al. (2014). Relationship between sunlight and the age of onset of bipolar disorder: an international multisite study. *Journal of Affective Disorders*, *167*, 104–111. https://doi.org/10.1016/j.jad.2014.05.032.

Bauer, M., Glenn, T., Alda, M., Andreassen, O. A., Angelopoulos, E., Ardau, R., et al. (2015). Influence of light exposure during early life on the age of onset of bipolar disorder. *Journal of Psychiatric Research*, *64*, 1–8. https://doi.org/10.1016/j.jpsychires.2015.03.013.

Baum, A. E., Akula, N., Cabanero, M., Cardona, I., Corona, W., Klemens, B., et al. (2008). A genome-wide association study implicates diacylglycerol kinase eta (DGKH) and several other genes in the etiology of bipolar disorder. *Molecular Psychiatry*, *13*(2), 197–207. https://doi.org/10.1038/sj.mp.4002012.

Belardinelli, C., Hatch, J. P., Olvera, R. L., Fonseca, M., Caetano, S. C., Nicoletti, M., et al. (2008). Family environment patterns in families with bipolar children. *Journal of Affective Disorders*, *107*(1–3), 299–305. https://doi.org/10.1016/j.jad.2007.08.011.

Benti, L., Manicavasagar, V., Proudfoot, J., & Parker, G. (2014). Identifying early indicators in bipolar disorder: a qualitative study. *The Psychiatric Quarterly*, *85*(2), 143–153. https://doi.org/10.1007/s11126-013-9279-x.

Blumberg, R. S. (2016). Environment and genes: what is the interaction? *Digestive Diseases*, *34*(1–2), 20–26. https://doi.org/10.1159/000442920.

Bootsman, F., Brouwer, R. M., Kemner, S. M., Schnack, H. G., van der Schot, A. C., Vonk, R., et al. (2015). Contribution of genes and unique environment to cross-sectional and longitudinal measures of subcortical volumes in bipolar disorder. *European Neuropsychopharmacology*, *25*(12), 2197–2209. https://doi.org/10.1016/j.euroneuro.2015.09.023.

Boyce, W. T. (2016). Differential susceptibility of the developing brain to contextual adversity and stress. *Neuropsychopharmacology*, *41*(1), 142–162. https://doi.org/10.1038/npp.2015.294.

Bradley, R. G., Binder, E. B., Epstein, M. P., Tang, Y., Nair, H. P., Liu, W., et al. (2008). Influence of child abuse on adult depression: moderation by the corticotropin-releasing hormone receptor gene. *Archives of General Psychiatry*, *65*(2), 190–200. https://doi.org/10.1001/archgenpsychiatry.2007.26.

Breen, M. E., Seifuddin, F., Zandi, P. P., Potash, J. B., & Willour, V. L. (2015). Investigating the role of early childhood abuse and HPA axis genes in suicide attempters with bipolar disorder. *Psychiatric Genetics*, *25*(3), 106–111. https://doi.org/10.1097/ypg.0000000000000082.

Brezo, J., Bureau, A., Merette, C., Jomphe, V., Barker, E. D., Vitaro, F., et al. (2010). Differences and similarities in the serotonergic diathesis for suicide attempts and mood disorders: a 22-year longitudinal gene-environment study. *Molecular Psychiatry*, *15*(8), 831–843. https://doi.org/10.1038/mp.2009.19.

Brietzke, E., Kauer Sant'anna, M., Jackowski, A., Grassi-Oliveira, R., Bucker, J., Zugman, A., et al. (2012). Impact of childhood stress on psychopathology. *Revista Brasileira de Psiquiatria*, *34*(4), 480–488.

Browne, R., Byrne, M., Mulryan, N., Scully, A., Morris, M., Kinsella, A., et al. (2000). Labour and delivery complications at birth and later mania. An Irish case register study. *The British Journal of Psychiatry*, *176*, 369–372.

Buka, S. L., Tsuang, M. T., & Lipsitt, L. P. (1993). Pregnancy/delivery complications and psychiatric diagnosis. A prospective study. *Archives of General Psychiatry, 50*(2), 151–156.

Canetta, S. E., Bao, Y., Co, M. D., Ennis, F. A., Cruz, J., Terajima, M., et al. (2014). Serological documentation of maternal influenza exposure and bipolar disorder in adult offspring. *The American Journal of Psychiatry, 171*(5), 557–563. https://doi.org/10.1176/appi. ajp.2013.13070943.

Chang, K. D., Blasey, C., Ketter, T. A., & Steiner, H. (2001). Family environment of children and adolescents with bipolar parents. *Bipolar Disorders, 3*(2), 73–78.

Chen, D. T., Jiang, X., Akula, N., Shugart, Y. Y., Wendland, J. R., Steele, C. J., et al. (2013). Genome-wide association study meta-analysis of European and Asian-ancestry samples identifies three novel loci associated with bipolar disorder. *Molecular Psychiatry, 18*(2), 195–205. https://doi.org/10.1038/mp.2011.157.

Chotai, J., Serretti, A., Lattuada, E., Lorenzi, C., & Lilli, R. (2003). Gene-environment interaction in psychiatric disorders as indicated by season of birth variations in tryptophan hydroxylase (TPH), serotonin transporter (5-HTTLPR) and dopamine receptor (DRD4) gene polymorphisms. *Psychiatry Research, 119*(1–2), 99–111.

Chudal, R., Brown, A. S., Gissler, M., Suominen, A., & Sourander, A. (2015). Is maternal smoking during pregnancy associated with bipolar disorder in offspring? *Journal of Affective Disorders, 171*, 132–136. https://doi.org/10.1016/j.jad.2014.09.030.

Chudal, R., Sourander, A., Polo-Kantola, P., Hinkka-Yli-Salomaki, S., Lehti, V., Sucksdorff, D., et al. (2014). Perinatal factors and the risk of bipolar disorder in Finland. *Journal of Affective Disorders, 155*, 75–80. https://doi.org/10.1016/j.jad.2013.10.026.

Cichon, S., Muhleisen, T. W., Degenhardt, F. A., Mattheisen, M., Miro, X., Strohmaier, J., et al. (2011). Genome-wide association study identifies genetic variation in neurocan as a susceptibility factor for bipolar disorder. *American Journal of Human Genetics, 88*(3), 372–381. https://doi.org/10.1016/j.ajhg.2011.01.017.

Dallaspezia, S., Poletti, S., Lorenzi, C., Pirovano, A., Colombo, C., & Benedetti, F. (2012). Influence of an interaction between lithium salts and a functional polymorphism in SLC1A2 on the history of illness in bipolar disorder. *Molecular Diagnosis & Therapy, 16*(5), 303–309. https://doi.org/10.1007/s40291-012-0004-5.

Dalle Molle, R., Fatemi, H., Dagher, A., Levitan, R. D., Silveira, P. P., & Dube, L. (2017). Gene and environment interaction: is the differential susceptibility hypothesis relevant for obesity? *Neuroscience and Biobehavioral Reviews, 73*, 326–339. https://doi.org/10.1016/j. neubiorev.2016.12.028.

De Pradier, M., Gorwood, P., Beaufils, B., Ades, J., & Dubertret, C. (2010). Influence of the serotonin transporter gene polymorphism, cannabis and childhood sexual abuse on phenotype of bipolar disorder: a preliminary study. *European Psychiatry, 25*(6), 323–327. https:// doi.org/10.1016/j.eurpsy.2009.10.002.

Debnath, M., Busson, M., Jamain, S., Etain, B., Hamdani, N., Oliveira, J., et al. (2013). The HLA-G low expressor genotype is associated with protection against bipolar disorder. *Human Immunology, 74*(5), 593–597. https://doi.org/10.1016/j.humimm.2012.11.032.

Del Grande, C., Galli, L., Schiavi, E., Dell'Osso, L., & Bruschi, F. (2017). Is toxoplasma gondii a trigger of bipolar disorder? *Pathogens, 6*(1), https://doi.org/10.3390/pathogens6010003.

Dickerson, F. B., Boronow, J. J., Stallings, C., Origoni, A. E., Cole, S., Leister, F., et al. (2006). The catechol O-methyltransferase Val158Met polymorphism and herpes simplex virus type 1 infection are risk factors for cognitive impairment in bipolar disorder: additive gene-environmental effects in a complex human psychiatric disorder. *Bipolar Disorders, 8*(2), 124–132. https://doi.org/10.1111/j.1399-5618.2006.00288.x.

Etain, B., Henry, C., Bellivier, F., Mathieu, F., & Leboyer, M. (2008). Beyond genetics: childhood affective trauma in bipolar disorder. *Bipolar Disorders*, *10*(8), 867–876. https://doi.org/10.1111/j.1399-5618.2008.00635.x.

Ferreira, G. S., Moreira, C. R., Kleinman, A., Nader, E. C., Gomes, B. C., Teixeira, A. M., et al. (2013). Dysfunctional family environment in affected versus unaffected offspring of parents with bipolar disorder. *The Australian and New Zealand Journal of Psychiatry*, *47*(11), 1051–1057. https://doi.org/10.1177/0004867413506754.

Fries, G. R., Li, Q., McAlpin, B., Rein, T., Walss-Bass, C., Soares, J. C., et al. (2016). The role of DNA methylation in the pathophysiology and treatment of bipolar disorder. *Neuroscience and Biobehavioral Reviews*, *68*, 474–488. https://doi.org/10.1016/j.neubiorev.2016.06.010.

Fries, G. R., Quevedo, J., Zeni, C. P., Kazimi, I. F., Zunta-Soares, G., Spiker, D. E., et al. (2017). Integrated transcriptome and methylome analysis in youth at high risk for bipolar disorder: a preliminary analysis. *Translational Psychiatry*, *7*(3), e1059, https://doi.org/10.1038/tp.2017.32.

Fries, G. R., Walss-Bass, C., Soares, J. C., & Quevedo, J. (2016). Non-genetic transgenerational transmission of bipolar disorder: targeting DNA methyltransferases. *Molecular Psychiatry*, *21*(12), 1653–1654. https://doi.org/10.1038/mp.2016.172.

Garno, J. L., Goldberg, J. F., Ramirez, P. M., & Ritzler, B. A. (2005). Impact of childhood abuse on the clinical course of bipolar disorder. *The British Journal of Psychiatry*, *186*, 121–125. https://doi.org/10.1192/bjp.186.2.121.

Goes, F. S. (2016). Genetics of bipolar disorder: recent update and future directions. *The Psychiatric Clinics of North America*, *39*(1), 139–155. https://doi.org/10.1016/j.psc.2015.10.004.

Gonzalez-Castro, T. B., Nicolini, H., Lanzagorta, N., Lopez-Narvaez, L., Genis, A., Pool Garcia, S., et al. (2015). The role of brain-derived neurotrophic factor (BDNF) Val66Met genetic polymorphism in bipolar disorder: a case-control study, comorbidities, and meta-analysis of 16,786 subjects. *Bipolar Disorders*, *17*(1), 27–38. https://doi.org/10.1111/bdi.12227.

Gonzalez-Pinto, A., Vega, P., Ibanez, B., Mosquera, F., Barbeito, S., Gutierrez, M., et al. (2008). Impact of cannabis and other drugs on age at onset of psychosis. *The Journal of Clinical Psychiatry*, *69*(8), 1210–1216.

Goodday, S., Levy, A., Flowerdew, G., Horrocks, J., Grof, P., Ellenbogen, M., et al. (2015). Early exposure to parental bipolar disorder and risk of mood disorder: the flourish Canadian prospective offspring cohort study. *Early Intervention in Psychiatry*, https://doi.org/10.1111/eip.12291.

Green, E. K., Hamshere, M., Forty, L., Gordon-Smith, K., Fraser, C., Russell, E., et al. (2013). Replication of bipolar disorder susceptibility alleles and identification of two novel genome-wide significant associations in a new bipolar disorder case-control sample. *Molecular Psychiatry*, *18*(12), 1302–1307. https://doi.org/10.1038/mp.2012.142.

Halldorsdottir, T., & Binder, E. B. (2017). Gene x environment interactions: from molecular mechanisms to behavior. *Annual Review of Psychology*, *68*, 215–241. https://doi.org/10.1146/annurev-psych-010416-044053.

Hosang, G. M., Fisher, H. L., Cohen-Woods, S., McGuffin, P., & Farmer, A. E. (2017). Stressful life events and catechol-O-methyl-transferase (COMT) gene in bipolar disorder. *Depression and Anxiety*, *34*(5), 419–426. https://doi.org/10.1002/da.22606.

Hosang, G. M., Uher, R., Keers, R., Cohen-Woods, S., Craig, I., Korszun, A., et al. (2010). Stressful life events and the brain-derived neurotrophic factor gene in bipolar disorder. *Journal of Affective Disorders*, *125*(1–3), 345–349. https://doi.org/10.1016/j.jad.2010.01.071.

Jawahar, M. C., Murgatroyd, C., Harrison, E. L., & Baune, B. T. (2015). Epigenetic alterations following early postnatal stress: a review on novel aetiological mechanisms of common psychiatric disorders. *Clinical Epigenetics*, *7*, 122. https://doi.org/10.1186/s13148-015-0156-3.

Jaworska-Andryszewska, P., & Rybakowski, J. (2016). Negative experiences in childhood and the development and course of bipolar disorder. *Psychiatria Polska*, *50*(5), 989–1000. https://doi.org/10.12740/pp/61159.

Karanovic, J., Ivkovic, M., Jovanovic, V. M., Svikovic, S., Pantovic-Stefanovic, M., Brkusanin, M., et al. (2017). Effect of childhood general traumas on suicide attempt depends on TPH2 and ADARB1 variants in psychiatric patients. *Journal of Neural Transmission*, *124*(5), 621–629. https://doi.org/10.1007/s00702-017-1677-z.

Kenneson, A., Funderburk, J. S., & Maisto, S. A. (2013). Substance use disorders increase the odds of subsequent mood disorders. *Drug and Alcohol Dependence*, *133*(2), 338–343. https://doi.org/10.1016/j.drugalcdep.2013.06.011.

Kinney, D. K., Yurgelun-Todd, D. A., Levy, D. L., Medoff, D., Lajonchere, C. M., & Radford-Paregol, M. (1993). Obstetrical complications in patients with bipolar disorder and their siblings. *Psychiatry Research*, *48*(1), 47–56.

Kinney, D. K., Yurgelun-Todd, D. A., Tohen, M., & Tramer, S. (1998). Pre- and perinatal complications and risk for bipolar disorder: a retrospective study. *Journal of Affective Disorders*, *50*(2–3), 117–124.

Kleinhaus, K., Harlap, S., Perrin, M., Manor, O., Margalit-Calderon, R., Opler, M., et al. (2013). Prenatal stress and affective disorders in a population birth cohort. *Bipolar Disorders*, *15*(1), 92–99. https://doi.org/10.1111/bdi.12015.

Klengel, T., Mehta, D., Anacker, C., Rex-Haffner, M., Pruessner, J. C., Pariante, C. M., et al. (2013). Allele-specific FKBP5 DNA demethylation mediates gene-childhood trauma interactions. *Nature Neuroscience*, *16*(1), 33–41. https://doi.org/10.1038/nn.3275.

Lagerberg, T. V., Sundet, K., Aminoff, S. R., Berg, A. O., Ringen, P. A., Andreassen, O. A., et al. (2011). Excessive cannabis use is associated with earlier age at onset in bipolar disorder. *European Archives of Psychiatry and Clinical Neuroscience*, *261*(6), 397–405. https://doi.org/10.1007/s00406-011-0188-4.

Leite, R. T., Nogueira Sde, O., do Nascimento, J. P., de Lima, L. S., da Nobrega, T. B., Virginio Mda, S., et al. (2015). The use of cannabis as a predictor of early onset of bipolar disorder and suicide attempts. *Neural Plasticity*, *2015*, 434127. https://doi.org/10.1155/2015/434127.

Leverich, G. S., McElroy, S. L., Suppes, T., Keck, P. E., Jr., Denicoff, K. D., Nolen, W. A., et al. (2002). Early physical and sexual abuse associated with an adverse course of bipolar illness. *Biological Psychiatry*, *51*(4), 288–297.

Lichtenstein, P., Yip, B. H., Bjork, C., Pawitan, Y., Cannon, T. D., Sullivan, P. F., et al. (2009). Common genetic determinants of schizophrenia and bipolar disorder in Swedish families: a population-based study. *Lancet*, *373*(9659), 234–239. https://doi.org/10.1016/s0140-6736(09)60072-6.

Ludwig, B., & Dwivedi, Y. (2016). Dissecting bipolar disorder complexity through epigenomic approach. *Molecular Psychiatry*, *21*(11), 1490–1498. https://doi.org/10.1038/mp.2016.123.

Machon, R. A., Mednick, S. A., & Huttunen, M. O. (1997). Adult major affective disorder after prenatal exposure to an influenza epidemic. *Archives of General Psychiatry*, *54*(4), 322–328.

Mackay, D. F., Anderson, J. J., Pell, J. P., Zammit, S., & Smith, D. J. (2017). Exposure to tobacco smoke in utero or during early childhood and risk of hypomania: prospective birth cohort study. *European Psychiatry*, *39*, 33–39. https://doi.org/10.1016/j.eurpsy.2016.06.001.

Marangoni, C., Hernandez, M., & Faedda, G. L. (2016). The role of environmental exposures as risk factors for bipolar disorder: a systematic review of longitudinal studies. *Journal of Affective Disorders*, *193*, 165–174. https://doi.org/10.1016/j.jad.2015.12.055.

Martins, S. S., Keyes, K. M., Storr, C. L., Zhu, H., & Chilcoat, H. D. (2009). Pathways between nonmedical opioid use/dependence and psychiatric disorders: results from the National Epidemiologic Survey on Alcohol and Related Conditions. *Drug and Alcohol Dependence*, *103*(1–2), 16–24. https://doi.org/10.1016/j.drugalcdep.2009.01.019.

Miller, S., Hallmayer, J., Wang, P. W., Hill, S. J., Johnson, S. L., & Ketter, T. A. (2013). Brain-derived neurotrophic factor val66met genotype and early life stress effects upon bipolar course. *Journal of Psychiatric Research*, *47*(2), 252–258. https://doi.org/10.1016/j.jpsychires.2012.10.015.

Mitchell, C., Schneper, L. M., & Notterman, D. A. (2016). DNA methylation, early life environment, and health outcomes. *Pediatric Research*, *79*(1–2), 212–219. https://doi.org/10.1038/pr.2015.193.

Mortensen, P. B., Pedersen, C. B., McGrath, J. J., Hougaard, D. M., Norgaard-Petersen, B., Mors, O., et al. (2011). Neonatal antibodies to infectious agents and risk of bipolar disorder: a population-based case-control study. *Bipolar Disorders*, *13*(7–8), 624–629. https://doi.org/10.1111/j.1399-5618.2011.00962.x.

Mortensen, P. B., Pedersen, C. B., Melbye, M., Mors, O., & Ewald, H. (2003). Individual and familial risk factors for bipolar affective disorders in Denmark. *Archives of General Psychiatry*, *60*(12), 1209–1215. https://doi.org/10.1001/archpsyc.60.12.1209.

Muhleisen, T. W., Leber, M., Schulze, T. G., Strohmaier, J., Degenhardt, F., Treutlein, J., et al. (2014). Genome-wide association study reveals two new risk loci for bipolar disorder. *Nature Communications*, *5*, 3339. https://doi.org/10.1038/ncomms4339.

Nosarti, C., Reichenberg, A., Murray, R. M., Cnattingius, S., Lambe, M. P., Yin, L., et al. (2012). Preterm birth and psychiatric disorders in young adult life. *Archives of General Psychiatry*, *69*(6), E1–8. https://doi.org/10.1001/archgenpsychiatry.2011.1374.

O'Neill, S. M., Curran, E. A., Dalman, C., Kenny, L. C., Kearney, P. M., Clarke, G., et al. (2016). Birth by caesarean section and the risk of adult psychosis: a population-based cohort study. *Schizophrenia Bulletin*, *42*(3), 633–641. https://doi.org/10.1093/schbul/sbv152.

Ogendahl, B. K., Agerbo, E., Byrne, M., Licht, R. W., Eaton, W. W., & Mortensen, P. B. (2006). Indicators of fetal growth and bipolar disorder: a Danish national register-based study. *Psychological Medicine*, *36*(9), 1219–1224. https://doi.org/10.1017/s0033291706008269.

Oliveira, J., Etain, B., Lajnef, M., Hamdani, N., Bennabi, M., Bengoufa, D., et al. (2015). Combined effect of TLR2 gene polymorphism and early life stress on the age at onset of bipolar disorders. *PLoS One*, *10*(3), e0119702https://doi.org/10.1371/journal.pone.0119702.

Oliveira, J., Kazma, R., Le Floch, E., Bennabi, M., Hamdani, N., Bengoufa, D., et al. (2016). Toxoplasma gondii exposure may modulate the influence of TLR2 genetic variation on bipolar disorder: a gene-environment interaction study. *International Journal of Bipolar Disorders*, *4*(1), 11. https://doi.org/10.1186/s40345-016-0052-6.

Ostiguy, C. S., Ellenbogen, M. A., Linnen, A. M., Walker, E. F., Hammen, C., & Hodgins, S. (2009). Chronic stress and stressful life events in the offspring of parents with bipolar disorder. *Journal of Affective Disorders*, *114*(1–3), 74–84. https://doi.org/10.1016/j.jad.2008.08.006.

Padmos, R. C., Van Baal, G. C., Vonk, R., Wijkhuijs, A. J., Kahn, R. S., Nolen, W. A., et al. (2009). Genetic and environmental influences on pro-inflammatory monocytes in bipolar disorder: a twin study. *Archives of General Psychiatry*, *66*(9), 957–965. https://doi.org/10.1001/archgenpsychiatry.2009.116.

Parboosing, R., Bao, Y., Shen, L., Schaefer, C. A., & Brown, A. S. (2013). Gestational influenza and bipolar disorder in adult offspring. *JAMA Psychiatry*, *70*(7), 677–685. https://doi.org/10.1001/jamapsychiatry.2013.896.

Park, M. H., Chang, K. D., Hallmayer, J., Howe, M. E., Kim, E., Hong, S. C., et al. (2015). Preliminary study of anxiety symptoms, family dysfunction, and the brain-derived neurotrophic factor (BDNF) Val66Met genotype in offspring of parents with bipolar disorder. *Journal of Psychiatric Research*, *61*, 81–88. https://doi.org/10.1016/j.jpsychires.2014.11.013.

Psychiatric GWAS Consortium Bipolar Disorder Working Group. (2011). Large-scale genome-wide association analysis of bipolar disorder identifies a new susceptibility locus near ODZ4. *Nature Genetics*, *43*(10), 977–983. https://doi.org/10.1038/ng.943.

Rasic, D., Hajek, T., Alda, M., & Uher, R. (2014). Risk of mental illness in offspring of parents with schizophrenia, bipolar disorder, and major depressive disorder: a meta-analysis of family high-risk studies. *Schizophrenia Bulletin*, *40*(1), 28–38. https://doi.org/10.1093/schbul/sbt114.

Romero, S., Delbello, M. P., Soutullo, C. A., Stanford, K., & Strakowski, S. M. (2005). Family environment in families with versus families without parental bipolar disorder: a preliminary comparison study. *Bipolar Disorders*, *7*(6), 617–622. https://doi.org/10.1111/j.1399-5618.2005.00270.x.

Scott, L. J., Muglia, P., Kong, X. Q., Guan, W., Flickinger, M., Upmanyu, R., et al. (2009). Genome-wide association and meta-analysis of bipolar disorder in individuals of European ancestry. *Proceedings of the National Academy of Sciences of the United States of America*, *106*(18), 7501–7506. https://doi.org/10.1073/pnas.0813386106.

Simanek, A. M., & Meier, H. C. (2015). Association between prenatal exposure to maternal infection and offspring mood disorders: a review of the literature. *Current Problems in Pediatric and Adolescent Health Care*, *45*(11), 325–364. https://doi.org/10.1016/j.cppeds.2015.06.008.

Singh, M. K., DelBello, M. P., Soutullo, C., Stanford, K. E., McDonough-Ryan, P., & Strakowski, S. M. (2007). Obstetrical complications in children at high risk for bipolar disorder. *Journal of Psychiatric Research*, *41*(8), 680–685. https://doi.org/10.1016/j.jpsychires.2006.02.009.

Sklar, P., Smoller, J. W., Fan, J., Ferreira, M. A., Perlis, R. H., Chambert, K., et al. (2008). Whole-genome association study of bipolar disorder. *Molecular Psychiatry*, *13*(6), 558–569. https://doi.org/10.1038/sj.mp.4002151.

Smith, E. N., Bloss, C. S., Badner, J. A., Barrett, T., Belmonte, P. L., Berrettini, W., et al. (2009). Genome-wide association study of bipolar disorder in European American and African American individuals. *Molecular Psychiatry*, *14*(8), 755–763. https://doi.org/10.1038/mp.2009.43.

Swartz, J. R., Hariri, A. R., & Williamson, D. E. (2017). An epigenetic mechanism links socioeconomic status to changes in depression-related brain function in high-risk adolescents. *Molecular Psychiatry*, *22*(2), 209–214. https://doi.org/10.1038/mp.2016.82.

Talati, A., Bao, Y., Kaufman, J., Shen, L., Schaefer, C. A., & Brown, A. S. (2013). Maternal smoking during pregnancy and bipolar disorder in offspring. *The American Journal of Psychiatry*, *170*(10), 1178–1185. https://doi.org/10.1176/appi.ajp.2013.12121500.

Uher, R. (2014). Gene-environment interactions in severe mental illness. *Frontiers in Psychiatry*, *5*, 48. https://doi.org/10.3389/fpsyt.2014.00048.

van der Schot, A. C., Vonk, R., Brans, R. G., van Haren, N. E., Koolschijn, P. C., Nuboer, V., et al. (2009). Influence of genes and environment on brain volumes in twin pairs concordant and discordant for bipolar disorder. *Archives of General Psychiatry*, *66*(2), 142–151. https://doi.org/10.1001/archgenpsychiatry.2008.541.

Vance, Y. H., Huntley Jones, S., Espie, J., Bentall, R., & Tai, S. (2008). Parental communication style and family relationships in children of bipolar parents. *The British Journal of Clinical Psychology*, *47*(Pt 3), 355–359. https://doi.org/10.1348/014466508x282824.

Vinberg, M., Trajkovska, V., Bennike, B., Knorr, U., Knudsen, G. M., & Kessing, L. V. (2009). The BDNF Val66Met polymorphism: relation to familiar risk of affective disorder, BDNF levels and salivary cortisol. *Psychoneuroendocrinology*, *34*(9), 1380–1389. https://doi.org/10.1016/j.psyneuen.2009.04.014.

Wellcome Trust Case Control Consortium. (2007). Genome-wide association study of 14,000 cases of seven common diseases and 3,000 shared controls. *Nature*, *447*(7145), 661–678. https://doi.org/10.1038/nature05911.

Wilens, T. E., Biederman, J., Abrantes, A. M., & Spencer, T. J. (1997). Clinical characteristics of psychiatrically referred adolescent outpatients with substance use disorder. *Journal of the American Academy of Child and Adolescent Psychiatry*, *36*(7), 941–947. https://doi.org/10.1097/00004583-199707000-00016.

Zeni, C. P., Mwangi, B., Cao, B., Hasan, K. M., Walss-Bass, C., ZuntasSoares, G., et al. (2016). Interaction between BDNF rs6265 met allele and low family cohesion is associated with smaller left hippocampal volume in pediatric bipolar disorder. *Journal of Affective Disorders*, *189*, 94–97. https://doi.org/10.1016/j.jad.2015.09.031.

Genetic risk in adult family members of patients with bipolar disorder

Colm McDonald

Centre for Neuroimaging and Cognitive Genomics (NICOG),
Clinical Science Institute, National University of Ireland, Galway, Ireland

CHAPTER OUTLINE

INTRODUCTION

Bipolar disorder has a strong genetic component with high monozygotic twin concordance and estimates of heritability from population-based twin samples reported as high as 93% (Kieseppa, Partonen, Haukka, Kaprio, & Lonnqvist, 2004). It also displays substantial genetic and phenotypic complexity, overlap of genetic risk with other psychiatric syndromes, notably schizophrenia, and strong evidence for polygenic risk, i.e., many genotypic risk alleles of small effect (Craddock & Sklar, 2013). Evidence for genes of large effect, such as structural genomic alterations or copy number variants, is limited in bipolar disorder—although rare mutations are reported in younger onset cases (Kerner, 2015)—and are clearly less commonly found in bipolar disorder than in schizophrenia (Bergen et al., 2012). The higher rate of mutations reported by molecular genetic studies in schizophrenia compared to bipolar

Bipolar Disorder Vulnerability. https://doi.org/10.1016/B978-0-12-812347-8.00005-1

disorder may be linked to the greater degree of neurodevelopmental disturbance and cognitive dysfunction that characterizes the former syndrome (Murray et al., 2004; Owen, O'Donovan, Thapar, & Craddock, 2011). In recent years, genome-wide association studies (GWAS), which comprise simultaneous analysis of very many genotypic variations across all chromosomes in large numbers of unrelated patients and controls, have provided replicable evidence for allelic variation in several genes associated with bipolar disorder, some of which also overlap with similar GWAS studies conducted in schizophrenia and unipolar depression (Craddock & Sklar, 2013; Harrison, 2016). Evidence for further genotypic variation associated with bipolar disorder continues to accumulate as sample sizes increase. Genes implicated to date by these studies include those associated with calcium signaling, as well as cell surface proteins, hormonal regulation, and glutamate receptor signaling (Harrison, 2016; Nurnberger Jr et al., 2014).

Despite the progress from molecular genetic studies in implicating these genes of small effect in the pathogenesis of bipolar disorder, a clear understanding of the complex role of genotypic variations, their interaction with environmental risk factors, and their impact on brain structure and functioning remains elusive. Clarifying the biological processes underlying the etiopathogenesis of bipolar disorder is likely to assist with the development of novel therapeutic targets, as well as establish the optimal use of current pharmacotherapy for individuals. As with other common complex neuropsychiatric disorders such as schizophrenia, efforts have been made in recent years to explore neurobiological or cognitive phenotypes as alternatives to the clinical syndrome in an effort to understand better how genetic effects contribute to pathogenesis.

The approach of identifying endophenotypes or intermediate phenotypes of mental illness has been in existence for the last three decades, initially in an effort to simplify the process of gene discovery by selecting phenotypes that may be more proximal to susceptibility gene action than the clinical syndrome. Such biological markers were proposed as useful if they fulfilled a number of criteria (Gershon & Goldin, 1986; Leboyer et al., 1998): (i) be heritable themselves; (ii) be associated with the illness in the general population; (iii) be state independent—i.e., be manifest whether or not the illness is active; (iv) co-segregate with the illness within families—i.e., among relatives who manifest the marker, the illness is more prevalent than among those relatives who do not; (v) be measurable in both affected and unaffected subjects; and (vi) be found more frequently among the biological relatives of patients than in healthy controls. The last of these criteria has been given substantial weight in practice and abnormalities in unaffected first-degree relatives of patients with mental illness in comparison to healthy volunteers are usually proposed as potential endophenotypes—even though other endophenotypic criteria may not be established for the biological marker under investigation in the specific study. However, the utilization of endophenotypes purely for the purposes of gene discovery in psychiatry has proved to be of limited value to date, often because the putative endophenotypes were no more heritable or clearly on the causal pathway than the clinical syndromes themselves, but largely because they are logistically more difficult to acquire in large sample sizes (Walters & Owen, 2007). Nevertheless it is clear that genotypic variation is likely to impact on brain structure and function, and that from such abnormalities the clinical syndromes of our current psychiatric diagnoses emerge. Therefore sustained interest continues in the identification of intermediate phenotypes, not just for the purposes of gene discovery, but also in order to study the biological associations of presumed risk alleles and illuminate potential mechanisms of illness development (Glahn et al., 2014; Rasetti & Weinberger, 2011). Indeed the Research Domain Criteria (RDoC) strategy adopted by the US National Institute of Mental Health, which aims to move away from traditional psychiatric diagnoses toward more biologically valid phenotypes, can be considered a version of the endophenotype approach (Glahn et al., 2014).

Several studies have compared measures of biological markers in the unaffected relatives of patients with bipolar disorder to those in control samples in order to assess their potential as intermediate phenotypes and identify the effects of susceptibility genes on neurobiological or cognitive measures. Studying unaffected individuals likely to be enriched with susceptibility genes for illness has the additional advantage of avoiding the potential confounds of illness progression and psychotropic medication usage that are present when investigating patients alone. Such study designs generally comprise either unaffected adult first-degree relatives, who have generally passed the likely age of illness onset in the third decade (siblings, parents, or co-twins), or else children/adolescents or young adults who have not yet developed bipolar disorder but are considered at high risk of doing so due to their genetic risk as offspring of patients and sometimes early symptoms of mental illness. The latter group are explored in detail elsewhere in the current publication and therefore the theme of this chapter will be to review selectively findings on the former group of adult unaffected relatives of patients with bipolar disorder, with a focus on studies investigating neuroimaging abnormalities.

STRUCTURAL NEUROIMAGING

Bipolar disorder itself has been repeatedly associated with neuroanatomical abnormalities, albeit with quite varied findings from cross-sectional studies to date, most likely related to clinical or methodological heterogeneity and differential medication use among research participants. Recent reviews and meta- and mega-analyses indicate the strongest evidence is for ventricular enlargement and reduced volume of the hippocampus and thalamus in patients (Hallahan et al., 2011; Hibar et al., 2016), as well as cortical thinning in the anterior cingulate and prefrontal regions (Hanford, Nazarov, Hall, & Sassi, 2016) and diffuse white matter microstructural abnormalities (Kempton, Geddes, Ettinger, Williams, & Grasby, 2008; Nortje, Stein, Radua, Mataix-Cols, & Horn, 2013).

A number of structural MRI studies have also reported neuroanatomical abnormalities in unaffected adult relatives of patients with bipolar disorder.

GRAY MATTER

McIntosh et al. (2004) reported anterior thalamic gray matter deficits in unaffected relatives of patients, as well as patients with bipolar disorder themselves, compared with healthy volunteers. McDonald et al. (2004) assessed a sample of multiply affected bipolar disorder families and reported that increasing genetic risk for bipolar disorder was associated with gray matter deficit in the right anterior cingulate and ventral striatum in relatives as well as patients. Elsewhere, increased gray matter in the insula and cerebellum in relatives has been reported (Frangou, 2011; Kempton et al., 2009), interpreted by the authors as indicative of resilience among unaffected but genetically at-risk relatives. Another study that included siblings of patients with euthymic bipolar disorder reported reduced volume of the left orbitofrontal cortex and cerebellum in relatives, interpreted by the authors as indicative of genetic risk, but increased volume of left dorsolateral prefrontal cortex in relatives, interpreted as indicative of resistance that could act to balance out the risk-enhancing neural abnormalities within the emotional regulation system associated with genetic vulnerability (Eker et al., 2014). Larger inferior frontal gyrus volume bilaterally and left parahippocampal gyrus, as well as reduced volume of the cerebellum, were reported in relatives of bipolar disorder patients by Saricicek et al. (2015). Larger right inferior frontal gyrus in unaffected relatives was reported by Hajek et al. (2013) in samples recruited from two international centers, and this region was also reported to reduce in size with progression of illness in patients who were not taking lithium. A study of relatives of bipolar I disorder patients by Matsuo et al. (2012) identified reduced left anterior insula gray matter volume in relatives as well as patients. Another study reported that a small sample of unaffected relatives shared right prefrontal cortex and bilateral globus pallidus deficits with affected patients (Sandoval et al., 2016). Matsubara et al. (2016) reported reduced left anterior cingulate gray matter in the unaffected relatives of bipolar disorder patients, an abnormality also found to a larger degree in patients, who additionally displayed left insula gray matter deficit.

Twin studies which include discordant monozygotic and dizygotic twin pairs are well placed to separate genetic and environmental contributions to biological markers as they include unaffected relatives of differing levels of genetic risk. Such studies have reported gray matter deficit in the right medial frontal gyrus, precentral gyrus, and insula in the unaffected co-twins of bipolar disorder patients (van der Schot et al., 2010), as well as genetic contributions to cortical thinning in the parahippocampal and orbitofrontal cortex (Hulshoff Pol et al., 2012) and to volume reduction of the thalamus, putamen, and nucleus accumbens (Bootsman et al., 2015).

Several negative studies have also been published which fail to identify abnormalities of gray matter in subcortical and cortical regions in the unaffected adult relatives of patients with bipolar disorder (McDonald et al., 2006; McIntosh et al., 2006; Nery et al., 2015; Nery, Monkul, & Lafer, 2013). These include the large Bipolar-Schizophrenia Network on Intermediate Phenotypes (B-SNIP) study, which included a cohort of unaffected relatives of patients with psychotic bipolar I disorder

(Ivleva et al., 2013), and a metaanalysis of studies examining individuals at enhanced genetic risk for bipolar disorder (including high-risk offspring) which examined sub-cortical structures and frontal lobes (Fusar-Poli, Howes, Bechdolf, & Borgwardt, 2012).

WHITE MATTER

An early indication for a genetic contribution toward white matter deficits in bipolar disorder came from the report that unaffected co-twins of patients display reduced global white matter in the left hemisphere (Kieseppa et al., 2003). This finding was later replicated in larger structural MRI studies of discordant bipolar twins, which reported a genetic contribution to white matter volume reduction and reduced frontal white matter density in bipolar disorder (van der Schot et al., 2009, 2010). Elsewhere volume of white matter hyperintensities in unaffected relatives was reported to be intermediate between healthy volunteers and patients (Tighe et al., 2012); and volume of white matter in the right medial frontal gyrus was reported to be reduced in unaffected relatives compared with controls (Matsuo et al., 2012). McDonald et al. (2004) reported white matter volume deficits in fronto-temporal and parietal regions with increasing genetic risk in unaffected relatives, as well as affected patients, from multiply affected bipolar disorder families, and proposed white matter deficit as endophenotypic across the psychosis spectrum since similar overlapping deficits were found in schizophrenia families. This apparent shared white matter deficit in genetically at-risk unaffected relatives crossing both schizophrenia and bipolar disorder was also reported in a large twin study of discordant pairs spanning both disorders (Hulshoff Pol et al., 2012). Francis et al. (2016) assessed callosal volumes in patients with psychotic disorders and their unaffected relatives. Similar to the findings in probands, unaffected relatives of patients with psychotic bipolar disorder displayed reduced volume of the anterior splenium compared with controls.

Hajek et al. (2015) used machine learning to distinguish unaffected young adult offspring of bipolar patients from controls and succeeded for white matter (but not gray matter) analyses, with the regions of greatest abnormality in relatives incorporating white matter of the inferior/middle frontal gyrus, inferior/middle temporal gyrus, and precuneus. Notably some other studies did not distinguish unaffected relatives from controls in terms of white matter hyperintensity load (Gunde et al., 2011) and regional white matter volume (McIntosh et al., 2005).

Several diffusion tensor imaging (DTI) analyses also report that white matter structural abnormalities are overrepresented in the unaffected relatives of patients with bipolar disorder. This magnetic resonance technique measures diffusion of water molecules within tissues and, since direction of diffusion is constrained by myelinated axons within white matter, impaired constraint of such diffusion (most commonly assessed using fractional anisotropy) can be compared between participant groups. Chaddock et al. (2009) reported distributed reduction of fractional anisotropy, including in the region of corpus callosum and fronto-temporal longitudinal tracts, with increasing genetic risk in families with bipolar disorder. A further study of this cohort by Emsell et al. (2014), which employed a tractography method to assess fractional anisotropy within specific tracts, did not identify abnormalities in unaffected relatives compared with controls at a group level, but reported that increasing genetic liability in patients was associated with reduced fractional anisotropy in the left uncinate fasciculus and superior longitudinal fasciculus.

Sprooten et al. (2011) identified widespread white matter integrity reductions, as indexed by lower fractional anisotropy measures, in a large sample of young unaffected relatives of patients with bipolar disorder. In a separate cohort of sibling pairs discordant for bipolar disorder, Sprooten et al. (2013) again identified fractional anisotropy reductions in unaffected relatives compared with controls incorporating the corpus callosum, posterior thalamic radiation, posterior corona radiata, and left superior longitudinal fasciculus. A later tractography study on this sample also identified fractional anisotropy reduction in relatives in the corpus callosum (Sprooten et al., 2016).

In the large B-SNIP project, Skudlarski et al. (2013) reported reduced fractional anisotropy in the unaffected relatives of psychotic bipolar patients which were similar to, though more attenuated than, those found in schizophrenia relatives and most prominent in the left posterior corona radiata and genu of the corpus callosum. Both monozygotic and dizygotic unaffected co-twins of patients with bipolar disorder were reported to display reduced fractional anisotropy in the anterior callosum, as did unaffected co-twins of depressed patients, indicating a potential endophenotype across affective disorders (Macoveanu et al., 2016). Mahon et al. (2013) reported reduced fractional anisotropy in the right temporal lobe in unaffected siblings of patients with bipolar disorder. Linke et al. (2013) reported reduced fractional anisotropy in the right anterior limb of the internal capsule and right uncinate fasciculus in unaffected relatives compared with controls. Saricicek et al. (2016) reported that, while patients with bipolar disorder displayed widespread fractional anisotropy reductions compared with controls, their unaffected siblings shared such fractional anisotropy deficits in the left posterior thalamic radiation, left sagittal stratum, and fornix.

In summary, there is a substantial and increasing literature reporting that neuroanatomical abnormalities are present in the unaffected adult relatives of patients with bipolar disorder. These studies use a variety of neuroimaging methods and clinical inclusion criteria and report somewhat varied results. Some studies report increased gray matter volume in relatives incorporating the frontal gyrus, i.e., neuroanatomical deviation which is in the opposite direction to that reported in patients, and tend to interpret this finding as a marker of resilience in these adult relatives who, despite carrying genetic risk, have not succumbed to illness. Several other studies report subtle gray matter deficits in relatives, i.e., in the same direction as that usually reported in patients but less marked in severity. These are then interpreted as the impact of susceptibility genes for illness manifest in neuroanatomy, with the assumption that patients themselves experience additional genetic and/or environmental risk factors which drive them across the threshold to develop frank illness. Apparent further support for this interpretation is provided by the location of the reported gray matter deficits which involve prefrontal cortex, anterior cingulate, and associated subcortical structures—regions which are neuroanatomically implicated in the circuitry of mood regulation.

Stronger evidence has emerged for a genetic contribution to white matter abnormalities from studies of bipolar disorder relatives, since most published studies have reported significant changes and these are consistent in their directionality—white matter volume deficit and reduced fractional anisotropy—with that repeatedly reported in patients. There is a lack of uniformity in the specific anatomical regions implicated by individual studies (as indeed there is for studies on patients themselves). However, abnormalities in unaffected relatives appear widespread, and seem to incorporate callosal and longitudinal tracts in frontal as well as posterior brain regions—areas implicated in emotional processing and regulation—and suggest that susceptibility genes for bipolar disorder may impact upon structural connectivity in networks subserving these functions.

FUNCTIONAL NEUROIMAGING

There is ample evidence that patients with bipolar disorder display neurophysiological abnormalities compared with healthy volunteers when measured by alterations in cerebral activation assessed using functional MRI, and that these abnormalities are found not only during episodes of mood exacerbation but also during euthymia (Langan & McDonald, 2009). Studies vary considerably in methodology and employ a range of cognitive and emotional stimuli designed to activate diverse neural networks, as well as examining the brain during resting state. Reviews of this literature support a model of functional abnormalities within the fronto-limbic system, and in particular disconnectivity between ventral prefrontal brain structures and limbic system structures including the amygdala, which impairs homeostatic control of mood regulation and reward processing (Strakowski et al., 2012). In these models of bipolar disorder, impaired cognitive control as evidenced by underactive prefrontal cortex is associated with overactivity in limbic system structures most prominently during emotional activation paradigms and manic mood exacerbation (Chen, Suckling, Lennox, Ooi, & Bullmore, 2011; Phillips & Swartz, 2014).

Several studies have reported that unaffected adult relatives of patients with bipolar disorder also display functional brain abnormalities compared with controls.

RESTING STATE

Analyses of resting state fMRI data is increasingly employed in neuropsychiatric research to examine for abnormalities in connectivity within and between neural networks. Using a resting state fMRI task and functional network connectivity analysis in the B-SNIP study, Meda et al. (2012) reported that unaffected relatives of bipolar disorder patients displayed reduced connectivity between two network pairs compared to controls, as did affected patients: (i) fronto-occipital regions (linked to visual perception and higher-order visual processing) and anterior default mode/prefrontal network (linked to social cognition); and (ii) meso-paralimbic network and sensory motor networks. A further study of the cohort, employing a within-network analysis, reported aberrant connectivity in fronto-occipital, fronto-thalamic-basal ganglia and sensory-motor networks in unaffected relatives of bipolar disorder patients, which was shared with relatives and patients across the psychosis spectrum, interpreted as representing a potential psychosis endophenotype (Khadka et al., 2013). Further analyses of low-frequency fluctuations in some of the cohort indicated that unaffected relatives of psychotic bipolar disorder patients displayed abnormal connectivity between precentral/postcentral gyrus and caudate nucleus, similar to probands (Lui et al., 2015). Another study did not identify abnormalities in local spontaneous fluctuations of neuronal firing in relatives of psychotic bipolar disorder patients, which were present in patients across the psychosis spectrum (Meda et al., 2015). A further study of the extended cohort exploring abnormalities of the default mode network using independent components analysis did not identify abnormalities in bipolar relatives, whereas unaffected relatives of schizophrenia patients were reported to display hypoconnectivity (Meda et al., 2014). In a study including unaffected siblings of bipolar disorder patients, Li et al. (2015) reported increased functional connectivity between dorsolateral prefrontal cortex and amygdala in unaffected relatives, which was further accentuated in patients.

COGNITIVE TASKS

In an fMRI study using the *n*-back working memory task in multiply affected families of patients with bipolar disorder and their unaffected relatives who participated in the Maudsley Family Study of Psychosis, Drapier et al. (2008) reported that relatives displayed increased activation in the left frontal pole/ventrolateral gyrus during the 1- and 2-back tasks compared with healthy volunteers. There was no difference in task performance between relatives and healthy controls, whereas patients performed more poorly and tended also to overactivate this region, suggesting a genetic contribution to inefficiency of this prefrontal region which was hyperactivated to maintain performance. In a study by Thermenos et al. (2010) employing a sequential letter 2-back task, unaffected relatives displayed increased activity (i.e., failure to suppress activation) in the left anterior insula and also in the left orbitofrontal cortex and superior parietal cortex compared with controls. In another study using the *n*-back task with unaffected siblings of patients with bipolar disorder (Alonso-Lana et al., 2016), relatives displayed increased activation (i.e., reduced deactivation) compared with controls in the medial prefrontal cortex, an abnormality which was also found to a more accentuated degree in patients.

In a further study of the Maudsley Family Study of Psychosis cohort which employed a verbal fluency task to probe neural networks underlying executive function, Allin et al. (2010) demonstrated hyperactivation (i.e., failure to suppress activation) in the retrospenial/posterior cingulate cortex and precuneus in unaffected relatives in both easy and hard versions of the task. Patients also failed to suppress this region in the easy version of the task. Relatives also displayed reduced activation of the medial frontal cortex while performing the hard condition. Using a similar task in a study that included a small sample of unaffected monozygotic co-twins of patients with bipolar disorder and an analysis that was confined to the inferior frontal cortex, Costafreda et al. (2009) did not find any abnormalities in bipolar disorder relatives, nor in patients, in contrast to the hyperactivation found in patients with schizophrenia and their co-twins.

Whalley et al. (2011) studied young unaffected adult relatives of bipolar disorder patients who participated in the Scottish Bipolar Family Study employing an extension of the verbal fluency paradigm: the Hayling sentence completion task. Relatives displayed increased activation in the left amygdala compared with controls. There was also increased activation in other regions associated with subclinical symptoms in relatives—in the ventral striatum with higher depressive symptom scores and the ventral prefrontal cortex with higher cyclothymic temperament scores. A 2-year follow-up study of this cohort (Whalley, Sussmann, et al., 2013) demonstrated that those unaffected relatives who later developed major depressive disorder (one-fifth of the sample) were more likely to have higher depressive and cyclothymic scores at baseline and also significantly hyperactivated the insula bilaterally during the sentence completion task, indicating that this abnormality in unaffected relatives could be a marker of those genetically at-risk subjects who will go on to develop a mood disorder.

When examining neural networks subserving sustained attention function by employing the continuous performance task paradigm with a sample of unaffected first-degree relatives of patients with bipolar disorder, Sepede et al. (2012) reported that, similar to patients, relatives displayed increased activation in the bilateral insula and middle cingulate cortex during errors in target recognition. Additionally, and in contrast to patients, relatives displayed increased activation of the left insula and bilateral inferior parietal lobule during correct target responses with a higher attention load.

Examining neural networks subserving inhibition processes by employing a Stroop Color Word test paradigm with unaffected first-degree relatives (a proportion of whom did have major depressive disorder), Pompei, Jogia, et al. (2011) reported that unaffected relatives, similar to patients, displayed reduced activity in the posterior and inferior parietal lobules, areas associated with selective attention processes, compared with controls. Additionally, hypoactivation of the left striatum was observed in relatives who had depressive illness, similar to patients with bipolar disorder and in contrast to those unaffected relatives with no mood disorder. A further exploration of this dataset using psychophysiological interaction analysis to assess functional connectivity of the inferior frontal gyrus (Frangou, 2011; Pompei, Dima, Rubia, Kumari, & Frangou, 2011) demonstrated that unaffected relatives displayed reduced connectivity between the ventrolateral prefrontal cortex and the anterior cingulate gyrus (similar to patients), as well as reduced connectivity between the ventrolateral prefrontal cortex and insula (in contrast to patients who had positive connectivity), and increased connectivity between the ventrolateral prefrontal cortex and dorsolateral prefrontal cortex.

An investigation of reward system abnormalities using a probabilistic reversal learning task in unaffected relatives of bipolar disorder patients was conducted by Linke et al. (2012). Similar to affected patients, unaffected relatives displayed regional overactivity in response to reward and reward-reversal contingencies in the right medial orbitofrontal cortex and the amygdala, interpreted as indicative of heightened sensitivity toward reward and deficient prediction error signal as endophenotypic markers. In contrast, when assessing the anticipation phase of reward processing, Kollmann, Scholz, Linke, Kirsch, and Wessa (2017) identified increased right anterior cingulate activation in euthymic patients with bipolar disorder compared with controls, but no abnormalities in their first-degree relatives. The authors interpreted this hyperactivation in patients as a consequence of the illness rather than a marker of genetic vulnerability.

In an investigation of the neural underpinnings of social cognition using a theory of mind task with bipolar disorder patients and their unaffected relatives, Willert et al. (2015) reported that patients displayed reduced bilateral activation of the temporoparietal junction, and reduced coupling between temporoparietal junction and medial prefrontal cortex on functional connectivity analysis, when compared with controls. Although not statistically significant, unaffected relatives tended toward reduced right temporoparietal activity and also displayed intermediate fronto-temporoparietal connectivity, which was interpreted as suggestive evidence for altered theory of mind processing as an intermediate phenotype that requires further investigation.

EMOTIONAL TASKS

In a further study of the Maudsley Family Study of Psychosis cohort that employed an implicit emotional face recognition task, Surguladze et al. (2010) demonstrated that unaffected relatives, similar to affected patients, displayed increased activity in the medial prefrontal cortex compared with controls in response to fearful and happy faces. Both relatives and patients also displayed increased activity in the left putamen in response to fearful faces and in the left amygdala in response to intensely happy faces. Roberts et al. (2013) investigated young adult relatives of bipolar disorder patients with a facial-emotion go/no-go task which sought to determine neural activity underpinning the inhibition of emotional material. In contrast to nonemotional inhibition, unaffected relatives displayed a lack of recruitment of the inferior frontal gyrus when inhibiting responses to fearful faces compared with controls, interpreted as indicating that impaired inhibitory function of the frontal cortex when exposed to emotional material may represent an endophenotypic marker of bipolar disorder.

Chan et al. (2016) employed an implicit facial emotion processing task incorporating angry and neutral faces in an investigation as part of the Scottish Bipolar Family Study. Young unaffected relatives who later developed depressive illness displayed reduced activation of the anterior cingulate, however unaffected relatives who remained well did not differ from controls, interpreted as indicating that this neural network abnormality was related to the imminent onset of illness rather than familial risk. Kanske, Heissler, Schonfelder, Forneck, and Wessa (2013) investigated the neural activity underpinning emotional distraction during mental arithmetic problems in patients with bipolar disorder, unaffected relatives, and healthy individuals with hypomanic personality traits. Whereas task-related activation of the prefrontal and parietal cortex was generally increased in each group under emotional activation, only patients had significant hyperactivity in the parietal region compared with controls; this was interpreted therefore as an illness effect rather than a genetic vulnerability marker. In another study examining the neural correlates of the emotion regulation strategies of reappraisal and distraction in unaffected relatives of bipolar I disorder patients, these authors reported that relatives as well as patients displayed impaired downregulation of the amygdala during reappraisal, but not during distraction (Kanske, Schonfelder, Forneck, & Wessa, 2015). Furthermore, negative connectivity between the amygdala and orbitofrontal cortex observed during reappraisal in controls was reversed in patients and relatives.

In summary, fMRI studies employing a wide range of methodologies have reported a number of subtle anomalies in the unaffected adult relatives of bipolar disorder patients which suggest endophenotypic effects within neural networks, including those subserving various cognitive and emotional processes. Although some studies reported negative findings, the majority of published studies to date have reported that unaffected adult relatives of bipolar disorder patients do display neurophysiological abnormalities as assessed by fMRI when compared with healthy volunteers.

Given the wide range of clinical samples and functional neuroimaging methodologies employed in these studies, it is unsurprising that several different neuroanatomical networks and regions are implicated, and further replication will be required to make definitive conclusions. The heterogeneity of clinical samples is underlined by the wide age range among the adults, with some likely to develop mood disorders over time—indeed longitudinal follow-up has enabled some researchers to separate, on the basis of cerebral activation abnormalities, those genetically at-risk individuals on a trajectory toward mood disorder from those genetically at-risk individuals who remained well (Chan et al., 2016; Whalley, Sussmann, et al., 2013). Furthermore, abnormalities found in genetically at-risk individuals that differ from those found in patients can be interpreted as compensatory mechanisms to support resilience rather than manifestations of genetic vulnerability: for example, increased connectivity between ventral and dorsolateral prefrontal cortex (Frangou, 2011) and hyperactivation of inefficient ventrolateral prefrontal regions in order to maintain behavioral performance on cognitive tasks (Drapier et al., 2008; Thermenos et al., 2010).

Despite these heterogeneities, some consistency has emerged to indicate that relatives are more likely to display functional disconnectivity as well as increased activation affecting limbic and paralimbic regions, with abnormal activation involving medial and ventrolateral prefrontal regions, and reduced activity in parietal regions. This pattern of abnormality largely reflects that reported in the heterogeneous literature on patients themselves (Phillips & Swartz, 2014; Piguet, Fodoulian, Aubry, Vuilleumier, & Houenou, 2015), and suggests that genotypic variation contributes to at least some of these functional abnormalities.

OTHER PHENOTYPIC MARKERS

In addition to the structural and functional neuroimaging findings described in detail above, several research groups have also investigated other measures in the unaffected relatives of bipolar disorder patients as potential endophenotypes. Most commonly, abnormalities in cognitive performance have been reported and were recently systematically reviewed by Miskowiak et al. (2017). The most robust support for deficits in nonemotional cognitive function in unaffected offspring and adult relatives are in the areas of verbal learning and memory, sustained attention, and executive function, which are similar to (though less pronounced than) those reported in the relatives of patients with schizophrenia. In relation to emotion-laden cognition, potentially endophenotypic abnormalities appear more specific to mood disorder and include impairments in facial emotion recognition and emotion regulation, increased reactivity to emotional stimuli, and increased interference of attentional resources by emotional stimuli (Miskowiak et al., 2017). Further support for social cognition abnormalities in the relatives of bipolar disorder patients comes from a metaanalysis of theory of mind and facial emotion recognition studies, which particularly implicated theory of mind deficits in relatives (Bora & Ozerdem, 2017).

Elsewhere, unaffected relatives have been reported to display impaired manual motor speed as assessed by finger tapping, with performance intermediate between controls and the more impaired affected patients (Correa-Ghisays et al., 2017). Unaffected siblings of bipolar disorder patients have been reported to display biomarkers of accelerated aging compared with controls, notably telomere shortening and increased levels of C−C motif chemokine 11 (Vasconcelos-Moreno et al., 2017).

Similar to remitted patients, unaffected siblings of bipolar disorder patients were reported to display reduced prepulse inhibition of the startle response compared with controls (Giakoumaki et al., 2007). Although prepulse inhibition did not differ from controls in the B-SNIP project, unaffected relatives of psychotic bipolar disorder patients were reported to display other neurophysiological abnormalities, similar to unaffected relatives of schizophrenia patients, including lower smooth pursuit eye movement maintenance and predictive pursuit gain, and lower auditory event-related potential theta/alpha and beta magnitudes to the first stimulus (Ivleva et al., 2014). In the Maudsley Family Study of Psychosis project, unaffected relatives of bipolar disorder patients displayed diminished P50 auditory evoked potential suppression and delayed P300 latency, which was similar in direction to but less pronounced in magnitude than that found in patients (Schulze et al., 2007, 2008). An extension of this approach to include unaffected co-twins of bipolar disorder probands provided further support for P300 amplitude reduction and latency delay as potentially endophenotypic for bipolar disorder (Hall et al., 2009).

ALLELIC VARIANTS AND NEUROIMAGING ENDOPHENOTYPES IN UNAFFECTED RELATIVES

As evidence mounts for genotypic variants associated with the clinical phenotype of bipolar disorder, research groups have sought to explore endophenotypes from the bottom-up direction of molecular genetics, i.e., using the genotype as a predictor variable, rather than the top-down direction of genetically at-risk individuals, i.e., comparing unaffected relatives with controls (Gurung & Prata, 2015). Most studies linking allelic variation with neuroimaging measures have been conducted in affected patients and healthy volunteers, but some have also included unaffected relatives in order to explore the impact of such genotypic variation on potential endophenotypes in genetically at-risk individuals free from any potential confounds related to illness or medication effects. Popular candidate genes for such studies come from genome-wide association studies of bipolar disorder, which have identified loci that display genome-wide significance for the illness, including CACNA1C, which encodes for the L-type calcium channel subunit, and ANK3, which encodes for the scaffold protein Ankyrin G (Craddock & Sklar, 2013; Harrison, 2016).

Jogia et al. (2011) examined the impact of the CACNA1C risk allele rs1006737 employing an fMRI study incorporating a facial affect recognition task in patients with bipolar disorder, their unaffected relatives, and controls. Risk allele carriers displayed increased right amygdala activation during fearful face recognition compared with neutral faces across participant groups, whereas only patients carrying the risk allele, and not relatives or controls, displayed reduced activation in the ventrolateral prefrontal cortex. This study supported the impact of the CACNA1C risk allele on amygdala and ventrolateral prefrontal cortex functioning during emotion processing. Radua et al. (2012) examined the impact of the same allele on effective connectivity between brain regions (medial prefrontal, left amygdala, and left putamen) previously identified as displaying increased activation in bipolar disorder patients and their relatives during emotional processing (Surguladze et al., 2010). The risk allele was associated with decreased outflow from the medial prefrontal gyrus to the left putamen across groups, which was most prominent in the patients. Erk et al. (2014)) completed an fMRI study incorporating an episodic memory task with relatives of patients with bipolar disorder, major depressive disorder, and schizophrenia. Carriers of the CACNA1C risk allele across relative groups displayed reduced activation of the hippocampus and perigenual anterior cingulate during episodic memory recall compared with the control group bearing no risk allele, indicating this genetic variant may have a role in dysfunctional regulation of limbic and prefrontal brain circuits in these individuals.

Delvecchio, Dima, and Frangou (2015) examined the impact of allelic variation in ANK3 in an fMRI study of working memory in patients with bipolar disorder and their unaffected first-degree relatives. In patients and their relatives, risk alleles at two loci were associated with hyperactivation of the right anterior cingulate cortex. In contrast, a mixed pattern of associations of each risk allele with either reduced temporal cortex activation or increased lateral prefrontal cortex activation was observed in healthy volunteers. The authors interpreted this as indicating that the effect of risk associated polymorphisms of ANK3 are modulated by risk status for the illness, and, in the presence of such risk, are associated with inability to suppress a key node within the anterior default mode network during working memory.

Outside of the neuroimaging field, Arts, Simons, and Os (2013) explored, in patients with bipolar disorder, their unaffected first-degree relatives, and controls, the effect of CACNA1C genotypic variation on several cognitive measures collected longitudinally, including verbal memory, sustained attention, selective attention, attentional span, and working memory. Risk allele carriers among patients displayed poorer performance on a composite measure of cognitive tasks, but this effect was not found in healthy volunteers or unaffected relatives. Ruberto et al. (2011) investigated, in patients with bipolar disorder, their unaffected first-degree relatives and controls, the impact of allelic variation in ANK3, and a series of cognitive tasks assessing general intellectual ability, memory, decision making, response inhibition, and sustained attention. Risk allele carriers displayed reduced sensitivity in target detection and increased errors of commission during sustained attention, regardless of subject group.

Notably, several other polymorphisms in individual genes proposed as having a role in the pathogenesis of bipolar disorder have also been assessed for association with functional and structural neuroimaging measures in studies that included unaffected relatives of patients with bipolar disorder with a mixture of positive and negative results, including microRNA MIR137 (Whalley, Papmeyer, Romaniuk, et al., 2012), ZNF804A (Sprooten et al., 2012), DISC1 (Whalley, Sussmann, et al., 2012), COMT (Lelli-Chiesa et al., 2011), and BDNF (Dutt et al., 2009).

VARIABLE GENETIC RISK AMONG UNAFFECTED RELATIVES

Unaffected first-degree relatives of patients with bipolar disorder are likely to vary in the extent to which they carry susceptibility genes for the illness. Rather than consider relatives as a homogenous group with the same risk, some studies on unaffected relatives have sought to build this variable risk into the study design. Hence the Maudsley Family Study of Psychosis and Scottish Bipolar Family studies specifically sampled relatives from multiply affected families with bipolar disorder on the assumption that these individuals would be more enriched with susceptibility genes (McDonald et al., 2006; McIntosh et al., 2004). These studies also used a proxy genetic liability scale, reflecting the extent to which families were more densely affected by the illness, as a predictor variable within some of the neuroimaging analyses (Chaddock et al., 2009; McDonald et al., 2004; McIntosh et al., 2006). Standard twin studies implement a variable risk design among unaffected relatives, whereby unaffected monozygotic twins are considered more likely to display genetically driven neurobiological abnormalities than unaffected dizygotic twins (Bootsman et al., 2015; Hulshoff Pol et al., 2012). Other studies combined twins and unaffected siblings in the same analysis in order to stratify genetic risk and assess heritability of endophenotypic traits using structural equation modeling (Hall et al., 2009). Employing this approach, Hall and colleagues also demonstrated that impaired P50 suppression in an auditory evoked potential task appeared a heritable endophenotype for psychotic bipolar disorder (Hall et al., 2008), whereas evoked gamma band responses were not endophenotypic despite displaying heritability (Hall et al., 2011). Similarly, orbitofrontal cortex hyperactivation during working memory processing was demonstrated by Sugihara et al. (2017) to be heritable as well as associated with genetic liability for bipolar disorder. These study designs, which attempt to relate genetic risk to potentially endophenotypic neurobiological markers, are based upon presumptions that certain groups of relatives are more likely to be enriched with susceptibility genes, and do not attempt to incorporate the emerging molecular genetic research on actual genotypic variation at an individual level.

POLYGENIC RISK SCORES

Given the likely polygenic nature of bipolar disorder and as molecular genetic studies have identified genotypic variants of small effect, a further approach to model variable genetic risk among individuals, including unaffected relatives of patients with bipolar disorder, is to employ polygenic risk scores. A polygenic risk score models the aggregate effect of single nucleotide polymorphisms (SNPs) associated with the illness in each individual. Genome-wide association studies test millions of SNPs for association with an illness and necessarily apply conservative thresholds for statistical significance that require very large sample sizes to detect genes of small effect. The amount of illness variation explained by each GWAS-supported SNP is very small, whereas there are likely to be a large number of SNPs underlying risk for an illness like bipolar disorder. A polygenic risk score incorporates information from both GWAS SNPs and SNPs that are not genome-wide significant, but that do meet nominal significance criteria (Dima & Breen, 2015). The polygenic risk score can thus confer substantially more statistical power than a single SNP in a phenotypic analysis and can be related to neurobiological variables to explore genetic effects in much smaller samples than are used for gene discovery in GWAS studies.

A bipolar disorder polygenic risk score was employed by Whalley, Papmeyer, Sprooten, et al. (2012) to explore genetically mediated cerebral activation in unaffected relatives of patients with bipolar disorder. Polygenic risk scores were calculated for each individual based on GWAS data from a large consortium of over 16,000 individuals. Unaffected young adults/adolescents who were considered high risk for bipolar disorder, since they had at least one first-degree or two second-degree relatives with bipolar disorder, had significantly higher polygenic risk scores than healthy volunteers. When performing a sentence completion task in the fMRI scanner, the polygenic risk score was significantly associated with increased activation in the anterior cingulate cortex and the right amygdala. Furthermore, the previously reported amygdala hyperactivation in relatives compared with controls (Whalley et al., 2011) was attenuated when controlling for the polygenic risk score, suggesting that these effects were mediated by polygenic load. This study suggested that profiling each unaffected individual on the basis of contributing risk alleles for bipolar disorder could be a more powerful approach to capturing phenotypic variation due to genetic factors than group comparisons between relatives and controls (Whalley, Papmeyer, Sprooten, et al., 2012). Using a similar approach with diffusion tensor imaging data acquired from the sample, Whalley, Sprooten, et al. (2013) demonstrated that polygenic risk scores for major depressive disorder (although not for bipolar disorder) were associated with reduced fractional anisotropy in an extensive region incorporating the right superior longitudinal fasciculus, inferior longitudinal fasciculus, and inferior fronto-occipital fasciculus. The authors interpreted this finding as indicating that polygenic contribution for mood disorder at an individual level is associated with impaired white matter integrity.

Dima, de Jong, Breen, and Frangou (2016) examined the effect of polygenic risk scores for bipolar disorder and task-related activation from an fMRI study incorporating a facial affect and a working memory processing task in patients with bipolar disorder, their unaffected first-degree relatives and healthy volunteers. In the facial affect recognition task, the polygenic risk score was associated with reduced activation in the visual cortex and there was no interaction with participant group. In the working memory 2-back task, the polygenic risk score was associated with increased activation in the ventromedial prefrontal cortex, and there was no interaction with participant group. This is an important region for emotion processing during the default state, and this report of failure to deactivate during working memory in association with increased genetic risk echoes the similar finding from group comparisons of unaffected relatives of patients with bipolar disorder and controls (Alonso-Lana et al., 2016).

A feature of the polygenic risk score is that it can be used to model variation of genetic risk for a disorder in healthy volunteers as well as unaffected relatives or patients and thus take advantage of existing datasets with large numbers of individuals who have had phenotypic and genetic data collected. Caseras, Tansey, Foley, and Linden (2015) examined the association between genetic risk scores for bipolar disorder or schizophrenia and measures of subcortical brain volumes in a large sample of healthy research participants. Decreased globus pallidus volume was associated with genetic risk for both bipolar disorder and schizophrenia, and decreased volume of the amygdala was associated with genetic risk for bipolar disorder (and not schizophrenia). In an fMRI study incorporating an emotional faces matching paradigm of patients with bipolar disorder and controls, Tesli et al. (2015) reported that polygenic risk for bipolar disorder was associated with increased activation in the right inferior frontal gyrus during negative face processing in the entire sample, echoing the more pronounced inferior frontal gyrus hyperactivation in young people with bipolar disorder in comparison to adults in response to emotional stimuli (Wegbreit et al., 2014). Wang et al. (2017) completed a functional connectivity study of a large number of healthy volunteers in China that examined the impact of a cross disorder polygenic risk score as well as scores for individual disorders. Polygenic risk for bipolar disorder was associated with reduced functional connectivity between the midbrain and insula bilaterally, and increased functional connectivity between the bilateral insula and the bilateral posterior cingulate, cuneus, and precuneus. The latter increased functional connectivity between the salience and default mode networks associated with increased genetic risk for bipolar disorder was interpreted as consistent with impaired switch functioning underlying cognitive and affective disturbance.

Some other studies investigating the association of polygenic risk scores for bipolar disorder with neuroimaging or cognitive variables in healthy volunteers have been less successful in detecting significant effects. A study employing imaging and genetic data from the large UK Biobank of middle-aged and older-age healthy volunteers found no significant association between polygenic risk scores for bipolar disorder and brain tissue volumes, regional subcortical volumes, general fractional anisotropy, or tract-specific diffusion measures (Reus et al., 2017). Nor was there any significant association between polygenic risk scores for major depressive disorder or schizophrenia and these neuroimaging measures. The authors speculated that the older age of this sample and the limited phenotypic variation that can be explained by polygenic risk scores may have undermined their ability to detect associations, and highlighted the need for independent replication of results from smaller samples. In another study relating polygenic risk scores for bipolar disorder (and several other psychiatric disorders) to executive functioning measures in a large sample of healthy twins, no significant associations were identified (Benca et al., 2017).

Outside the neuroimaging field, polygenic risk scores for bipolar disorder have been linked to various other traits, including reduced P300 amplitude in patients (Hall et al., 2015), increased risk of alcohol and substance misuse (Reginsson et al., 2017), higher creativity (Power et al., 2015), and higher hair cortisol concentration (Streit et al., 2016) in healthy volunteers.

Although a potentially powerful approach to link advances in molecular genetics with endophenotypic markers of bipolar disorder, the polygenic risk score carries its own weaknesses. While substantially more powerful than single SNPs, the current polygenic risk score explains only about 5% of the variance of bipolar disorder. Furthermore it is unclear at present how polygenic risk scores are optimally defined—SNPs that meet a certain significance threshold in discovery samples are binned together; however, adding SNPs can increase noise as well as signal and the threshold P value that optimizes signal-to-noise ratio is likely to vary with the phenotypic measure and sample size. As more SNPs are identified using the clinical phenotype, it is likely that the power of the polygenic risk approach to identify endophenotypic markers will also be enhanced.

CONCLUSION

Bipolar disorder is a heritable, complex illness of likely polygenic risk. The clinical treatments to date have largely been identified serendipitously and the development of more efficacious and better-tolerated pharmacotherapy is hampered by the lack of understanding of the underlying biology of the illness. A clearer picture of the mechanisms of gene action and the identification of intermediate phenotypes between genetic risk and the development of clinical symptoms would facilitate biomarkers for treatment response and illness progression to be established, paving the way for the development of novel therapeutic interventions and the optimal use of current ones.

The study of genetically at-risk adult unaffected relatives of bipolar disorder patients is a valuable tool in the search for neuroimaging endophenotypes linking genotypic variation and the established clinical syndrome of the illness. Unaffected relatives carry the research benefits of enrichment with bipolar disorder susceptibility genes and freedom from the effects of illness and psychotropic medication, which can confound studies of affected patients—including studies examining the effects of allelic variation in specific genes linked to the disorder or of polygenic rick scores emerging from genome-wide association studies.

This review of abnormalities associated with genetic risk in family members of patients with bipolar disorder focused on unaffected adult relatives and on neuroimaging findings, as well as recent studies linking polygenic risk with potential endophenotypes. What has emerged from the literature to date is a clear rejection of the null hypothesis, i.e., that neuroimaging measures do not differ between unaffected relatives and controls. However, there is a lack of clear consistency among the published studies to date. This is unsurprising given the sources of clinical and methodological heterogeneity, especially among the varied techniques employed in functional neuroimaging studies, as well as variable genetic risk among samples of relatives which likely alters their expression of endophenotypic markers. Nevertheless, a number of replicated findings have emerged and abnormalities identified in relatives tend to be in the same direction and more subtle than those identified in patients (with some notable exceptions where changes are in the opposite direction and interpreted as resilience markers): this lends convergent validity to the findings.

Structural neuroimaging studies to date have reported subtle abnormalities among genetically at-risk unaffected relatives including gray matter deficits in the anterior limbic system, insula, prefrontal cortex, and anterior cingulate, and increased volume of the inferior frontal gyrus. More consistent evidence for neuroanatomical abnormalities has come from studies of white matter through both structural magnetic resonance and diffusion tensor imaging methodologies, with most of the studies conducted reporting abnormalities in relatives which are consistent in direction. These include diffuse white matter volume deficits and reduced fractional anisotropy incorporating the major transverse and longitudinal white matter tracts and extending beyond the anterior limbic system. Such studies support the hypothesis that susceptibility genes for bipolar disorder impact on structural connectivity within the brain.

A wide range of functional magnetic resonance imaging studies have been conducted which include resting state functional connectivity and experimental designs incorporating working memory, verbal fluency, sustained attention, selective attention, reward processing, social cognition, and emotional face recognition tasks. Despite the range of neural networks stimulated by these tasks and consequent varied findings, certain consistent abnormalities are reported in relatives, with most evidence pointing toward functional disconnectivity, increased activation in limbic and paralimbic regions, abnormal activation in medial and prefrontal cortex, and reduced activation in parietal regions. The validity of these abnormalities as potential endophenotypic markers is supported by the similar directionality to that found in patients and the well-established role of these networks in emotional regulation. Outside the neuroimaging field there is additional evidence for abnormalities in unaffected relatives compared with healthy volunteers, most notably deficits of emotional laden cognition and social cognition.

As molecular genetics studies with the diagnostic phenotype of bipolar disorder advance, allelic variation in proposed susceptibility genes have been associated with neuroimaging data in unaffected relatives as a bottom-up approach to identifying endophenotypes. Allelic variation in CACNA1C and ANK3 has been linked to dysfunctional limbic and prefrontal circuitry in relatives. Polygenic risk score variation is a potentially more powerful technique than single nucleotide polymorphism variation in detecting neuroimaging endophenotypes. Studies utilizing this measure support a genetic contribution to limbic and ventromedial prefrontal cortex hyperactivation as endophenotypic markers of bipolar disorder. The role of unaffected relatives in assessing allelic variation or polygenic risk score effects may be especially valuable in that they carry a genetic milieu more similar to patients, which could better account for gene-gene or gene-environment interactions than healthy volunteer samples. The study of unaffected relatives of patients with bipolar disorder will thus continue to be informative in both bottom-up and top-down approaches to identifying endophenotypic markers, with larger multimodal studies, international collaborative consortia, and the study of relatives from a high genetic risk environment likely to provide future fruitful advances.

REFERENCES

Allin, M. P., Marshall, N., Schulze, K., Walshe, M., Hall, M. H., Picchioni, M., et al. (2010). A functional MRI study of verbal fluency in adults with bipolar disorder and their unaffected relatives. *Psychological Medicine*, *40*(12), 2025–2035. pii: S0033291710000127. https://doi.org/10.1017/S0033291710000127.

Alonso-Lana, S., Valenti, M., Romaguera, A., Sarri, C., Sarro, S., Rodriguez-Martinez, A., et al. (2016). Brain functional changes in first-degree relatives of patients with bipolar disorder: evidence for default mode network dysfunction. *Psychological Medicine*, *46*(12), 2513–2521. https://doi.org/10.1017/S0033291716001148.

Arts, B., Simons, C. J., & Os, J. (2013). Evidence for the impact of the CACNA1C risk allele rs1006737 on 2-year cognitive functioning in bipolar disorder. *Psychiatric Genetics*, *23*(1), 41–42. https://doi.org/10.1097/YPG.0b013e328358641c.

Benca, C. E., Derringer, J. L., Corley, R. P., Young, S. E., Keller, M. C., Hewitt, J. K., et al. (2017). Predicting cognitive executive functioning with polygenic risk scores for psychiatric disorders. *Behavior Genetics*, *47*(1), 11–24. https://doi.org/10.1007/s10519-016-9814-2.

Bergen, S. E., O'Dushlaine, C. T., Ripke, S., Lee, P. H., Ruderfer, D. M., Akterin, S., et al. (2012). Genome-wide association study in a Swedish population yields support for greater CNV and MHC involvement in schizophrenia compared with bipolar disorder. *Molecular Psychiatry*, *17*(9), 880–886. https://doi.org/10.1038/mp.2012.73.

Bootsman, F., Brouwer, R. M., Kemner, S. M., Schnack, H. G., van der Schot, A. C., Vonk, R., et al. (2015). Contribution of genes and unique environment to cross-sectional and longitudinal measures of subcortical volumes in bipolar disorder. *European Neuropsychopharmacology*, *25*(12), 2197–2209. https://doi.org/10.1016/j.euroneuro.2015.09.023.

Bora, E., & Ozerdem, A. (2017). Social cognition in first-degree relatives of patients with bipolar disorder: a meta-analysis. *European Neuropsychopharmacology*, *27*(4), 293–300. https://doi.org/10.1016/j.euroneuro.2017.02.009.

Caseras, X., Tansey, K. E., Foley, S., & Linden, D. (2015). Association between genetic risk scoring for schizophrenia and bipolar disorder with regional subcortical volumes. *Translational Psychiatry, 5*, e692, https://doi.org/10.1038/tp.2015.195.

Chaddock, C. A., Barker, G. J., Marshall, N., Schulze, K., Hall, M. H., Fern, A., et al. (2009). White matter microstructural impairments and genetic liability to familial bipolar I disorder. *The British Journal of Psychiatry, 194*(6), 527–534. pii: 194/6/527. https://doi.org/10.1192/bjp.bp.107.047498.

Chan, S. W., Sussmann, J. E., Romaniuk, L., Stewart, T., Lawrie, S. M., Hall, J., et al. (2016). Deactivation in anterior cingulate cortex during facial processing in young individuals with high familial risk and early development of depression: fMRI findings from the Scottish Bipolar Family Study. *Journal of Child Psychology and Psychiatry, 57*(11), 1277–1286. https://doi.org/10.1111/jcpp.12591.

Chen, C. H., Suckling, J., Lennox, B. R., Ooi, C., & Bullmore, E. T. (2011). A quantitative meta-analysis of fMRI studies in bipolar disorder. *Bipolar Disorders, 13*(1), 1–15. https://doi.org/10.1111/j.1399-5618.2011.00893.x.

Correa-Ghisays, P., Balanza-Martinez, V., Selva-Vera, G., Vila-Frances, J., Soria-Olivas, E., Vivas-Lalinde, J., et al. (2017). Manual motor speed dysfunction as a neurocognitive endophenotype in euthymic bipolar disorder patients and their healthy relatives. Evidence from a 5-year follow-up study. *Journal of Affective Disorders, 215*, 156–162. https://doi.org/10.1016/j.jad.2017.03.041.

Costafreda, S. G., Fu, C. H., Picchioni, M., Kane, F., McDonald, C., Prata, D. P., et al. (2009). Increased inferior frontal activation during word generation: a marker of genetic risk for schizophrenia but not bipolar disorder? *Human Brain Mapping, 30*(10), 3287–3298. https://doi.org/10.1002/hbm.20749.

Craddock, N., & Sklar, P. (2013). Genetics of bipolar disorder. *Lancet, 381*(9878), 1654–1662. https://doi.org/10.1016/S0140-6736(13)60855-7.

Delvecchio, G., Dima, D., & Frangou, S. (2015). The effect of ANK3 bipolar-risk polymorphisms on the working memory circuitry differs between loci and according to risk-status for bipolar disorder. *American Journal of Medical Genetics. Part B, Neuropsychiatric Genetics, 168B*(3), 188–196. https://doi.org/10.1002/ajmg.b.32294.

Dima, D., & Breen, G. (2015). Polygenic risk scores in imaging genetics: usefulness and applications. *Journal of Psychopharmacology, 29*(8), 867–871. https://doi.org/10.1177/0269881115584470.

Dima, D., de Jong, S., Breen, G., & Frangou, S. (2016). The polygenic risk for bipolar disorder influences brain regional function relating to visual and default state processing of emotional information. *Neuroimage Clinical, 12*, 838–844. https://doi.org/10.1016/j.nicl.2016.10.022.

Drapier, D., Surguladze, S., Marshall, N., Schulze, K., Fern, A., Hall, M. H., et al. (2008). Genetic liability for bipolar disorder is characterized by excess frontal activation in response to a working memory task. *Biological Psychiatry, 64*(6), 513–520.

Dutt, A., McDonald, C., Dempster, E., Prata, D., Shaikh, M., Williams, I., et al. (2009). The effect of COMT, BDNF, 5-HTT, NRG1 and DTNBP1 genes on hippocampal and lateral ventricular volume in psychosis. *Psychological Medicine, 39*(11), 1783–1797. https://doi.org/10.1017/S0033291709990316.

Eker, C., Simsek, F., Yilmazer, E. E., Kitis, O., Cinar, C., Eker, O. D., et al. (2014). Brain regions associated with risk and resistance for bipolar I disorder: a voxel-based MRI study of patients with bipolar disorder and their healthy siblings. *Bipolar Disorders, 16*(3), 249–261. https://doi.org/10.1111/bdi.12181.

Emsell, L., Chaddock, C., Forde, N., Van Hecke, W., Barker, G. J., Leemans, A., et al. (2014). White matter microstructural abnormalities in families multiply affected with bipolar I disorder: a diffusion tensor tractography study. *Psychological Medicine*, *44*(10), 2139–2150. https://doi.org/10.1017/S0033291713002845.

Erk, S., Meyer-Lindenberg, A., Schmierer, P., Mohnke, S., Grimm, O., Garbusow, M., et al. (2014). Hippocampal and frontolimbic function as intermediate phenotype for psychosis: evidence from healthy relatives and a common risk variant in CACNA1C. *Biological Psychiatry*, *76*(6), 466–475. https://doi.org/10.1016/j.biopsych.2013.11.025.

Francis, A. N., Mothi, S. S., Mathew, I. T., Tandon, N., Clementz, B., Pearlson, G. D., et al. (2016). Callosal abnormalities across the psychosis dimension: bipolar schizophrenia network on intermediate phenotypes. *Biological Psychiatry*, *80*(8), 627–635. https://doi.org/10.1016/j.biopsych.2015.12.026.

Frangou, S. (2011). Brain structural and functional correlates of resilience to bipolar disorder. *Frontiers in Human Neuroscience*, *5*, 184. https://doi.org/10.3389/fnhum.2011.00184.

Fusar-Poli, P., Howes, O., Bechdolf, A., & Borgwardt, S. (2012). Mapping vulnerability to bipolar disorder: a systematic review and meta-analysis of neuroimaging studies. *Journal of Psychiatry & Neuroscience*, *37*(3), 170–184. https://doi.org/10.1503/jpn.110061.

Gershon, E. S., & Goldin, L. R. (1986). Clinical methods in psychiatric genetics. I. Robustness of genetic marker investigative strategies. *Acta Psychiatrica Scandinavica*, *74*(2), 113–118.

Giakoumaki, S. G., Roussos, P., Rogdaki, M., Karli, C., Bitsios, P., & Frangou, S. (2007). Evidence of disrupted prepulse inhibition in unaffected siblings of bipolar disorder patients. *Biological Psychiatry*, *62*(12), 1418–1422. https://doi.org/10.1016/j.biopsych.2006.12.002.

Glahn, D. C., Knowles, E. E., McKay, D. R., Sprooten, E., Raventos, H., Blangero, J., et al. (2014). Arguments for the sake of endophenotypes: examining common misconceptions about the use of endophenotypes in psychiatric genetics. *American Journal of Medical Genetics. Part B, Neuropsychiatric Genetics*, *165B*(2), 122–130. https://doi.org/10.1002/ajmg.b.32221.

Gunde, E., Novak, T., Kopecek, M., Schmidt, M., Propper, L., Stopkova, P., et al. (2011). White matter hyperintensities in affected and unaffected late teenage and early adulthood offspring of bipolar parents: a two-center high-risk study. *Journal of Psychiatric Research*, *45*(1), 76–82. https://doi.org/10.1016/j.jpsychires.2010.04.019.

Gurung, R., & Prata, D. P. (2015). What is the impact of genome-wide supported risk variants for schizophrenia and bipolar disorder on brain structure and function? A systematic review. *Psychological Medicine*, *45*(12), 2461–2480. https://doi.org/10.1017/S0033291715000537.

Hajek, T., Cooke, C., Kopecek, M., Novak, T., Hoschl, C., & Alda, M. (2015). Using structural MRI to identify individuals at genetic risk for bipolar disorders: a 2-cohort, machine learning study. *Journal of Psychiatry & Neuroscience*, *40*(5), 316–324.

Hajek, T., Cullis, J., Novak, T., Kopecek, M., Blagdon, R., Propper, L., et al. (2013). Brain structural signature of familial predisposition for bipolar disorder: replicable evidence for involvement of the right inferior frontal gyrus. *Biological Psychiatry*, *73*(2), 144–152. https://doi.org/10.1016/j.biopsych.2012.06.015.

Hall, M. H., Chen, C. Y., Cohen, B. M., Spencer, K. M., Levy, D. L., Ongur, D., et al. (2015). Genomewide association analyses of electrophysiological endophenotypes for schizophrenia and psychotic bipolar disorders: a preliminary report. *American Journal of Medical Genetics. Part B, Neuropsychiatric Genetics*, *168B*(3), 151–161. https://doi.org/10.1002/ajmg.b.32298.

Hall, M. H., Schulze, K., Rijsdijk, F., Kalidindi, S., McDonald, C., Bramon, E., et al. (2009). Are auditory P300 and duration MMN heritable and putative endophenotypes of psychotic bipolar disorder? A Maudsley Bipolar Twin and Family Study. *Psychological Medicine*, *39*(8), 1277–1287. pii: S0033291709005261. https://doi.org/10.1017/S0033291709005261.

Hall, M. H., Schulze, K., Sham, P., Kalidindi, S., McDonald, C., Bramon, E., et al. (2008). Further evidence for shared genetic effects between psychotic bipolar disorder and P50 suppression: a combined twin and family study. *American Journal of Medical Genetics. Part B, Neuropsychiatric Genetics, 147B*(5), 619–627. https://doi.org/10.1002/ajmg.b.30653.

Hall, M. H., Spencer, K. M., Schulze, K., McDonald, C., Kalidindi, S., Kravariti, E., et al. (2011). The genetic and environmental influences of event-related gamma oscillations on bipolar disorder. *Bipolar Disorders, 13*(3), 260–271. https://doi.org/10.1111/j.1399-5618.2011.00925.x.

Hallahan, B., Newell, J., Soares, J. C., Brambilla, P., Strakowski, S. M., Fleck, D. E., et al. (2011). Structural magnetic resonance imaging in bipolar disorder: an international collaborative mega-analysis of individual adult patient data. *Biological Psychiatry, 69*(4), 326–335. pii: S0006-3223(10)00913-3. https://doi.org/10.1016/j.biopsych.2010.08.029.

Hanford, L. C., Nazarov, A., Hall, G. B., & Sassi, R. B. (2016). Cortical thickness in bipolar disorder: a systematic review. *Bipolar Disorders, 18*(1), 4–18. https://doi.org/10.1111/bdi.12362.

Harrison, P. J. (2016). Molecular neurobiological clues to the pathogenesis of bipolar disorder. *Current Opinion in Neurobiology, 36*, 1–6. https://doi.org/10.1016/j.conb.2015.07.002.

Hibar, D. P., Westlye, L. T., van Erp, T. G., Rasmussen, J., Leonardo, C. D., Faskowitz, J., et al. (2016). Subcortical volumetric abnormalities in bipolar disorder. *Molecular Psychiatry, 21*(12), 1710–1716. https://doi.org/10.1038/mp.2015.227.

Hulshoff Pol, H. E., van Baal, G. C., Schnack, H. G., Brans, R. G., van der Schot, A. C., Brouwer, R. M., et al. (2012). Overlapping and segregating structural brain abnormalities in twins with schizophrenia or bipolar disorder. *Archives of General Psychiatry, 69*(4), 349–359. pii: 69/4/349. https://doi.org/10.1001/archgenpsychiatry.2011.1615.

Ivleva, E. I., Bidesi, A. S., Keshavan, M. S., Pearlson, G. D., Meda, S. A., Dodig, D., et al. (2013). Gray matter volume as an intermediate phenotype for psychosis: bipolar-schizophrenia network on intermediate phenotypes (B-SNIP). *The American Journal of Psychiatry, 170*(11), 1285–1296. https://doi.org/10.1176/appi.ajp.2013.13010126.

Ivleva, E. I., Moates, A. F., Hamm, J. P., Bernstein, I. H., O'Neill, H. B., Cole, D., et al. (2014). Smooth pursuit eye movement, prepulse inhibition, and auditory paired stimuli processing endophenotypes across the schizophrenia-bipolar disorder psychosis dimension. *Schizophrenia Bulletin, 40*(3), 642–652. https://doi.org/10.1093/schbul/sbt047.

Jogia, J., Ruberto, G., Lelli-Chiesa, G., Vassos, E., Maieru, M., Tatarelli, R., et al. (2011). The impact of the CACNA1C gene polymorphism on frontolimbic function in bipolar disorder. *Molecular Psychiatry, 16*(11), 1070–1071. https://doi.org/10.1038/mp.2011.49.

Kanske, P., Heissler, J., Schonfelder, S., Forneck, J., & Wessa, M. (2013). Neural correlates of emotional distractibility in bipolar disorder patients, unaffected relatives, and individuals with hypomanic personality. *The American Journal of Psychiatry, 170*(12), 1487–1496. https://doi.org/10.1176/appi.ajp.2013.12081044.

Kanske, P., Schonfelder, S., Forneck, J., & Wessa, M. (2015). Impaired regulation of emotion: neural correlates of reappraisal and distraction in bipolar disorder and unaffected relatives. *Translational Psychiatry, 5*, e497. https://doi.org/10.1038/tp.2014.137.

Kempton, M. J., Geddes, J. R., Ettinger, U., Williams, S. C., & Grasby, P. M. (2008). Metaanalysis, database, and meta-regression of 98 structural imaging studies in bipolar disorder. *Archives of General Psychiatry, 65*(9), 1017–1032.

Kempton, M. J., Haldane, M., Jogia, J., Grasby, P. M., Collier, D., & Frangou, S. (2009). Dissociable brain structural changes associated with predisposition, resilience, and disease expression in bipolar disorder. *The Journal of Neuroscience, 29*(35), 10863–10868. https://doi.org/10.1523/JNEUROSCI.2204-09.2009.

Kerner, B. (2015). Toward a deeper understanding of the genetics of bipolar disorder. *Frontiers in Psychiatry, 6*, 105. https://doi.org/10.3389/fpsyt.2015.00105.

Khadka, S., Meda, S. A., Stevens, M. C., Glahn, D. C., Calhoun, V. D., Sweeney, J. A., et al. (2013). Is aberrant functional connectivity a psychosis endophenotype? A resting state functional magnetic resonance imaging study. *Biological Psychiatry, 74*(6), 458–466. https://doi.org/10.1016/j.biopsych.2013.04.024.

Kieseppa, T., Partonen, T., Haukka, J., Kaprio, J., & Lonnqvist, J. (2004). High concordance of bipolar I disorder in a nationwide sample of twins. *The American Journal of Psychiatry, 161*(10), 1814–1821. https://doi.org/10.1176/ajp.161.10.1814.

Kieseppa, T., van Erp, T. G., Haukka, J., Partonen, T., Cannon, T. D., Poutanen, V. P., et al. (2003). Reduced left hemispheric white matter volume in twins with bipolar I disorder. *Biological Psychiatry, 54*(9), 896–905.

Kollmann, B., Scholz, V., Linke, J., Kirsch, P., & Wessa, M. (2017). Reward anticipation revisited- evidence from an fMRI study in euthymic bipolar I patients and healthy first-degree relatives. *Journal of Affective Disorders, 219*, 178–186. https://doi.org/10.1016/j.jad.2017.04.044.

Langan, C., & McDonald, C. (2009). Neurobiological trait abnormalities in bipolar disorder. *Molecular Psychiatry, 14*(9), 833–846. pii: mp200939. https://doi.org/10.1038/mp.2009.39.

Leboyer, M., Bellivier, F., Nosten-Bertrand, M., Jouvent, R., Pauls, D., & Mallet, J. (1998). Psychiatric genetics: search for phenotypes. *Trends in Neurosciences, 21*(3), 102–105.

Lelli-Chiesa, G., Kempton, M. J., Jogia, J., Tatarelli, R., Girardi, P., Powell, J., et al. (2011). The impact of the Val158Met catechol-O-methyltransferase genotype on neural correlates of sad facial affect processing in patients with bipolar disorder and their relatives. *Psychological Medicine, 41*(4), 779–788. https://doi.org/10.1017/S0033291710001431.

Li, C. T., Tu, P. C., Hsieh, J. C., Lee, H. C., Bai, Y. M., Tsai, C. F., et al. (2015). Functional dysconnection in the prefrontal-amygdala circuitry in unaffected siblings of patients with bipolar I disorder. *Bipolar Disorders, 17*(6), 626–635. https://doi.org/10.1111/bdi.12321.

Linke, J., King, A. V., Poupon, C., Hennerici, M. G., Gass, A., & Wessa, M. (2013). Impaired anatomical connectivity and related executive functions: differentiating vulnerability and disease marker in bipolar disorder. *Biological Psychiatry, 74*(12), 908–916. https://doi.org/10.1016/j.biopsych.2013.04.010.

Linke, J., King, A. V., Rietschel, M., Strohmaier, J., Hennerici, M., Gass, A., et al. (2012). Increased medial orbitofrontal and amygdala activation: evidence for a systems-level endophenotype of bipolar I disorder. *The American Journal of Psychiatry, 169*(3), 316–325. https://doi.org/10.1176/appi.ajp.2011.11050711.

Lui, S., Yao, L., Xiao, Y., Keedy, S. K., Reilly, J. L., Keefe, R. S., et al. (2015). Resting-state brain function in schizophrenia and psychotic bipolar probands and their first-degree relatives. *Psychological Medicine, 45*(1), 97–108. https://doi.org/10.1017/S003329171400110X.

Macoveanu, J., Vinberg, M., Madsen, K., Kessing, L. V., Siebner, H. R., & Baare, W. (2016). Unaffected twins discordant for affective disorders show changes in anterior callosal white matter microstructure. *Acta Psychiatrica Scandinavica, 134*(5), 441–451. https://doi.org/10.1111/acps.12638.

Mahon, K., Burdick, K. E., Ikuta, T., Braga, R. J., Gruner, P., Malhotra, A. K., et al. (2013). Abnormal temporal lobe white matter as a biomarker for genetic risk of bipolar disorder. *Biological Psychiatry, 73*(2), 177–182. https://doi.org/10.1016/j.biopsych.2012.07.033.

Matsubara, T., Matsuo, K., Harada, K., Nakano, M., Nakashima, M., Watanuki, T., et al. (2016). Distinct and shared Endophenotypes of neural substrates in bipolar and major depressive disorders. *PLoS One, 11*(12), e0168493. https://doi.org/10.1371/journal.pone.0168493.

Matsuo, K., Kopecek, M., Nicoletti, M. A., Hatch, J. P., Watanabe, Y., Nery, F. G., et al. (2012). New structural brain imaging endophenotype in bipolar disorder. *Molecular Psychiatry*, *17*(4), 412–420. https://doi.org/10.1038/mp.2011.3.

McDonald, C., Bullmore, E. T., Sham, P. C., Chitnis, X., Wickham, H., Bramon, E., et al. (2004). Association of genetic risks for schizophrenia and bipolar disorder with specific and generic brain structural endophenotypes. *Archives of General Psychiatry*, *61*(10), 974–984.

McDonald, C., Marshall, N., Sham, P. C., Bullmore, E. T., Schulze, K., Chapple, B., et al. (2006). Regional brain morphometry in patients with schizophrenia or bipolar disorder and their unaffected relatives. *The American Journal of Psychiatry*, *163*(3), 478–487.

McIntosh, A. M., Job, D. E., Moorhead, T. W., Harrison, L. K., Forrester, K., Lawrie, S. M., et al. (2004). Voxel-based morphometry of patients with schizophrenia or bipolar disorder and their unaffected relatives. *Biological Psychiatry*, *56*(8), 544–552.

McIntosh, A. M., Job, D. E., Moorhead, T. W., Harrison, L. K., Lawrie, S. M., & Johnstone, E. C. (2005). White matter density in patients with schizophrenia, bipolar disorder and their unaffected relatives. *Biological Psychiatry*, *58*(3), 254–257.

McIntosh, A. M., Job, D. E., Moorhead, W. J., Harrison, L. K., Whalley, H. C., Johnstone, E. C., et al. (2006). Genetic liability to schizophrenia or bipolar disorder and its relationship to brain structure. *American Journal of Medical Genetics. Part B, Neuropsychiatric Genetics*, *141*(1), 76–83.

Meda, S. A., Gill, A., Stevens, M. C., Lorenzoni, R. P., Glahn, D. C., Calhoun, V. D., et al. (2012). Differences in resting-state functional magnetic resonance imaging functional network connectivity between schizophrenia and psychotic bipolar probands and their unaffected first-degree relatives. *Biological Psychiatry*, *71*(10), 881–889. https://doi.org/10.1016/j.biopsych.2012.01.025.

Meda, S. A., Ruano, G., Windemuth, A., O'Neil, K., Berwise, C., Dunn, S. M., et al. (2014). Multivariate analysis reveals genetic associations of the resting default mode network in psychotic bipolar disorder and schizophrenia. *Proceedings of the National Academy of Sciences of the United States of America*, *111*(19), E2066–2075. https://doi.org/10.1073/pnas.1313093111.

Meda, S. A., Wang, Z., Ivleva, E. I., Poudyal, G., Keshavan, M. S., Tamminga, C. A., et al. (2015). Frequency-specific neural signatures of spontaneous low-frequency resting state fluctuations in psychosis: evidence from bipolar-schizophrenia network on intermediate phenotypes (B-SNIP) consortium. *Schizophrenia Bulletin*, *41*(6), 1336–1348. https://doi.org/10.1093/schbul/sbv064.

Miskowiak, K. W., Kjaerstad, H. L., Meluken, I., Petersen, J. Z., Maciel, B. R., Kohler, C. A., et al. (2017). The search for neuroimaging and cognitive endophenotypes: a critical systematic review of studies involving unaffected first-degree relatives of individuals with bipolar disorder. *Neuroscience and Biobehavioral Reviews*, *73*, 1–22. https://doi.org/10.1016/j.neubiorev.2016.12.011.

Murray, R. M., Sham, P., Van Os, J., Zanelli, J., Cannon, M., & McDonald, C. (2004). A developmental model for similarities and dissimilarities between schizophrenia and bipolar disorder. *Schizophrenia Research*, *71*(2–3), 405–416.

Nery, F. G., Gigante, A. D., Amaral, J. A., Fernandes, F. B., Berutti, M., Almeida, K. M., et al. (2015). Gray matter volumes in patients with bipolar disorder and their first-degree relatives. *Psychiatry Research*, *234*(2), 188–193. https://doi.org/10.1016/j.pscychresns.2015.09.005.

Nery, F. G., Monkul, E. S., & Lafer, B. (2013). Gray matter abnormalities as brain structural vulnerability factors for bipolar disorder: a review of neuroimaging studies of individuals at high genetic risk for bipolar disorder. *The Australian and New Zealand Journal of Psychiatry*, *47*(12), 1124–1135. https://doi.org/10.1177/0004867413496482.

Nortje, G., Stein, D. J., Radua, J., Mataix-Cols, D., & Horn, N. (2013). Systematic review and voxel-based meta-analysis of diffusion tensor imaging studies in bipolar disorder. *Journal of Affective Disorders, 150*(2), 192–200. https://doi.org/10.1016/j.jad.2013.05.034.

Nurnberger, J. I., Jr., Koller, D. L., Jung, J., Edenberg, H. J., Foroud, T., Guella, I., et al. (2014). Identification of pathways for bipolar disorder: a meta-analysis. *JAMA Psychiatry, 71*(6), 657–664. https://doi.org/10.1001/jamapsychiatry.2014.176.

Owen, M. J., O'Donovan, M. C., Thapar, A., & Craddock, N. (2011). Neurodevelopmental hypothesis of schizophrenia. *The British Journal of Psychiatry, 198*(3), 173–175. https://doi.org/10.1192/bjp.bp.110.084384.

Phillips, M. L., & Swartz, H. A. (2014). A critical appraisal of neuroimaging studies of bipolar disorder: toward a new conceptualization of underlying neural circuitry and a road map for future research. *The American Journal of Psychiatry, 171*(8), 829–843. https://doi.org/10.1176/appi.ajp.2014.13081008.

Piguet, C., Fodoulian, L., Aubry, J. M., Vuilleumier, P., & Houenou, J. (2015). Bipolar disorder: functional neuroimaging markers in relatives. *Neuroscience and Biobehavioral Reviews, 57*, 284–296. https://doi.org/10.1016/j.neubiorev.2015.08.015.

Pompei, F., Dima, D., Rubia, K., Kumari, V., & Frangou, S. (2011). Dissociable functional connectivity changes during the Stroop task relating to risk, resilience and disease expression in bipolar disorder. *NeuroImage, 57*(2), 576–582. https://doi.org/10.1016/j.neuroimage.2011.04.055.

Pompei, F., Jogia, J., Tatarelli, R., Girardi, P., Rubia, K., Kumari, V., et al. (2011). Familial and disease specific abnormalities in the neural correlates of the Stroop task in bipolar disorder. *NeuroImage, 56*(3), 1677–1684. https://doi.org/10.1016/j.neuroimage.2011.02.052.

Power, R. A., Steinberg, S., Bjornsdottir, G., Rietveld, C. A., Abdellaoui, A., Nivard, M. M., et al. (2015). Polygenic risk scores for schizophrenia and bipolar disorder predict creativity. *Nature Neuroscience, 18*(7), 953–955. https://doi.org/10.1038/nn.4040.

Radua, J., Surguladze, S. A., Marshall, N., Walshe, M., Bramon, E., Collier, D. A., et al. (2012). The impact of CACNA1C allelic variation on effective connectivity during emotional processing in bipolar disorder. *Molecular Psychiatry*, https://doi.org/10.1038/mp.2012.61. pii: mp201261.

Rasetti, R., & Weinberger, D. R. (2011). Intermediate phenotypes in psychiatric disorders. *Current Opinion in Genetics & Development, 21*(3), 340–348. https://doi.org/10.1016/j.gde.2011.02.003.

Reginsson, G. W., Ingason, A., Euesden, J., Bjornsdottir, G., Olafsson, S., Sigurdsson, E., et al. (2017). Polygenic risk scores for schizophrenia and bipolar disorder associate with addiction. *Addiction Biology*, https://doi.org/10.1111/adb.12496.

Reus, L. M., Shen, X., Gibson, J., Wigmore, E., Ligthart, L., Adams, M. J., et al. (2017). Association of polygenic risk for major psychiatric illness with subcortical volumes and white matter integrity in UK biobank. *Scientific Reports, 7*, 42140. https://doi.org/10.1038/srep42140.

Roberts, G., Green, M. J., Breakspear, M., McCormack, C., Frankland, A., Wright, A., et al. (2013). Reduced inferior frontal gyrus activation during response inhibition to emotional stimuli in youth at high risk of bipolar disorder. *Biological Psychiatry, 74*(1), 55–61. https://doi.org/10.1016/j.biopsych.2012.11.004.

Ruberto, G., Vassos, E., Lewis, C. M., Tatarelli, R., Girardi, P., Collier, D., et al. (2011). The cognitive impact of the ANK3 risk variant for bipolar disorder: initial evidence of selectivity to signal detection during sustained attention. *PLoS One, 6*(1), e16671. https://doi.org/10.1371/journal.pone.0016671.

Sandoval, H., Soares, J. C., Mwangi, B., Asonye, S., Alvarado, L. A., Zavala, J., et al. (2016). Confirmation of MRI anatomical measurements as endophenotypic markers for bipolar disorder in a new sample from the NIMH genetics of bipolar disorder in Latino populations study. *Psychiatry Research, 247*, 34–41. https://doi.org/10.1016/j.pscychresns.2015.11.004.

Saricicek, A., Yalin, N., Hidiroglu, C., Cavusoglu, B., Tas, C., Ceylan, D., et al. (2015). Neuroanatomical correlates of genetic risk for bipolar disorder: a voxel-based morphometry study in bipolar type I patients and healthy first degree relatives. *Journal of Affective Disorders, 186*, 110–118. https://doi.org/10.1016/j.jad.2015.06.055.

Saricicek, A., Zorlu, N., Yalin, N., Hidiroglu, C., Cavusoglu, B., Ceylan, D., et al. (2016). Abnormal white matter integrity as a structural endophenotype for bipolar disorder. *Psychological Medicine, 46*(7), 1547–1558. https://doi.org/10.1017/S0033291716000180.

Schulze, K. K., Hall, M. H., McDonald, C., Marshall, N., Walshe, M., Murray, R. M., et al. (2007). P50 auditory evoked potential suppression in bipolar disorder patients with psychotic features and their unaffected relatives. *Biological Psychiatry, 62*(2), 121–128. https://doi.org/10.1016/j.biopsych.2006.08.006.

Schulze, K. K., Hall, M. H., McDonald, C., Marshall, N., Walshe, M., Murray, R. M., et al. (2008). Auditory P300 in patients with bipolar disorder and their unaffected relatives. *Bipolar Disorders, 10*(3), 377–386. https://doi.org/10.1111/j.1399-5618.2007.00527.x.

Sepede, G., De Berardis, D., Campanella, D., Perrucci, M. G., Ferretti, A., Serroni, N., et al. (2012). Impaired sustained attention in euthymic bipolar disorder patients and non-affected relatives: an fMRI study. *Bipolar Disorders, 14*(7), 764–779. https://doi.org/10.1111/bdi.12007.

Skudlarski, P., Schretlen, D. J., Thaker, G. K., Stevens, M. C., Keshavan, M. S., Sweeney, J. A., et al. (2013). Diffusion tensor imaging white matter endophenotypes in patients with schizophrenia or psychotic bipolar disorder and their relatives. *The American Journal of Psychiatry, 170*(8), 886–898. https://doi.org/10.1176/appi.ajp.2013.12111448.

Sprooten, E., Barrett, J., McKay, D. R., Knowles, E. E., Mathias, S. R., Winkler, A. M., et al. (2016). A comprehensive tractography study of patients with bipolar disorder and their unaffected siblings. *Human Brain Mapping, 37*(10), 3474–3485. https://doi.org/10.1002/hbm.23253.

Sprooten, E., Brumbaugh, M. S., Knowles, E. E., McKay, D. R., Lewis, J., Barrett, J., et al. (2013). Reduced white matter integrity in sibling pairs discordant for bipolar disorder. *The American Journal of Psychiatry, 170*(11), 1317–1325. https://doi.org/10.1176/appi.ajp.2013.12111462.

Sprooten, E., McIntosh, A. M., Lawrie, S. M., Hall, J., Sussmann, J. E., Dahmen, N., et al. (2012). An investigation of a genomewide supported psychosis variant in ZNF804A and white matter integrity in the human brain. *Magnetic Resonance Imaging, 30*(10), 1373–1380. https://doi.org/10.1016/j.mri.2012.05.013.

Sprooten, E., Sussmann, J. E., Clugston, A., Peel, A., McKirdy, J., Moorhead, T. W., et al. (2011). White matter integrity in individuals at high genetic risk of bipolar disorder. *Biological Psychiatry, 70*(4), 350–356. https://doi.org/10.1016/j.biopsych.2011.01.021.

Strakowski, S. M., Adler, C. M., Almeida, J., Altshuler, L. L., Blumberg, H. P., Chang, K. D., et al. (2012). The functional neuroanatomy of bipolar disorder: a consensus model. *Bipolar Disorders, 14*(4), 313–325. https://doi.org/10.1111/j.1399-5618.2012.01022.x.

Streit, F., Memic, A., Hasandedic, L., Rietschel, L., Frank, J., Lang, M., et al. (2016). Perceived stress and hair cortisol: differences in bipolar disorder and schizophrenia. *Psychoneuroendocrinology, 69*, 26–34. https://doi.org/10.1016/j.psyneuen.2016.03.010.

Sugihara, G., Kane, F., Picchioni, M. M., Chaddock, C. A., Kravariti, E., Kalidindi, S., et al. (2017). Effects of risk for bipolar disorder on brain function: a twin and family study. *European Neuropsychopharmacology, 27,* 494–503. https://doi.org/10.1016/j.euroneuro.2017.03.001.

Surguladze, S. A., Marshall, N., Schulze, K., Hall, M. H., Walshe, M., Bramon, E., et al. (2010). Exaggerated neural response to emotional faces in patients with bipolar disorder and their first-degree relatives. *NeuroImage, 53*(1), 58–64. pii: S1053-8119(10)00821-9. https://doi.org/10.1016/j.neuroimage.2010.05.069.

Tesli, M., Kauppi, K., Bettella, F., Brandt, C. L., Kaufmann, T., Espeseth, T., et al. (2015). Altered brain activation during emotional face processing in relation to both diagnosis and polygenic risk of bipolar disorder. *PLoS One, 10*(7), e0134202. https://doi.org/10.1371/journal.pone.0134202.

Thermenos, H. W., Goldstein, J. M., Milanovic, S. M., Whitfield-Gabrieli, S., Makris, N., Laviolette, P., et al. (2010). An fMRI study of working memory in persons with bipolar disorder or at genetic risk for bipolar disorder. *American Journal of Medical Genetics. Part B, Neuropsychiatric Genetics, 153B*(1), 120–131. https://doi.org/10.1002/ajmg.b.30964.

Tighe, S. K., Reading, S. A., Rivkin, P., Caffo, B., Schweizer, B., Pearlson, G., et al. (2012). Total white matter hyperintensity volume in bipolar disorder patients and their healthy relatives. *Bipolar Disorders, 14*(8), 888–893. https://doi.org/10.1111/bdi.12019.

van der Schot, A. C., Vonk, R., Brans, R. G., van Haren, N. E., Koolschijn, P. C., Nuboer, V., et al. (2009). Influence of genes and environment on brain volumes in twin pairs concordant and discordant for bipolar disorder. *Archives of General Psychiatry, 66*(2), 142–151.

van der Schot, A. C., Vonk, R., Brouwer, R. M., van Baal, G. C., Brans, R. G., van Haren, N. E., et al. (2010). Genetic and environmental influences on focal brain density in bipolar disorder. *Brain, 133*(10), 3080–3092. https://doi.org/10.1093/brain/awq236.

Vasconcelos-Moreno, M. P., Fries, G. R., Gubert, C., Dos Santos, B. T., Fijtman, A., Sartori, J., et al. (2017). Telomere length, oxidative stress, inflammation and BDNF levels in siblings of patients with bipolar disorder: implications for accelerated cellular aging. *International Journal of Neuropsychopharmacology,* https://doi.org/10.1093/ijnp/pyx001.

Walters, J. T., & Owen, M. J. (2007). Endophenotypes in psychiatric genetics. *Molecular Psychiatry, 12*(10), 886–890. https://doi.org/10.1038/sj.mp.4002068.

Wang, T., Zhang, X., Li, A., Zhu, M., Liu, S., Qin, W., et al. (2017). Polygenic risk for five psychiatric disorders and cross-disorder and disorder-specific neural connectivity in two independent populations. *Neuroimage Clinical, 14,* 441–449. https://doi.org/10.1016/j.nicl.2017.02.011.

Wegbreit, E., Cushman, G. K., Puzia, M. E., Weissman, A. B., Kim, K. L., Laird, A. R., et al. (2014). Developmental meta-analyses of the functional neural correlates of bipolar disorder. *JAMA Psychiatry, 71*(8), 926–935. https://doi.org/10.1001/jamapsychiatry.2014.660.

Whalley, H. C., Papmeyer, M., Romaniuk, L., Sprooten, E., Johnstone, E. C., Hall, J., et al. (2012). Impact of a microRNA MIR137 susceptibility variant on brain function in people at high genetic risk of schizophrenia or bipolar disorder. *Neuropsychopharmacology, 37*(12), 2720–2729. https://doi.org/10.1038/npp.2012.137.

Whalley, H. C., Papmeyer, M., Sprooten, E., Romaniuk, L., Blackwood, D. H., Glahn, D. C., et al. (2012). The influence of polygenic risk for bipolar disorder on neural activation assessed using fMRI. *Translational Psychiatry, 2,* e130. https://doi.org/10.1038/tp.2012.60.

Whalley, H. C., Sprooten, E., Hackett, S., Hall, L., Blackwood, D. H., Glahn, D. C., et al. (2013). Polygenic risk and white matter integrity in individuals at high risk of mood disorder. *Biological Psychiatry, 74*(4), 280–286. https://doi.org/10.1016/j.biopsych.2013.01.027.

Whalley, H. C., Sussmann, J. E., Chakirova, G., Mukerjee, P., Peel, A., McKirdy, J., et al. (2011). The neural basis of familial risk and temperamental variation in individuals at high risk of bipolar disorder. *Biological Psychiatry*, *70*(4), 343–349. https://doi.org/10.1016/j. biopsych.2011.04.007.

Whalley, H. C., Sussmann, J. E., Johnstone, M., Romaniuk, L., Redpath, H., Chakirova, G., et al. (2012). Effects of a mis-sense DISC1 variant on brain activation in two cohorts at high risk of bipolar disorder or schizophrenia. *American Journal of Medical Genetics. Part B, Neuropsychiatric Genetics*, *159B*(3), 343–353. https://doi.org/10.1002/ajmg.b.32035.

Whalley, H. C., Sussmann, J. E., Romaniuk, L., Stewart, T., Papmeyer, M., Sprooten, E., et al. (2013). Prediction of depression in individuals at high familial risk of mood disorders using functional magnetic resonance imaging. *PLoS One*, *8*(3), e57357https://doi.org/10.1371/journal.pone.0057357.

Willert, A., Mohnke, S., Erk, S., Schnell, K., Romanczuk-Seiferth, N., Quinlivan, E., et al. (2015). Alterations in neural theory of mind processing in euthymic patients with bipolar disorder and unaffected relatives. *Bipolar Disorders*, *17*(8), 880–891. https://doi.org/10.1111/bdi.12352.

Neurobiological markers of stress in youth at risk for bipolar disorder

6

Manpreet K. Singh

Division of Child and Adolescent Psychiatry, Stanford University School of Medicine, Stanford, CA, United States

CHAPTER OUTLINE

INTRODUCTION

Bipolar disorder (BD) is a serious, chronic, and highly familial psychiatric disorder with an onset typically during adolescence. In youth, bipolar disorder is associated with a high risk of suicide attempts and self-injurious behaviors (Goldstein et al., 2012), high rates of co-occurring psychiatric disorders (Frías, Palma, & Farriols, 2014), family dysfunction (Ferreira et al., 2013), academic problems (Pavuluri et al., 2006), and substance abuse (Goldstein & Bukstein, 2010). While the occurrence of mania defines bipolar I disorder, exposure to stress often precedes and complicates the onset and course of mania (Agnew-Blais & Danese, 2016; Jaworska-Andryszewska & Rybakowski, 2016). Numerous studies have demonstrated that despite advances in identifying cases of bipolar disorder in young people, as well as improvements in the pharmacological management of this disorder, problems related to repeated stress exposures, symptom relapse, medication nonadherence, and comorbid disorders such as substance abuse continue to affect the long-term morbidity and mortality of

Bipolar Disorder Vulnerability. https://doi.org/10.1016/B978-0-12-812347-8.00006-3

individuals with BD (Okkels, Trabjerg, Arendt, & Pedersen, 2017; Post et al., 2010). Importantly, youth offspring of parents with BD are exposed to stress as early as in infancy (Johnson et al., 2014). Despite the adverse impact of bipolar disorder on families and society and its obvious significant public health relevance, the neurophysiological impact of stress in BD is yet to be fully realized.

Neuroscience offers several tools to investigate the early stress-mediated risk factors for the pathogenesis of bipolar disorder. Function in the hypothalamus-pituitary-adrenal (HPA) axis provides an evaluation of consequent physiological effects of stress exposure. Neuroimaging studies have demonstrated that bipolar disorder is a brain-based disorder, and that there are likely to be early stress-mediated neurobiological cascades that are critical to the onset and progression of BD (Apter et al., 2017). Multimodal magnetic resonance imaging (MRI) provides a safe, noninvasive tool that is ideally suited for simultaneously identifying aberrant brain structure, function, and neurochemistry in bipolar disorder. With these in vivo assessments, we can bridge a clinical assessment of stress exposure with biologically mediated neuroendocrine abnormalities to advance our understanding of the pathophysiology of bipolar disorder.

Bipolar disorder is characterized by dysfunction in the regulation of emotion and cognition that is governed by complex gene and environmental interactions. Disturbances in core emotional and cognitive functions examined in the context of stress may provide a basis for understanding the origins of symptom manifestations related to stress exposure in youth at risk for BD. In addition, investigating the neuroendocrine sources of these core features may be more informative than symptom expression alone, particularly when controversies regarding diagnostic criteria of bipolar disorder in high-risk youth inevitably arise. In this chapter, we review clinical, endocrine, structural, functional, and neurochemical assessments of stress in youth with a familial risk for developing bipolar disorder. After reviewing each assessment as it applies to youth at familial risk for developing bipolar disorder, we will conclude by illustrating how, taken together, these studies suggest a model for bipolar disorder that is rooted in abnormalities in stress regulation. Finally, we will propose areas of future study that will further explain stress correlates of mania onset and progression.

CLINICAL CONSEQUENCES OF STRESS IN BIPOLAR RISK

Bipolar disorder (BD) is a highly familial disorder. Twin and family studies have reported a 59%–87% heritability of BD (Smoller & Finn, 2003), reflecting the high level of risk for first-degree relatives of probands with BD to develop the disorder themselves (Wozniak et al., 2012). Children of parents with BD are especially vulnerable to developing mood problems at an early age and more commonly have a severe course of illness (Axelson et al., 2015; Hafeman et al., 2016), signifying the importance of elucidating factors that predict the early onset of BD. Indeed, assessing familial risk can improve identification of bipolar symptom onset in youth (Algorta et al., 2013), and combining this with a biological assessment is likely to increase the accuracy of predicting outcomes in high-risk youth (Brietzke, Mansur, Soczynska, Kapczinski, et al., 2012), while also providing insights about the timing and mechanisms of risk for developing BD. Importantly, offspring of parents with BD have greater exposure to stressful life events compared to offspring of healthy controls, with greater frequency and severity of such events being associated with the presence of Axis I psychiatric disorders (Ostiguy et al., 2009; Pan et al., 2017).

An early exposure to adversity through exposure to parental illness or neglect is likely mediated through increased stress reactivity that may manifest biologically and behaviorally (Duffy, Jones, Goodday, & Bentall, 2015). For example, offspring of parents with BD are biologically sensitive to stress (Ostiguy et al., 2011), and through exposure to significant adversity or negative life events (Ostiguy et al., 2009), and by living in family environments that have low levels of cohesion and organization and high levels of conflict and chaos (Belardinelli et al., 2008; Chang, Blasey, Ketter, & Steiner, 2001; Romero et al., 2005), even unaffected siblings of individuals with BD have been found to show greater levels of impulsivity compared to healthy controls (Lombardo et al., 2012). This behavioral dyscontrol appears to be compounded by a chaotic family environment and poor psychosocial functioning (Bella et al., 2011; Jones & Bentall, 2008). Importantly, however, it is not known whether exposure to such dysfunctional family settings is associated with disruptions in neural circuitry in these high-risk youth.

Not all youth at familial risk for BD go on to develop BD, suggesting that some may be resilient to the mechanisms by which BD is transmitted (Frangou, 2009). Examining healthy offspring of parents with BD provides a unique opportunity to evaluate simultaneously factors associated with risk and resilience, and to examine associations between neural network function and specific intrinsic (e.g., poor inhibitory control) or environmental (e.g., family chaos), as opposed to acute illness-associated (Schneider et al., 2012) factors that precede symptom onset.

ENDOCRINE CONSEQUENCES OF STRESS IN BIPOLAR RISK

Bipolar disorder is frequently associated with hypothalamic-pituitary-adrenal (HPA) axis dysregulation. Whether biologically or psychologically mediated, stress may be responsible for the initiation and progression of a mania diathesis (Muneer, 2016). Indeed, there is a complex interaction between stress and brain development, but most prior studies, when metaanalytically combined, suggest that BD is associated with significantly increased levels of cortisol (basal and postdexamethasone) and ACTH, but not of CRH (Belvederi Murri et al., 2016). One recent study demonstrated a relative *hypo*cortisolism in adult chronic BD, and that youth with BD are more likely to show relative *hyper*cortisolism in the face of stress (Maripuu et al., 2017). This difference may be explained by a variety of theoretical frameworks such as diathesis stress or allostatic load (Brietzke, Mansur, Soczynska, Powell, et al., 2012). Regardless, the deleterious effects of stress on neuroendocrine function motivate the development of early intervention and prevention strategies to reverse the consequences of stress, including ensuing mania.

Stress reactivity related to bipolar risk can be acute, like exposures to specific stressful life events, and chronic, such as repeated exposure to parental manic and depressive episodes or recurrent neglect. Indeed, glucocorticoids are important mediators of neuroprogression toward mania, because when they are chronically in excess, they can incite a number of dysfunctional cellular processes in the mitochondria, lead to oxidative stress and inflammation, and prevent mechanisms of neuroprotection (Grande et al., 2012). Key regions of the brain such as the hippocampus provide negative feedback to the HPA axis such that dysfunction in that loop may account for neuroendocrine abnormalities found in BD. The hippocampus is among several brain regions involved in both emotional processing and the regulation of stress. Its critical role has been demonstrated across species, with studies demonstrating impairments in hippocampal cellular processes such as long-term potentiation, neurogenesis, and dendritic remodeling in pathological disease states and after a corticosteroid challenge (Brown, Rush, & McEwen, 1999).

Stress may be measured in three important ways: endogenously over a basal diurnal cycle, in response to a psychological stressor (e.g., Trier), or by being demonstrated as a consequence of a pharmacological challenge (e.g., corticosteroid). Depending on the reliability of the measurement, basal salivary cortisol may be highly variable (Goodday et al., 2016) and in some instances has not distinguished risk groups from fully syndromal individuals with BD (Deshauer et al., 2006). However, several studies have now demonstrated that youth at risk for BD have higher basal cortisol levels after waking and throughout the day compared to typically developing controls (Ellenbogen et al., 2006, 2010; Ellenbogen, Hodgins, & Walker, 2004; Ostiguy et al., 2011). When prospectively followed for a few years in late adolescence, elevated daytime cortisol persisted (Ellenbogen et al., 2010) and predicted the development of an affective disorder in high-risk offspring (Ellenbogen, Hodgins, Linnen, & Ostiguy, 2011). In contrast, in response to a Trier Social Stress Test (TSST) in laboratory psychological stressor, youth offspring of parents with BD did not separate from healthy controls and showed normal reactivity in response to a psychological stressor (Ellenbogen et al., 2006).

Other factors might contribute to maladaptive stress reactivity in youth at risk for BD. For example, parenting style contributes to stress response in high-risk youth, such that low levels of structure provided by parents in middle childhood predicted heightened cortisol upon awakening and during a TSST (Ellenbogen & Hodgins, 2009). Childhood trauma appears to have a differential effect on daytime cortisol levels among bipolar offspring compared to healthy controls when presented with a pharmacological challenge with dexamethasone, independent of lifetime or current psychiatric diagnoses, or stressful life events (Schreuder et al., 2016). Such childhood trauma-related changes in daytime HPA axis activity thus appear to be a specific trait in bipolar offspring who have increased risk for mood disorders compared to healthy individuals. Collectively, these findings implicate a unique vulnerability toward dysfunctional stress response in BD offspring.

NEURAL CONSEQUENCES OF STRESS IN BIPOLAR RISK

The negative impact of stress on brain development is apparently the highest among youth who are the most disadvantaged (Noble et al., 2015). As described above, and in previous chapters, youth offspring of parents with BD are commonly exposed to such disadvantage due to parental illness. In these parents, BD is principally characterized by disruptions in prefrontal and subcortical neural networks (Green, Cahill, & Malhi, 2007; Houenou et al., 2011; Leow et al., 2013). Findings from task-based fMRI studies in youth who have developed BD have demonstrated either over- or underactivation of prefrontal emotion regulatory regions (Pavuluri et al., 2012) or limbic hyperactivity (Garrett et al., 2012; Pavuluri, Passarotti, Harral, & Sweeney, 2009; Rich et al., 2006). It is unclear from these studies whether the observed neural dysfunction precedes or is a consequence of bipolar illness. Contemporary studies have suggested that brain regions implicated in the bipolar syndrome are showing evidence of potential maladaptive and adaptive responses to stress preceding the onset of mood or other psychiatric symptoms. Such neural findings are summarized according to neuroimaging modality below.

STRUCTURAL FINDINGS IN YOUTH AT RISK FOR BIPOLAR DISORDER

Few studies have examined neurobiological endophenotypes associated with risk for bipolar disorder independent of common confounds associated with illness onset including comorbidities, medication exposure, or substance use. Studies of youth at risk for bipolar disorder not only identify potential endophenotypic neural abnormalities that may exist prior to illness onset, but also may be useful to identify markers associated with resilience in youth. High-risk offspring may be followed longitudinally to assess how these markers progress with illness development.

Studies in youth at risk for bipolar disorder may be categorized based on the level of symptoms present at the time of assessment. Some youth offspring of parents with bipolar disorder already show phenotypic expression of mood symptoms that may not meet duration or severity threshold criteria for fully syndromal bipolar disorder, while other youth at risk for bipolar disorder may be symptom-free at the time of assessment. In one study, 22 youth of parents with bipolar disorder had amygdala, hippocampus, and thalamus volumes measured by manual tracing; of note, although they did not meet criteria for bipolar disorder, these young people exhibited ADHD and moderate mood symptoms at the time of assessment. No significant volumetric differences were found in any of these brain regions as compared to youth of healthy parents (Karchemskiy et al., 2011), suggesting that any morphometric abnormalities in these structures found in subjects with bipolar disorder may occur only after more prolonged illness rather than as a preexisting risk factor. This study may have been underpowered to differentiate at-risk youth most likely to progress to bipolar disorder from healthy subjects, as conversion rates in at-risk youth are relatively low. Similarly, another structural neuroimaging study of at-risk youth found no structural abnormalities differentiating symptomatic high-risk offspring of parents with bipolar disorder from youth of healthy parents (Singh et al., 2008). However, asymptomatic or a healthy subset of offspring of parents with bipolar disorder in this study showed trends for increased prefrontal cortical volumes (Singh et al., 2008), suggesting an abnormality in typical neuronal proliferation, in pruning of prefrontal cortical circuits, or representing a compensatory neuroprotective effect. In another study, increases of gray matter volume in the parahippocampal gyri were found in healthy offspring of parents with bipolar disorder compared to offspring of healthy parents (Ladouceur et al., 2008). These latter two studies suggested either the presence of trait-related structural changes predating illness onset, an abnormality in the healthy neuronal proliferation or pruning, or compensatory effects on neural circuits to prevent the onset of mood symptoms in youth at risk for bipolar disorder. Moreover, increases in prefrontal and parahippocampal regions and lack of structural differences in other regions from typically developing youth may represent features of resilience rather than risk in a subset of at-risk youth who do not develop symptoms. Longitudinal studies are needed to confirm whether volumetric changes represent risk or resilience factors (or neither) for the development of bipolar disorder.

Cortisol reactivity to stress has been linked to white matter microstructure in early childhood, and this relation is moderated by positive early caregiving (Sheikh et al., 2014). White matter structural abnormalities related to bipolar risk appear to be both focal to white matter integrity of various limbic regions of the brain and diffuse, suggesting nonselective or network-mediated downstream responses of stress. Where gray matter volume differences have largely been absent in healthy offspring of BD parents, decreases of white matter volume in the prefrontal, occipital, and parietal lobes were recently reported in bipolar offspring prior to any psychiatric symptom development (Nery et al., 2017). Diffusion tensor imaging (DTI) provides a more detailed assessment of white matter microstructure, and compared to healthy controls, bipolar and at-risk groups have shown reduced fractional anisotrophy (FA) in bilateral superior longitudinal fasciculus (SLF), suggesting trait-related white matter deficits (Frazier et al., 2007). More recent studies of symptomatic high-risk youth demonstrated diffuse white matter tract abnormalities in the cingulum, superior fronto-occipital fasciculus (SFOF), and SLF in offspring of parents with BD (Roybal et al., 2015), and widespread FA alterations in association and projection tracts in youth with subthreshold manic symptoms (Paillère Martinot et al., 2014). Another DTI study in 20 healthy offspring with a parent diagnosed with bipolar disorder and 25 healthy control offspring of healthy parents showed altered white matter development in the corpus callosum and temporal associative tracts (Versace et al., 2010). Given that this sample was without any medical or psychiatric diagnosis, these findings may represent an endophenotype unrelated to bipolar disorder development or potential early resiliency markers or risk factors for future bipolar disorder development. DTI analyses conducted in conjunction with neuroendocrine assessments and over time are needed to understand these results in the context of bipolar risk as a consequence of stress.

TASK-BASED FUNCTIONAL MRI STUDIES IN YOUTH AT RISK FOR BIPOLAR DISORDER

Evidence from youth at risk for BD who have not fully developed BD suggest certain patterns of aberrant neural activations in key prefrontal and limbic regions including the ventromedial and ventrolateral (VLPFC) prefrontal cortices, the striatum, and the amygdala (Kim et al., 2012; Mourão-Miranda et al., 2012; Olsavsky et al., 2012; Roberts et al., 2012) and impairments in neurocognitive performance (Belleau et al., 2013; Brotman et al., 2008; Glahn et al., 2010; Whitney et al., 2013) that may precede BD onset. Most of these studies have been limited, however, by symptom or illness-related confounds such as mood state, comorbidities, and medication exposure (Chang et al., 2004; Garrett et al., 2012; Passarotti et al., 2012; Pavuluri et al., 2012; Rich et al., 2006; Singh et al., 2013), and have reported activations in discrete brain regions rather than the functional interactions among them. Nevertheless, taken together, these studies raise the possibility that disruption of connections among different neural regions that constitute large-scale networks (Mesulam, 1990) may represent early risk factors for developing BD.

Recent studies of still healthy youth at risk for BD have identified neurofunctional risk factors that support a well-characterized dysfunctional stress response among these youth, which appears to precede the onset of any psychiatric symptoms (Ellenbogen et al., 2006, 2010, 2011, 2012). Relative to control populations, healthy offspring of parents with BD have aberrant neural responses to face processing (Olsavsky et al., 2012; Wiggins et al., 2017) and monetary rewards (Singh, Kelley, et al., 2014), and dysfunctional emotion processing during sustained attention (Welge et al., 2016). Although these findings are inferred to be trait markers unrelated to any symptom development, it is challenging to disentangle whether performance represents a vulnerability toward mania or a compensatory marker of neuroprotection in the face of external stress to prevent symptoms from developing.

RESTING-STATE FUNCTIONAL CONNECTIVITY IN YOUTH AT RISK FOR BIPOLAR DISORDER

Functional MRI studies typically use paradigms that target specific brain regions and functions associated with cognition and emotion. Task-independent spontaneous resting-state functional connectivity (RSFC) has been used to better understand brain activity and relations among prefrontal and subcortical structures at rest. Unlike task-activation studies, RSFC can directly assess neural systems without contamination by differences in performance that might inflate individual variability across the study group. RSFC is determined by correlating this activity temporally and such patterns are believed to reflect synchronous interaction among brain regions. One particular resting-state network in healthy subjects is the default mode network (DMN), which collectively involves the anterior cingulate (ACC) and a large portion of medial prefrontal cortex (MPFC) extending inferiorly into the orbitofrontal cortex (OFC). This network may be involved with internally generated thought and is inhibited when attending to external stimuli requiring attention and cognition. Younger children have weaker connectivity in these areas. It is likely that the DMN becomes established during adolescence or young adulthood (Fair et al., 2008).

Youth at risk for BD have altered patterns of intrinsic functional connectivity both within brain networks and between key prefrontal and subcortical brain regions. Compared to controls, high-risk offspring have increased connectivity between the ventrolateral prefrontal cortex (VLPFC) and executive control networks, and decreased functional connectivity between the left VLPFC and left caudate (Singh, Chang, et al., 2014). Importantly, older and higher-functioning high-risk offspring showed stronger connectivity in the VLPFC region of left ECN, suggesting a potential neuroprotective mechanism that may prevent the onset of mood symptoms for these youth as they pass through adolescence. However, older high-risk youth also had stronger connectivity between the left VLPFC and caudate than did their younger peers, but had weaker connectivity between these regions with more family chaos. It is possible that low fronto-striatal connectivity in high-risk youth represents a vulnerability marker for developing BD that is compounded by a chaotic family environment. This is the first study to demonstrate the impact of family stress on neural connectivity in youth at risk for BD. The association between a disorganized family structure and disrupted fronto-striatal connectivity in high-risk youth motivates early family intervention to prevent the onset or progression of psychiatric symptoms in youth at familial risk for psychopathology (Miklowitz et al., 2017).

MAGNETIC RESONANCE SPECTROSCOPY

Data demonstrating macroscopic structural and functional changes in selective brain regions in bipolar disorder imply underlying cellular and molecular dysfunction that requires further investigation. Magnetic resonance spectroscopy (MRS) is a noninvasive neuroimaging method that yields molecular-level biochemical data to examine neuronal function quantitatively. Most MRS studies in youth with bipolar disorder have employed proton (^{1}H-MRS) acquisitions focused primarily on key prefrontal cortical regions. For example, studies of bipolar youth have shown altered medial and dorsolateral prefrontal concentrations of *N*-acetyl aspartate (NAA) and phosphocreatine/creatine (PCr/Cr), healthy nerve cell markers putatively involved in maintaining fluid balance, energy production, and myelin formation in the brain. In addition, increases and decreases in prefrontal myo-inositol (mI) levels, a marker for cellular metabolism and second messenger signaling pathways, have also been found in youth with bipolar disorder. Some, but not all, prior studies have demonstrated that alterations in neurometabolite concentrations may precede the onset of bipolar disorder, and are described below.

N-Acetyl aspartate (NAA) decreases have been observed in pediatric bipolar disorder in prefrontal regions, including the dorsolateral prefrontal cortex (DLPFC) and medial prefrontal cortex. The relation of these neurometabolic findings to stress exposure has not yet been directly examined. However, these studies do suggest abnormal dendridic arborization and neuropil in the neurodevelopmental milieu of mania, loss of neuronal matter due to age-dependent processes, and disease-related fluid shifts in prefrontal regions of the brain after the development of fully syndromal mania or long duration of bipolar illness (Gallelli et al., 2005).

Increased orbitofrontal myo-inositol (mI) has been demonstrated in offspring of parents with bipolar disorder who were scanned while they had moderate mood symptoms not meeting criteria for bipolar I or II disorder (Cecil, DelBello, Sellars, & Strakowski, 2003). Another neurometabolite measurable by ^{1}H-MRS in the prefrontal cortex that may be associated with abnormal mood regulation is the excitatory neurotransmitter glutamate. In a recent study of offspring of parents with bipolar disorder, decreases in ACC glutamate concentrations and trends for decreases in glutamate levels relative to creatine concentrations were found, but only in a subset of youth who had developed syndromal mania (Singh et al., 2010). This suggests that for high-risk offspring, altered glutamatergic functioning may represent a marker for a more fully symptomatic clinical course of mania rather than for familial risk alone.

Studies on the neural aspects of emotion have discovered that, in addition to regulating motor coordination, balance, and speech, the cerebellum may play an important role in regulating emotion. The cerebellum has rich bidirectional connections to key regions in the cerebral cortex that modulate emotion. Symptomatic offspring of parents with bipolar disorder who themselves have mood disorders other than bipolar I or II disorder have demonstrated decreased concentrations of N-acetyl aspartate (NAA) and phosphocreatine/creatine (PCr/Cr) in the cerebellar vermis (Cecil et al., 2003). Cerebellar vermis mI and choline deficits were also recently observed in youth at familial risk for bipolar disorder who had a nonbipolar mood disorder compared with healthy subjects (Singh et al., 2011). Myo-inositol reductions in this latter study may be associated with altered cellular signaling via second messenger pathways, regulation of neuronal osmolarity, and metabolism of membrane-bound phospholipids. Similarly, reductions in choline could reflect disruption in cell membrane synthesis, maintenance, and repair. As these two studies sampled youth at risk for bipolar disorder who did not meet criteria for bipolar I or II disorder, these findings suggest that metabolite alterations in the cerebellum may be occurring even prior to the development of a fully symptomatic clinical course of bipolar disorder.

In summary, although researchers may disagree about which brain regions are the most likely candidates of vulnerability for the onset and persistence of bipolar symptoms, most would agree that altered interactions between prefrontal and subcortical regions of the brain appear to contribute to dysfunctional regulation of emotion and cognitive processes in the face of stress. Moreover, altered interactions among these regions may be a developmental phenomenon. For example, in typical neurodevelopment, subcortical gray matter development precedes prefrontal development and pruning, and there appears to be a shift in functional dependence from subcortical to prefrontal structures around puberty (Thompson et al., 2000). Consequently, puberty may represent a critical period when such functional shifts are vulnerable to alterations and the development of psychopathology (Sinclair, Webster, Wong, & Weickert, 2011). It is therefore not surprising that the most common onset of bipolar disorder occurs in adolescence.

NEUROBIOLOGICAL MARKERS OF RESILIENCE

Although bipolar disorder in parents confer vulnerability for a broad range of problems for children, we know far less about factors that boost resilience, or that increase the capacity of youth to avoid maladaptive consequences of having a parent with BD. Nonhuman animal studies have elucidated neural and molecular mechanisms of resilience (Karatsoreos & McEwen, 2011), but limited studies in humans have examined the mechanisms by which resilience shapes health into adulthood (Russo et al., 2012; Smieskova, Fusar-Poli, Riecher-Rössler, & Borgwardt, 2012). Indeed, there are individual differences in biological sensitivity to negative and positive experiences (Beauchaine et al., 2011), determined in part by factors such as social context (Silk et al., 2007), genetic endowment, physical environment and resources (Suchdev et al., 2017), and temperament (Southwick & Charney, 2012). Moreover, resilience is mediated by adaptive changes in neural circuits that regulate reward, emotion reactivity, and social behavior, and involve numerous molecular pathways that shape the functioning of these neural circuits to mediate successful coping with stress (Feder, Nestler, & Charney, 2009). However, the neurobiology of resilience or the enduring effects of key adaptive systems have rarely been examined in youth with a familial risk for BD (Neigh et al., 2013). In one study of healthy offspring of parents with BD, offspring with stronger ventrolateral prefrontal cortex connectivity could more accurately and cost-effectively predict long-term outcomes for high-risk youth. If we knew which factors confer resilience versus risk (Greenberg, 2006), we could leverage resilient factors to develop better and lasting interventions (Beardslee et al., 1992; Beardslee, Gladstone, Wright, & Cooper, 2003; Everly Jr, Smith, & Lating, 2009; Leve, Fisher, & Chamberlain, 2009; Lieberman, Perkins, & Fredrik Jarskog, 2007). Critical analyses are needed to examine the biological correlates of resilience in offspring of parents with BD that comprehensively assess neural, neuroendocrine, and cognitive factors that predict better functioning in high-risk youth.

NEUROENDOCRINE MODEL OF STRESS IN RISK FOR BIPOLAR DISORDER

Taking into account genetic and epigenetic factors, this review is guided by a theoretical model of stress-related processes underlying bipolar disorder onset and progression. Stress-related factors preceding or relating to illness onset may be considered etiological and should be differentiated from factors that may be associated with the course or progression of illness. For example, genetic, inflammatory, and stress diatheses of bipolar disorder may all be etiological and expressed phenotypically as dysregulation of emotion: hyperarousal, hypersensitivity, difficulty with affect decoding or labeling, or problems with "repair" of emotional responses. With time, chronic illness-related stress may incite more stress and inflammatory states, which then may have a direct neurotoxic effect, contributing to the development and progression of mood episodes. Moreover, the combination of genetic vulnerability, inflammation, temperamental phenotypes, and disturbances in mood regulation converge in the expression of early subthreshold forms of bipolar disorder: persistent hypomania or depression, episodic irritability, rapid mood fluctuation, and low social functioning. These pathological processes, layered with a family history of bipolar disorder, may elevate the liability for progression to bipolar disorder and to neurobiological sequelae.

Over time, untreated or repeated mood episodes may place individuals with bipolar disorder at risk for neurobiological sequelae manifested as structural and functional brain impairment. One stress model that may explain the onset of bipolar disorder in youth is an abnormality in prefrontal-subcortical connectivity that may result in failure of normal prefrontal regulation of subcortical processes associated with emotion. This failure may in turn result in the appearance of disease-specific anatomic abnormalities, thereby creating vulnerabilities in other brain regions involved in emotion including the ventrolateral prefrontal cortex (VLPFC), anterior cingulate cortex (ACC), thalamus, striatum, hippocampus, and cerebellar vermis. Abnormal emotion regulation and cognition in bipolar disorder may arise from aberrant reciprocal connections between amygdala and dorsal and ventral prefrontal areas as described by others in mood disorders in general. Comparisons of pediatric and adult bipolar risk studies of neuroendocrine responses to stress add a developmental component to this model, suggesting that hypercortisolemia and limbic hyperactivity (e.g., in the amygdala) may occur early in childhood, eventually becoming associated with hypocortisolemia and decreased prefrontal activation and control over amygdala activity in adulthood. As bipolar disorder progresses, prefrontal abnormalities become more important, and neurodegenerative sequelae of the illness prevail.

Based on the studies reviewed in this chapter, a neurobiological framework for stress-mediated development of bipolar disorder may be postulated. Literature on adults with bipolar disorder points to a neurodegenerative model that has created an imperative for early intervention to prevent neuronal loss and to regenerate neural tissue to restore function. In youth with bipolar disorder, we have the capacity to lessen the impact of common confounds associated with chronic course of illness, including medication exposure and longstanding comorbidities such as substance dependence, and we can better examine etiological factors associated with the onset and progression of bipolar disorder. With poor coping skills, environmental stress, and inborn genetic vulnerabilities, children may develop mood dysregulation and begin to exhibit dysfunction in cognitive, emotional, and psychosocial domains. Pathological reactions to stress can be detected acutely by clinical symptoms, and over time, by examining brain structural and functional changes, which interrupt healthy development of networks of brain structures that are important for mood regulation. Without adequate intervention, children then eventually develop a full manic episode and display further functional connectivity abnormalities between nodes in mood regulation networks; e.g., inverse functional connectivity between amygdala and DLPFC is lost, or task-dependent excessive medial prefrontal cortex (MPFC) activation is observed. With repeated mood episodes into adulthood, further neurodegeneration of prefrontal areas might occur, leading to rapid cycling, treatment resistance, and more severe episodes, causing increasing morbidity and mortality.

Studies in youth at risk for bipolar disorder indicate that most structural and neurochemical changes do not occur prior to the onset of syndromal mania, suggesting the possibility of preventing disease onset and progression if an underlying mechanism can be identified. Furthermore, a relative lack of morphometric abnormalities in healthy youth at high risk for bipolar disorder points to the probability that structural abnormalities in BD are likely a consequence of illness. Studies in healthy offspring of parents with bipolar disorder suggest clues for resilience factors that may be important in preventing the onset of illness, including enlargement of prefrontal and limbic structures, and increased intrinsic connectivity between the emotion regulating VLPFC and executive control networks. Together, these studies in youth at risk for bipolar disorder are identifying many important early neuroendocrine findings that may be amenable to early intervention to prevent or ameliorate the natural course of bipolar disorder. Larger studies are needed to test these suggestions.

CONCLUSION

Convergence in clinical, endocrine, structural, functional, and neurochemical abnormalities preceding the onset of BD support the hypothesis that stress-mediated neuroendocrine pathways underlie the pathophysiology of bipolar disorder. Available studies to date, however, have been limited by lack of consistent and comprehensive assays of stress, which are then correlated with neural and clinical assessments. Moreover, once symptoms in high-risk youth become established, confounding illness-related variables make it difficult to elucidate the interactions between stress and symptoms in the neurodevelopment of bipolar disorder. Use of other measurements and technologies, such as serum cortisol and genetic markers of stress, are of great interest to the field. Advances in genetic and imaging technology may enable us to understand the role of genetics in bipolar disorder development.

Other MRI techniques, such as cortical thickness measurements or advanced brain mapping approaches, have not yet been related to other measures of stress in youth at risk for bipolar disorder. Replication of the extant literature is also needed to resolve inconsistencies across groups in the presence, absence, and directionality of neuroendocrine findings. Functional and resting-state connectivity studies are important to provide a network-based understanding of this complex disorder. In addition, paradigms that highlight core symptoms of mania that are related to acute or chronic exposures to stress could expand how we conceptualize the etiopathogenesis of bipolar disorder. Studies that stage and prospectively follow youth from healthy, to subthreshold, to fully syndromal forms of bipolar disorder would aid in characterizing risk more accurately, and improving predictive power toward conversion. In summary, it is crucial that we find systematic ways of integrating clinical, endocrine, and neuroimaging data across multiple modalities with the use of multivariate statistical approaches and conduct meaningful longitudinal studies in at-risk populations.

Despite recent progress in understanding the neurobiological basis of bipolar disorder, future longitudinal controlled studies are needed to examine the effects of stress on the development of mania, in order to identify neurobiomarkers that establish rational prevention and treatment strategies in youth at risk for developing bipolar disorder, and to determine the neurodevelopmental and genetic mechanisms that contribute to illness onset. With additional investigations, the complex etiopathophysiology of bipolar disorder may be clarified so that youth at risk for bipolar disorder may be more accurately diagnosed and effectively treated.

REFERENCES

Agnew-Blais, J., & Danese, A. (2016). Childhood maltreatment and Unfavourable clinical outcomes in bipolar disorder: a systematic review and meta-analysis. *The Lancet Psychiatry*, *3*(4), 342–349.

Algorta, G. P., et al. (2013). An inexpensive family index of risk for mood issues improves identification of pediatric bipolar disorder. *Psychological Assessment*, *25*(1), 12–22.

Apter, G., et al. (2017). Update on mental health of infants and children of parents affected with mental health issues. *Current Psychiatry Reports*, *19*(10), 72.

Axelson, D., et al. (2015). Diagnostic precursors to bipolar disorder in offspring of parents with bipolar disorder: a longitudinal study. *The American Journal of Psychiatry*, *172*(7), 638–646.

Beardslee, W. R., Gladstone, T. R. G., Wright, E. J., & Cooper, A. B. (2003). A family-based approach to the prevention of depressive symptoms in children at risk: evidence of parental and child change. *Pediatrics*, *112*(2), e119–131.

Beardslee, W. R., et al. (1992). Initial findings on preventive intervention for families with parental affective disorders. *The American Journal of Psychiatry*, *149*(10), 1335–1340.

Beauchaine, T. P., et al. (2011). The effects of allostatic load on neural systems subserving motivation, mood regulation, and social affiliation. *Development and Psychopathology*, *23*(4), 975–999.

Belardinelli, C., et al. (2008). Family environment patterns in families with bipolar children. *Journal of Affective Disorders*, *107*(1–3), 299–305.

Bella, T., et al. (2011). Psychosocial functioning in offspring of parents with bipolar disorder. *Journal of Affective Disorders*, *133*(1–2), 204–211.

Belleau, E. L., et al. (2013). Aberrant executive attention in unaffected youth at familial risk for mood disorders. *Journal of Affective Disorders*, *147*(1–3), 397–400.

Belvederi Murri, M., et al. (2016). The HPA Axis in bipolar disorder: systematic review and meta-analysis. *Psychoneuroendocrinology*, *63*, 327–342.

Brietzke, E., Mansur, R. B., Soczynska, J. K., Kapczinski, F., et al. (2012). Towards a multifactorial approach for prediction of bipolar disorder in at risk populations. *Journal of Affective Disorders*, *140*(1), 82–91.

Brietzke, E., Mansur, R. B., Soczynska, J., Powell, A. M., et al. (2012). A theoretical framework informing research about the role of stress in the pathophysiology of bipolar disorder. *Progress in Neuro-Psychopharmacology & Biological Psychiatry*, *39*(1), 1–8.

Brotman, M. A., et al. (2008). Risk for bipolar disorder is associated with face-processing deficits across emotions. *Journal of the American Academy of Child and Adolescent Psychiatry*, *47*(12), 1455–1461.

Brown, E. S., Rush, A. J., & McEwen, B. S. (1999). Hippocampal remodeling and damage by corticosteroids: implications for mood disorders. *Neuropsychopharmacology: Official Publication of the American College of Neuropsychopharmacology*, *21*(4), 474–484.

Cecil, K. M., DelBello, M. P., Sellars, M. C., & Strakowski, S. M. (2003). Proton magnetic resonance spectroscopy of the frontal lobe and cerebellar vermis in children with a mood disorder and a familial risk for bipolar disorders. *Journal of Child and Adolescent Psychopharmacology*, *13*(4), 545–555.

Chang, K. D., Blasey, C., Ketter, T. A., & Steiner, H. (2001). Family environment of children and adolescents with bipolar parents. *Bipolar Disorders*, *3*(2), 73–78.

Chang, K., et al. (2004). Anomalous prefrontal-subcortical activation in familial pediatric bipolar disorder: a functional magnetic resonance imaging investigation. *Archives of General Psychiatry*, *61*(8), 781–792.

Deshauer, D., et al. (2006). Salivary cortisol secretion in remitted bipolar patients and offspring of bipolar parents. *Bipolar Disorders*, *8*(4), 345–349.

Duffy, A., Jones, S., Goodday, S., & Bentall, R. (2015). Candidate risks indicators for bipolar disorder: early intervention opportunities in high-risk youth. *The International Journal of Neuropsychopharmacology*, *19*(1).

Ellenbogen, M. A., & Hodgins, S. (2009). Structure provided by parents in middle childhood predicts cortisol reactivity in adolescence among the offspring of parents with bipolar disorder and controls. *Psychoneuroendocrinology*, *34*(5), 773–785.

Ellenbogen, M. A., Hodgins, S., Linnen, A.-M., & Ostiguy, C. S. (2011). Elevated daytime cortisol levels: a biomarker of subsequent major affective disorder? *Journal of Affective Disorders*, *132*(1–2), 265–269.

Ellenbogen, M. A., Hodgins, S., & Walker, C.-D. (2004). High levels of cortisol among adolescent offspring of parents with bipolar disorder: a pilot study. *Psychoneuroendocrinology*, *29*(1), 99–106.

Ellenbogen, M. A., et al. (2006). Daytime cortisol and stress reactivity in the offspring of parents with bipolar disorder. *Psychoneuroendocrinology*, *31*(10), 1164–1180.

Ellenbogen, M. A., et al. (2010). High cortisol levels in the offspring of parents with bipolar disorder during two weeks of daily sampling. *Bipolar Disorders*, *12*(1), 77–86.

Ellenbogen, M. A., et al. (2012). Salivary cortisol and interpersonal functioning: an event-contingent recording study in the offspring of parents with bipolar disorder. *Psychoneuroendocrinology*, *38*(7), 997–1006.

Everly, G. S., Jr., Smith, K. J., & Lating, J. M. (2009). A rationale for cognitively-based resilience and psychological first aid (PFA) training: a structural modeling analysis. *International Journal of Emergency Mental Health*, *11*(4), 249–262.

Fair, D. A., et al. (2008). The maturing architecture of the Brain's default network. *Proceedings of the National Academy of Sciences of the United States of America*, *105*(10), 4028–4032.

Feder, A., Nestler, E. J., & Charney, D. S. (2009). Psychobiology and molecular genetics of resilience. *Nature Reviews. Neuroscience*, *10*(6), 446–457.

Ferreira, G. S., et al. (2013). Dysfunctional family environment in affected versus unaffected offspring of parents with bipolar disorder. *The Australian and New Zealand Journal of Psychiatry*, *47*(11), 1051–1057.

Frangou, S. (2009). Risk and resilience in bipolar disorder: rationale and design of the vulnerability to bipolar disorders study (VIBES). *Biochemical Society Transactions*, *37*(Pt 5), 1085–1089.

Frazier, J. A., et al. (2007). White matter abnormalities in children with and at risk for bipolar disorder. *Bipolar Disorders*, *9*(8), 799–809.

Frías, Á., Palma, C., & Farriols, N. (2014). Comorbidity in pediatric bipolar disorder: prevalence, clinical impact, etiology and treatment. *Journal of Affective Disorders*, *174C*, 378–389.

Gallelli, K. A., et al. (2005). *N*-Acetylaspartate levels in bipolar offspring with and at high-risk for bipolar disorder. *Bipolar Disorders*, *7*(6), 589–597.

Garrett, A. S., et al. (2012). Abnormal amygdala and prefrontal cortex activation to facial expressions in pediatric bipolar disorder. *Journal of the American Academy of Child and Adolescent Psychiatry*, *51*(8), 821–831.

Glahn, D. C., et al. (2010). Neurocognitive endophenotypes for bipolar disorder identified in multiplex multigenerational families. *Archives of General Psychiatry*, *67*(2), 168–177.

Goldstein, B. I., & Bukstein, O. G. (2010). Comorbid substance use disorders among youth with bipolar disorder: opportunities for early identification and prevention. *The Journal of Clinical Psychiatry*, *71*(3), 348–358.

Goldstein, T. R., et al. (2012). Predictors of prospectively examined suicide attempts among youth with bipolar disorder. *Archives of General Psychiatry*, *69*(11), 1113–1122.

Goodday, S. M., et al. (2016). Repeated salivary daytime cortisol and onset of mood episodes in offspring of bipolar parents. *International Journal of Bipolar Disorders*, *4*(1), 12.

Grande, I., et al. (2012). Mediators of allostasis and systemic toxicity in bipolar disorder. *Physiology & Behavior*, *106*(1), 46–50.

Green, M. J., Cahill, C. M., & Malhi, G. S. (2007). The cognitive and neurophysiological basis of emotion dysregulation in bipolar disorder. *Journal of Affective Disorders*, *103*(1–3), 29–42.

Greenberg, M. T. (2006). Promoting resilience in children and youth: preventive interventions and their Interface with neuroscience. *Annals of the New York Academy of Sciences*, *1094*, 139–150.

Hafeman, D. M., et al. (2016). Toward the definition of a bipolar Prodrome: dimensional predictors of bipolar Spectrum disorders in at-risk youths. *The American Journal of Psychiatry*, *173*(7), 695–704.

Houenou, J., et al. (2011). Neuroimaging-based markers of bipolar disorder: evidence from two meta-analyses. *Journal of Affective Disorders*, *132*(3), 344–355.

Jaworska-Andryszewska, P., & Rybakowski, J. (2016). Negative experiences in childhood and the development and course of bipolar disorder. *Psychiatria Polska*, *50*(5), 989–1000.

Johnson, K. C., et al. (2014). Physiological regulation in infants of women with a mood disorder: examining associations with maternal symptoms and stress. *Journal of Child Psychology and Psychiatry, and Allied Disciplines*, *55*(2), 191–198.

Jones, S. H., & Bentall, R. P. (2008). A review of potential cognitive and environmental risk markers in children of bipolar parents. *Clinical Psychology Review*, *28*(7), 1083–1095.

Karatsoreos, I. N., & McEwen, B. S. (2011). Psychobiological allostasis: resistance, resilience and vulnerability. *Trends in Cognitive Sciences*, *15*(12), 576–584.

Karchemskiy, A., et al. (2011). Amygdalar, hippocampal, and thalamic volumes in youth at high risk for development of bipolar disorder. *Psychiatry Research*, *194*(3), 319–325.

Kim, P., et al. (2012). Neural correlates of cognitive flexibility in children at risk for bipolar disorder. *Journal of Psychiatric Research*, *46*(1), 22–30.

Ladouceur, C. D., et al. (2008). Subcortical gray matter volume abnormalities in healthy bipolar offspring: potential neuroanatomical risk marker for bipolar disorder? *Journal of the American Academy of Child and Adolescent Psychiatry*, *47*(5), 532–539.

Leow, A., et al. (2013). Impaired inter-hemispheric integration in bipolar disorder revealed with brain network analyses. *Biological Psychiatry*, *73*(2), 183–193.

Leve, L. D., Fisher, P. A., & Chamberlain, P. (2009). Multidimensional treatment Foster Care as a preventive intervention to promote resiliency among youth in the child welfare system. *Journal of Personality*, *77*(6), 1869–1902.

Lieberman, J. A., Perkins, D. O., & Fredrik Jarskog, L. (2007). Neuroprotection: a therapeutic strategy to prevent deterioration associated with schizophrenia. *CNS Spectrums, 12*(3 Suppl 4), 1–13. quiz 14.

Lombardo, L. E., et al. (2012). Trait impulsivity as an Endophenotype for bipolar I disorder. *Bipolar Disorders, 14*(5), 565–570.

Maripuu, M., et al. (2017). Hyper- and hypocortisolism in bipolar disorder—a beneficial influence of lithium on the HPA-axis? *Journal of Affective Disorders, 213*, 161–167.

Mesulam, M. M. (1990). Large-scale neurocognitive networks and distributed processing for attention, language, and memory. *Annals of Neurology, 28*(5), 597–613.

Miklowitz, D. J., et al. (2017). Early intervention for youth at high risk for bipolar disorder: a multisite randomized trial of family-focused treatment. *Early Intervention in Psychiatry,* 1–9. https://doi.org/10.1111/eip.12463.

Mourão-Miranda, J., et al. (2012). Pattern recognition analyses of brain activation elicited by happy and neutral faces in unipolar and bipolar depression. *Bipolar Disorders, 14*(4), 451–460.

Muneer, A. (2016). The neurobiology of bipolar disorder: an integrated approach. *Chonnam Medical Journal, 52*(1), 18–37.

Neigh, G. N., et al. (2013). Translational reciprocity: Bridging the gap between preclinical studies and clinical treatment of stress effects on the adolescent brain. *Neuroscience, 249,* 139–153.

Nery, F. G., et al. (2017). White matter volumes in youth offspring of bipolar parents. *Journal of Affective Disorders, 209,* 246–253.

Noble, K. G., et al. (2015). Family income, parental education and brain structure in children and adolescents. *Nature Neuroscience, 18*(5), 773–778.

Okkels, N., Trabjerg, B., Arendt, M., & Pedersen, C. B. (2017). Traumatic stress disorders and risk of subsequent schizophrenia Spectrum disorder or bipolar disorder: a Nationwide cohort study. *Schizophrenia Bulletin, 43*(1), 180–186.

Olsavsky, A. K., et al. (2012). Amygdala hyperactivation during face emotion processing in unaffected youth at risk for bipolar disorder. *Journal of the American Academy of Child and Adolescent Psychiatry, 51*(3), 294–303.

Ostiguy, C. S., et al. (2009). Chronic stress and stressful life events in the offspring of parents with bipolar disorder. *Journal of Affective Disorders, 114*(1–3), 74–84.

Ostiguy, C. S., et al. (2011). Sensitivity to stress among the offspring of parents with bipolar disorder: a study of daytime cortisol levels. *Psychological Medicine, 41*(11), 2447–2457.

Paillère Martinot, M.-L., et al. (2014). White-matter microstructure and gray-matter volumes in adolescents with subthreshold bipolar symptoms. *Molecular Psychiatry, 19*(4), 462–470.

Pan, L. A., et al. (2017). The relationship between stressful life events and Axis I diagnoses among adolescent offspring of Probands with bipolar and non-bipolar psychiatric disorders and healthy controls: the Pittsburgh bipolar offspring study (BIOS). *The Journal of Clinical Psychiatry, 78*(3), e234–243.

Passarotti, A. M., et al. (2012). Reduced functional connectivity of prefrontal regions and amygdala within affect and working memory networks in pediatric bipolar disorder. *Brain Connectivity, 2*(6), 320–334.

Pavuluri, M. N., Passarotti, A. M., Harral, E. M., & Sweeney, J. A. (2009). An fMRI study of the neural correlates of incidental versus directed emotion processing in pediatric bipolar disorder. *Journal of the American Academy of Child and Adolescent Psychiatry, 48*(3), 308–319.

Pavuluri, M. N., et al. (2006). Impact of neurocognitive function on academic difficulties in pediatric bipolar disorder: a clinical translation. *Biological Psychiatry, 60*(9), 951–956.

Pavuluri, M. N., et al. (2012). Pharmacotherapy impacts functional connectivity among affective circuits during response inhibition in pediatric mania. *Behavioural Brain Research*, *226*(2), 493–503.

Post, R. M., et al. (2010). Early-onset bipolar disorder and treatment delay are risk factors for poor outcome in adulthood. *The Journal of Clinical Psychiatry*, *71*(7), 864–872.

Rich, B. A., et al. (2006). Limbic Hyperactivation during processing of neutral facial expressions in children with bipolar disorder. *Proceedings of the National Academy of Sciences of the United States of America*, *103*(23), 8900–8905.

Roberts, G., et al. (2012). Reduced inferior frontal gyrus activation during response inhibition to emotional stimuli in youth at high risk of bipolar disorder. *Biological Psychiatry*, *74*(1), 55–61. https://doi.org/10.1016/j.biopsych.2012.11.004.

Romero, S., et al. (2005). Family environment in families with versus families without parental bipolar disorder: A preliminary comparison study. *Bipolar Disorders*, *7*(6), 617–622.

Roybal, D. J., et al. (2015). Widespread white matter tract aberrations in youth with familial risk for bipolar disorder. *Psychiatry Research*, *232*(2), 184–192.

Russo, S. J., et al. (2012). Neurobiology of resilience. *Nature Neuroscience*.

Schneider, M. R., et al. (2012). Neuroprogression in bipolar disorder. *Bipolar Disorders*, *14*(4), 356–374.

Schreuder, M. M., et al. (2016). Childhood trauma and HPA axis functionality in offspring of bipolar parents. *Psychoneuroendocrinology*, *74*, 316–323.

Sheikh, H. I., et al. (2014). Links between white matter microstructure and cortisol reactivity to stress in early childhood: evidence for moderation by parenting. *NeuroImage: Clinical*, *6*, 77–85.

Silk, J. S., et al. (2007). Resilience among children and adolescents at risk for depression: mediation and moderation across social and neurobiological contexts. *Development and Psychopathology*, *19*(3), 841–865.

Sinclair, D., Webster, M. J., Wong, J., & Weickert, C. S. (2011). Dynamic molecular and anatomical changes in the glucocorticoid receptor in human cortical development. *Molecular Psychiatry*, *16*(5), 504–515.

Singh, M. K., Chang, K. D., et al. (2014). Early signs of anomalous neural functional connectivity in healthy offspring of parents with bipolar disorder. *Bipolar Disorders*, *16*(7), 678–689. https://doi.org/10.1111/bdi.12221.

Singh, M. K., Kelley, R. G., et al. (2014). Reward processing in healthy offspring of parents with bipolar disorder. *JAMA Psychiatry*, *52*(1), 68–83. https://doi.org/10.1016/j.jaac.2012.10.004.

Singh, M. K., et al. (2008). Neuroanatomical characterization of child offspring of bipolar parents. *Journal of the American Academy of Child and Adolescent Psychiatry*, *47*(5), 526–531.

Singh, M., et al. (2010). Brain glutamatergic characteristics of pediatric offspring of parents with bipolar disorder. *Psychiatry Research*, *182*(2), 165–171.

Singh, M. K., et al. (2011). Neurochemical deficits in the cerebellar vermis in child offspring of parents with bipolar disorder. *Bipolar Disorders*, *13*(2), 189–197.

Singh, M. K., et al. (2013). Reward processing in adolescents with bipolar I disorder. *Journal of the American Academy of Child and Adolescent Psychiatry*, *52*(1), 68–83.

Smieskova, R., Fusar-Poli, P., Riecher-Rössler, A., & Borgwardt, S. (2012). Neuroimaging and resilience factors—staging of the at-risk mental state? *Current Pharmaceutical Design*, *18*(4), 416–421.

Smoller, J. W., & Finn, C. T. (2003). Family, twin, and adoption studies of bipolar disorder. *American Journal of Medical Genetics. Part C, Seminars in Medical Genetics*, *123C*(1), 48–58.

Southwick, S. M., & Charney, D. S. (2012). The science of resilience: implications for the prevention and treatment of depression. *Science (New York, NY), 338*(6103), 79–82.

Suchdev, P. S., et al. (2017). Assessment of neurodevelopment, nutrition, and inflammation from fetal life to adolescence in low-resource settings. *Pediatrics, 139*(Suppl. 1), S23–37.

Thompson, P. M., et al. (2000). Growth patterns in the developing brain detected by using continuum mechanical tensor maps. *Nature, 404*(6774), 190–193.

Versace, A., et al. (2010). Altered development of white matter in youth at high familial risk for bipolar disorder: A diffusion tensor imaging study. *Journal of the American Academy of Child and Adolescent Psychiatry, 49*(12), 1249–1259. 1259.e1.

Welge, J. A., et al. (2016). Neurofunctional differences among youth with and at varying risk for developing mania. *Journal of the American Academy of Child and Adolescent Psychiatry, 55*(11), 980–989.

Whitney, J., et al. (2013). Socio-emotional processing and functioning of youth at high risk for bipolar disorder. *Journal of Affective Disorders, 148*(1), 112–117. https://doi.org/10.1016/j.jad.2012.08.016.

Wiggins, J. L., et al. (2017). Neural markers in pediatric bipolar disorder and familial risk for bipolar disorder. *Journal of the American Academy of Child and Adolescent Psychiatry, 56*(1), 67–78.

Wozniak, J., et al. (2012). Further evidence for robust familiality of pediatric bipolar I disorder: results from a very large controlled family study of pediatric bipolar I disorder and a meta-analysis. *The Journal of Clinical Psychiatry, 73*(10), 1328–1334.

Neuroimaging findings in youth at risk for bipolar disorder

Fabiano G. Nery, Melissa P. DelBello

Division of Bipolar Disorder Research, Department of Psychiatry and Behavioral Neuroscience, University of Cincinnati College of Medicine, Cincinnati, OH, United States

CHAPTER OUTLINE

INTRODUCTION

Because bipolar disorder is highly heritable, a great interest exists in studying individuals at genetic risk for bipolar disorder in order to understand the early manifestations of the disease (DelBello & Geller, 2001; Nery, Monkul, & Lafer, 2013). It is well-established that children of parents with bipolar disorder (bipolar offspring) have an increased risk of developing mood disorders in general, and bipolar disorder in particular (Birmaher et al., 2009; DelBello & Geller, 2001; Duffy et al., 2014; Gottesman, Laursen, Bertelsen, & Mortensen, 2010). In addition, the first symptoms of bipolar disorder often manifest during adolescence or young adulthood, which makes studying young individuals at risk for bipolar disorder an ideal population in which to investigate the neurobiology of illness risk or very early manifestations (DelBello & Geller, 2001). Moreover, studying young individuals prior to the onset

Bipolar Disorder Vulnerability. https://doi.org/10.1016/B978-0-12-812347-8.00007-5

115

of illness minimizes common confounding factors in neurobiological studies of patients with bipolar disorder, including medication effects and burden of illness. Therefore, the study of children, adolescents, and young adults at increased familial risk for bipolar disorder (also called at-risk) is an unique opportunity to: (1) investigate neurobiological abnormalities related to genetic risk that could be potential genetic markers of predisposition to the disorder; (2) identify early neurobiological manifestations that might be the expression of early prodromal symptoms, helping to understand the chain of neurobiological events that leads to a mood episode; and (3) understand neurobiological abnormalities in bipolar disorder that are free of confounding effects, such as the burden of chronic illness.

Bipolar disorder is a disorder characterized by a dysfunction of emotion regulation and attentional processes (Goodwin, Jamison, & Ghaemi, 2007). Expert consensus considers that this dysfunction is caused by structural, neurochemical, and functional abnormalities within emotional control networks that include the ventrolateral and ventromedial prefrontal cortices, thalamus, amygdala, and striatum (Strakowski et al., 2012; Strakowski, Delbello, & Adler, 2005). Several investigators have hypothesized that abnormal brain development, particularly in prefrontal white matter and cortical pruning, leads to abnormal connectivity in the emotional control network, impairing the at-risk youth's ability to regulate emotion and attention properly, which in turn gives rise to early symptoms of the disorder, and to the predisposition to develop enduring and severe mood states (Schneider, DelBello, McNamara, Strakowski, & Adler, 2012). In fact, cognitive abnormalities, such as difficulties in tasks of executive functioning, memory, and attention, exist in unaffected youth bipolar offspring, which suggests a preexistent fronto-limbic dysfunction that may contribute to the development of mood disorders in those youth at familial risk (Gotlib, Traill, Montoya, Joormann, & Chang, 2005; Ladouceur et al., 2013; Meyer et al., 2004; Schneider et al., 2012). These abovementioned hypotheses have been put forward based on mounting evidence from substantial research using several modalities of magnetic resonance imaging (MRI) techniques over the past 20 years. The interest in studying youth at risk for bipolar disorder lies in the assumption that there is a continuity between the anatomy and physiology of being "at-risk" and having the illness (Gottesman & Gould, 2003; Hasler, Drevets, Gould, Gottesman, & Manji, 2006). In this hypothesis, the "at-risk" condition would be characterized by brain abnormalities that are similar to those seen in patients with fully established conditions and that could be the direct expression of genetic risk or intermediate expressions of genetic risk and environmental interactions (Gottesman & Gould, 2003; Hasler et al., 2006). In addition, brain abnormalities that are potentially identified in those most likely to become ill (by virtue of familial risk) can give important clues to the initial path of disease.

With these considerations in mind, over the next sections, we will review findings from imaging studies performed in children and adolescents at increased familial risk for bipolar disorder (offspring or first-degree relative of a patient diagnosed with bipolar disorder). We will review studies that utilized magnetic resonance imaging to examine group differences in regional brain structure, neurochemistry, and functional activation between at-risk youth and normal developing children and adolescents. We will summarize findings, discuss current perspectives, and propose future directions.

STRUCTURAL IMAGING STUDIES (BOX 1)
GRAY MATTER FINDINGS

One of the first ROI imaging studies investigated differences in prefrontal cortical, striatal, amygdala, and thalamic volumes in 21 at-risk children (ages 8–12 years old) and 24 healthy controls (Singh, Delbello, Adler, Stanford, & Strakowski, 2008). Prefrontal cortical volumes included all intracranial tissue anterior to the genu of the corpus callosum, so also included white matter. Forty-three percent of the at-risk children had some nonbipolar mood disorder at time of the study. There were no group differences in regional volumes of prefrontal cortex, amygdala, thalamus, and striatum. A more recent ROI study compared 29 bipolar offspring and 17 healthy controls in regards to amygdala volumes and social anxiety scores, and investigated correlations between anxiety scores and amygdala volumes (Park et al., 2015). Participants were aged between 9 and 17 years old, were offspring of parents with bipolar disorder type I or II, and had subthreshold affective symptoms. Bipolar offspring exhibited higher depressive, manic, and social anxiety symptomatology than healthy controls. Amygdala volumes were not significantly different between the bipolar offspring and controls, and between bipolar offspring with and without high social anxiety scores. Among the bipolar offspring with high social anxiety, bilateral

---

BOX 1 STRUCTURAL MAGNETIC RESONANCE IMAGING

Structural imaging studies use T1-weighed magnetic resonance imaging (MRI) to investigate, beyond individual variations, differences in volume, shape, size, or thickness of gray and white matter in brain areas that might be attributable to psychiatric disorders or pathological phenomena associated with these disorders (Devlin & Poldrack, 2007).

For *gray matter*, two basic approaches are used: manual tracing of region-of-interest (ROI) (the manual delineation of brain structure across multiple slices of the MRI scan, in a stereotaxic space, and its posterior segmentation in gray and white matter, with calculation of volume in cm^3) (Giuliani, Calhoun, Pearlson, Francis, & Buchanan, 2005), or voxelwise approach (automated methods that search for differences in density, volume, or thickness of gray matter or white matter, after normalizing the MRI images to a standard template, and comparing groups voxel-by-voxel (Foland-Ross et al., 2011; Good et al., 2001). More recent approaches (e.g., FreeSurfer) use automated methods to extract regional brain volumes (e.g., amygdala volumes) to perform ROI-based analyses. The main advantages of ROI over voxelwise are its reproducibility and validity, and its main disadvantage is that it is very time consuming (Giuliani et al., 2005).

For *white matter*, a popular imaging technique is diffusion tensor imaging (DTI), in which the diffusion properties of water molecules along the axon myelin sheath are used to reconstruct contrast images, and rebuild axonal tract and fasciculi. DTI is also used to obtain measures of water diffusivity along the axons, such as fractional anisotropy, which gives an indirect measure of white matter integrity (Assaf & Pasternak, 2008; Neil, 2008).

amygdala volumes were negatively correlated with social anxiety scores. Another ROI study extracted hippocampal volumes and investigated associations with inhibited temperament in bipolar offspring (Kim et al., 2017). The authors compared 24 bipolar offspring with a history of major depression or ADHD, and moderate manic or depressive symptoms, 21 bipolar offspring without any psychopathology, and 24 healthy controls, all between 8 and 17 years old. The authors found a positive correlation between hippocampal volumes and inhibited temperament scores in the bipolar offspring with psychopathology, but not in the healthy bipolar offspring and in the healthy control groups. Group effects on hippocampal volumes were not studied. Of note, both studies found associations between amygdala or hippocampal volumes and psychopathology, but not with at-risk condition per se. More details of these studies are displayed in Table 1.

Most of the other studies in at-risk children and adolescents have used voxelwise approaches, usually voxel-based morphometry (VBM). Ladouceur et al. (2008) studied 20 unaffected bipolar offspring and 22 healthy controls (all with ages between 8 and 17 years old) and found that bipolar offspring exhibited increased gray matter volumes in the left parahippocampal gyrus compared with controls (Ladouceur et al., 2008). Another VBM study, using a two-center, replication design, compared 50 bipolar offspring with no lifetime history of psychiatric disorders (unaffected offspring), 36 bipolar offspring with personal history of major depression or mania (affected offspring), and 49 age and sex group-matched healthy controls (Hajek et al., 2013). All subjects were between 15 and 25 years old. Both bipolar offspring groups exhibited a larger right inferior frontal gyrus (Brodmann area 47) compared with the controls. In the second sample of this same study, increased gray matter volumes in the right inferior frontal gyrus sample were also found in patients with bipolar disorder in early stages of their illness (Hajek et al., 2013). Another VBM study compared 31 bipolar offspring and 20 healthy controls, group-matched for age and sex (Hanford, Hall, Minuzzi, & Sassi, 2016). All participants were adolescents with average age of 13 years. Bipolar offspring presented with decreased gray matter volumes in the right inferior orbitofrontal cortex, middle frontal, and bilateral superior and middle temporal gyrus compared with healthy controls. When the bipolar offspring group was divided according to presence or absence of lifetime psychiatric diagnoses (other than bipolar disorder), both subgroups with or without lifetime psychiatric diagnoses presented with decreased gray matter volumes in these same brain regions. Hanford, Sassi, Minuzzi, and Hall (2016) also used cortical thickness analysis in this same sample and found that bipolar offspring had cortical thinning in temporal, supramarginal, and middle frontal areas compared with healthy controls (Hanford, Sassi, et al., 2016). Symptomatic and asymptomatic bipolar offspring had cortical thinning of the right inferior temporal and supramarginal regions, but the symptomatic subgroup had cortical thinning in the superior frontal, somatosensory related cortices, and superior temporal and insular cortices (Hanford, Sassi, et al., 2016), highlighting the need of longitudinal studies to observe the association of these deficits with the onset of a bipolar disorder diagnosis.

Table 1 Summary of structural gray matter findings in imaging studies of children and adolescents at risk for bipolar disorder

Authors	Sample characteristics	Method	Results	Comments
ROI analyses				
Singh et al. (2008)	21 bipolar offspring 24 healthy controls Ages 8–12 years old	ROI manual tracing of prefrontal cortex, amygdala, thalamus, striatum	No differences between groups	All parents were bipolar type I Prefrontal cortex included white matter
Park et al. (2015)	29 bipolar offspring 17 healthy controls Ages 9–17 years old	Extraction of amygdala volumes using FreeSurfer Analyses of association between amygdala volumes and anxiety scores	No difference in amygdala volumes between groups Among highly anxious bipolar offspring, bilateral amygdala volumes decreased as anxiety increased	Bipolar parents were type I or II Bipolar offspring had subthreshold affective symptoms
Kim et al. (2017)	24 bipolar offspring with history of major depressive disorder or ADHD 21 bipolar offspring without psychopathology 24 healthy controls Ages 8–17 years old	Extraction of hippocampal volumes using FreeSurfer Analyses of association between hippocampal volumes and temperament	Positive correlation between hippocampal volumes and inhibited temperament scores in bipolar offspring without psychopathology	Parents had bipolar type I or II
Voxelwise analyses				
Ladouceur et al. (2008)	20 bipolar offspring without psychopathology 22 healthy controls Ages 8–17 years old	VBM	Increased gray matter volumes in the left parahippocampal gyrus in bipolar offspring	

Continued

Table 1 Summary of structural gray matter findings in imaging studies of children and adolescents at risk for bipolar disorder—cont'd

Authors	Sample characteristics	Method	Results	Comments
Hajek et al. (2013)	50 bipolar offspring without psychopathology 36 bipolar offspring with history of depression or mania 49 healthy controls Ages 15–25 years old	VBM	Increased right inferior frontal gyrus in both groups of bipolar offspring	Increased right inferior frontal gyrus also found in a replication sample of this study in bipolar patients in early stages of their illness
Hanford, Hall, et al. (2016)	31 bipolar offspring 20 healthy controls Ages 8–16 years old	VBM	Decreased gray matter volumes in right inferior orbitofrontal cortex, middle frontal, bilateral superior, and middle temporal gyrus in bipolar offspring	Bipolar offspring with or without psychopathology had decreased gray matter in the same brain regions
Hanford, Sassi, et al. (2016)	31 bipolar offspring 20 healthy controls Ages 8–16 years old	Cortical thickness analyses obtained by FreeSurfer	Cortical thinning in right inferior temporal and supramarginal regions in bipolar offspring	Same sample as in Hanford, Hall, et al. (2016) Bipolar offspring with psychopathology had cortical thinning in superior frontal, somatosensory areas, and superior and insular cortices
Sugranyes et al. (2015)	77 bipolar offspring 83 healthy controls Ages 6–17 years old	VBM	No differences between groups	
Nery et al. (2017)	47 bipolar offspring without psychopathology 68 bipolar offspring with psychopathology 57 healthy controls	VBM	No differences between groups	Bipolar offspring had childhood-onset psychopathology (e.g., ADHD, separation anxiety), but no history of mania, major depression, or psychosis

Abbreviations: ADHD, attention-deficit/hyperactivity disorder; ROI, region-of-interest; VBM, voxel-based morphometry.

In contrast, recent VBM studies using larger samples of bipolar offspring have not found differences in gray matter volumes between bipolar offspring and healthy controls (Nery et al., 2017; Sugranyes et al., 2015). Sugranyes et al. (2015) compared 77 bipolar offspring (aged 6–17 years old) to 83 healthy controls and found no differences in gray matter between the groups (Sugranyes et al., 2015). Our group performed a large VBM imaging study comparing 47 healthy bipolar offspring, 68 symptomatic bipolar offspring, and 57 healthy controls, all aged 9–20 years old. The symptomatic bipolar offspring had varied childhood-onset psychopathology (mostly attention deficit/hyperactivity disorder, but no bipolar disorder or major depressive disorder). No differences in gray matter volumes were found between groups (Nery et al., 2017).

WHITE MATTER FINDINGS

In contrast to studies of gray matter abnormalities, findings of white matter abnormalities in individuals at risk for bipolar disorder (including adults) have been more consistent (Nery et al., 2013) (Table 2). An earlier study using diffusion tensor imaging (DTI) in a sample of 10 children with bipolar disorder, 8 healthy controls, and 7 at-risk children (57% with a sibling and 43% with a parent with bipolar disorder) found that the at-risk children had reduced fractional anisotropy, a measure of white matter integrity, in the bilateral longitudinal fasciculi, when compared with controls (Frazier et al., 2007). Another DTI study investigated developmental abnormalities of white matter in 20 youth at risk for bipolar disorder (Versace et al., 2010). A linear decrease in fractional anisotropy and an increase in radial diffusivity with age in the left corpus callosum was found in bipolar offspring, which was different from healthy controls, where a linear decrease in fractional anisotropy and in radial diffusivity with age was found in the left corpus callosum, and in the right inferior longitudinal fasciculi. This study did not report group differences between bipolar offspring and healthy control offspring. No differences in fractional anisotropy and radial diffusivity was seen in a DTI study comparing 18 healthy offspring of parents with bipolar disorder, aged 6–17 years old, and healthy controls (Teixeira et al., 2014). Most recently, Roybal et al. (2015) used two complementary DTI approaches to investigate white matter findings in 25 children and adolescents at increased risk for bipolar disorder (defined by having symptomatic mood dysregulation and a parent with bipolar disorder) (Roybal et al., 2015). They found that the at-risk youth had significant and widespread abnormal measures of white matter integrity in the bilateral uncinated fasciculus, cingulum, cingulate, superior fronto-occipital fasciculus, superior and inferior longitudinal fasciculi, and corpus callosum (Roybal et al., 2015). Finally, in our VBM study of children and adolescent bipolar offspring cited above, we found that healthy bipolar offspring exhibited decreased white matter volumes in large areas encompassing the right frontal, temporal, and parietal lobes, and in the left temporal and parietal lobes, compared with healthy controls (Nery et al., 2017). Surprisingly, symptomatic bipolar offspring (those with psychiatric diagnoses other than mood disorders or psychotic disorders) had no differences in white matter volumes compared with healthy controls.

Table 2 Summary of white matter findings in imaging studies of children and adolescents at increased risk for bipolar disorder

Authors	Sample characteristics	Methods	Results	Comments
Frazier et al. (2007)	10 children with bipolar disorder 7 at-risk children 8 healthy controls Ages 4–12 years old	DTI	Reduced fractional anisotropy in the bilateral longitudinal fasciculi compared with controls	
Versace et al. (2010)	27 at-risk children 25 controls Ages 8–17 years old	DTI	Linear decrease in fractional anisotropy and increase in radial diffusivity with age in left corpus callosum and right inferior longitudinal fasciculi in at-risk	No associations of age and fractional anisotropy in controls
Teixeira et al. (2014)	18 bipolar offspring 20 healthy controls Ages 6–17 years old	DTI	No differences in fractional anisotropy and radial diffusivity between groups	
Roybal et al. (2015)	25 at-risk children 16 healthy controls Ages 10–18 years old	DTI	Increased bilateral uncinated fasciculus, cingulum, cingulate, superior fronto-occipital fasciculus, superior and inferior longitudinal fasciculi, and corpus callosum	
Nery et al. (2017)	47 bipolar offspring without psychopathology 68 bipolar offspring with psychopathology 57 healthy controls Ages 9–20 years old	VBM	Decreased white matter volumes in right frontal, temporal, and parietal lobes, and in left temporal and parietal lobes in bipolar offspring without psychopathology	No differences in bipolar offspring with psychopathology compared with controls

Abbreviations: DTI, diffusion tensor imaging.

SUMMARY OF STRUCTURAL IMAGING STUDIES

In summary, it is still unclear whether gray matter abnormalities are an expression of familial risk for bipolar disorder. Most of the studies that focus on children and adolescents, with ages spanning over late childhood and adolescence, found no evidence for abnormal brain structure on gray matter, with the exception of three studies (Hanford, Hall, et al., 2016; Hanford, Sassi, et al., 2016; Ladouceur et al., 2008). Most recent studies with larger samples of youth bipolar offspring yielded negative results (Nery et al., 2017; Sugranyes et al., 2015). Of particular relevance among the positive findings is the replication of abnormal (increased or decreased) gray matter volumes in the right ventrolateral prefrontal cortex, a key area in attention and emotion regulation and in the neurophysiological model of bipolar disorder (Phillips, Ladouceur, & Drevets, 2008; Phillips & Swartz, 2014; Strakowski et al., 2012). There are several explanations for the discrepancies among studies. First, studies with positive findings had small sample sizes, and particular sample characteristics (limiting the generalizability of the findings) or possible type I errors may explain the difficulty with replicating findings. Second, former structural imaging techniques or current VBM approaches may lack the sensitivity to detect subtle or small group differences between youth bipolar offspring and healthy controls. Third, most studies are cross-sectional. Therefore, it is unclear whether abnormal findings in regional gray matter volumes or cortical thickness are an early expression of a disease process that has not yet fully manifested, are a correlate of protection against the disease, or are a pure genetic risk marker. A fourth possibility is that abnormal gray matter as a marker of familial risk might interact with brain developmental processes, and manifest later in the disease process. This could reconcile negative findings of gray matter in children and adolescents (Nery et al., 2017; Singh et al., 2008; Sugranyes et al., 2015) with findings of larger inferior frontal gyrus in bipolar offspring samples whose age spans from late adolescence to young adulthood (Hajek et al., 2013).

The few studies of white matter abnormalities in children and adolescents at risk for bipolar disorder point to abnormalities in the prefrontal areas, cingulum, corpus callosum, and white matter tracts that connect frontal, temporal, and parietal lobes. This is largely consistent with several studies in adults at risk for bipolar disorder, who are beyond the peak of highest risk to develop the illness (Nery et al., 2017). As explained above for gray matter, the exact biological meaning of these findings is unknown. To the best of our knowledge, no prospective study has investigated these regional white matter abnormalities as predictive or risk factors for the late development of bipolar disorder. However, these findings lend support to the theory that predisposition to bipolar disorder may develop from an abnormal structural connectivity between prefrontal-temporal-subcortical areas which might lead to abnormal brain functioning. It is likely that abnormal white matter is not the exclusive etiological factor in the causation of disease, and the concomitant presence of other risk factors might contribute to the pathophysiological cascade.

SPECTROSCOPY IMAGING STUDIES (BOX 2)

One of the first ^{1}H-MRS studies was a small study in 9 youth with a mood disorder and who were offspring of parents with bipolar disorder and 10 healthy controls, matched for age (8–12 years old) and sex (Cecil, DelBello, Sellars, & Strakowski, 2003). Metabolites were acquired in the medial prefrontal cortex, medial frontal white matter, and cerebellar vermis. Bipolar offspring with a mood disorder had a 16% elevation in myo-inositol concentrations in the medial frontal cortex compared with controls, and a nonsignificant decrease in NAA and PCr+Cr concentrations in the cerebellar vermis compared with controls. A larger study investigated prefrontal glutamate concentrations in the anterior cingulate cortex of 60 children and adolescents, aged 9–18 years old (Singh et al., 2010). Of those 60, 20 were bipolar offspring with a personal history of mania, 20 were bipolar offspring with subthreshold manic symptoms, and 20 were controls. The authors found that bipolar offspring with a history of mania (hence with a diagnosis of bipolar disorder) had reduced glutamate concentrations in their anterior cingulate cortex compared with the other groups. No other group differences were found in other metabolites (NAA, PCr+Cr, and myo-inositol). In another study, Singh et al. (2011) investigated neurometabolites in the cerebellar vermis of 22 youth at familial risk for bipolar disorder (with nonbipolar mood symptoms) and 25 healthy controls (Singh et al., 2011). They found reduced myo-inositol and choline levels in the cerebellar vermis of bipolar offspring compared with controls. There was no difference in NAA, PCr+Cr, or glutamate concentrations. Others have found normal concentrations of metabolites in bipolar offspring compared with healthy controls. In a comparison of metabolite levels in the left and right dorsolateral prefrontal cortex of 60 offspring of parents with bipolar disorder I or II, no differences were found in ratios of NAA to creatine between groups (Gallelli et al., 2005). Hajek et al. (2008) studied at-risk (mostly offspring, but also siblings) individuals aged 15–30 years old, with two samples, a primary with 14 affected and 15 unaffected at-risk, and an extended, with 19 affected and 21 unaffected, and including parents with bipolar II, and 31 age- and sex-matched controls (Hajek et al., 2008). The authors found similar concentrations of NAA, PCr+Cr, choline-containing compounds, and myo-inositol in the bilateral dorsal and ventromedial prefrontal cortices

BOX 2 MAGNETIC RESONANCE SPECTROSCOPY (MRS)

Magnetic resonance spectroscopy (MRS) is an imaging technique that allows the in vivo measurements of brain metabolites. The most common modality of MRS uses the magnetic properties of the hydrogen (H) proton in the MR scanner to obtain indirect measurements of N-acetyl aspartate (NAA), a putative marker of neuronal integrity and neuronal metabolism, choline-containing compounds, a marker of cell membrane turnover and breakdown, phosphocreatine and creatine (PCr+Cr), a marker of cell energy metabolism, myo-inositol, involved in the phosphatidylinositol pathway and a marker of glial proliferation, and glutamate, the major excitatory neurotransmitter in the brain (Cecil, 2013). Other modalities of MRS use phosphorus to obtain indirect measurements of phosphorus-containing compounds involved in neuronal energetics, such as phosphocreatine and adenosine triphosphate (Yuksel et al., 2015).

between bipolar offspring and controls. Finally, Singh et al. (2013) investigated progression of neurochemical abnormalities with time in their original sample of bipolar offspring previously reported in 2010 (Singh et al., 2013). They followed 64 child and adolescent offspring of bipolar parents, of whom 36 had bipolar disorder and 28 had subsyndromal mania, and 28 controls. They examined differences in dorsolateral prefrontal cortex metabolites levels, at baseline, and at 5-year follow-up. No differences at baseline or at follow-up were found in metabolite concentrations between groups. Details of MRS studies in bipolar offspring are displayed in Table 3.

SUMMARY OF SPECTROSCOPY IMAGING FINDINGS

Unfortunately, most of the ^{1}H-MRS studies in at-risk children and adolescents are limited by the relatively small sample sizes, by the cross-sectional designs, and by the fact that often the bipolar offspring were already affected by subsyndromal symptoms or by bipolar disorder. Therefore, the understanding of whether the few positive findings represent a marker of familial risk or a marker of incipient psychopathology is uncertain.

FUNCTIONAL IMAGING STUDIES (BOX 3)
TASK-BASED fMRI FINDINGS

Greater amygdala activation was found in a study that used an emotional face processing (happy and fearful faces) task to compare 32 youth with bipolar disorder, 13 healthy bipolar offspring, and 56 healthy controls, between 8 and 18 years old (Olsavsky et al., 2012). Both bipolar and at-risk youth exhibited greater amygdala activation compared with healthy controls when rating fearful faces, but not happy faces. Using an emotional working memory paradigm, Ladouceur et al. (2013) compared 15 healthy bipolar offspring, to 16 healthy controls. The imaging analyses focused on regions involved with emotional processing and voluntary emotion regulation. They found that bipolar offspring exhibited increased right ventrolateral prefrontal cortical activation in response to positive emotional distracters, and reduced connectivity between the ventrolateral prefrontal cortex and amygdala for positive and negative emotional distracters. Another study investigated the functional activation of brain areas involved with emotional face processing in 29 bipolar offspring, 29 offspring of nonbipolar parents, and 23 age- and sex-matched healthy controls (Manelis et al., 2015). There was greater right amygdala activation during emotional faces versus shapes in both offspring of bipolar and nonbipolar parents, suggesting that it might be a marker of risk for affective disorders in general. Functional connectivity analyses showed that there was an increased connectivity between the right amygdala and the anterior cingulate cortex when processing emotional faces versus shapes, and a negative connectivity between the right amygdala and the left ventrolateral prefrontal cortex when processing happy faces in bipolar offspring compared with the other two groups. Using a task of sustained attention with emotional distracters, a recent study

Table 3 Summary of neurochemical findings in imaging studies of children and adolescents at risk for bipolar disorder

Authors	Sample characteristics	Methods	Results	Comments
Cecil et al. (2003)	9 bipolar offspring with mood disorders 10 healthy controls Ages 8–12 years old	Voxels in medial prefrontal cortex, frontal white matter, cerebellar vermis	16% elevation in myo-inositol levels nonsignificant decreased in N-acetyl aspartate and creatine and phosphocreatine levels in cerebellar vermis	Bipolar parents had bipolar type I or II
Gallelli et al. (2005)	60 bipolar offspring 26 healthy controls Ages 9–18 years old	Bilateral dorsolateral prefrontal cortex	No group differences in N-acetyl aspartate/creatine ratios	Offspring included 32 diagnosed with bipolar disorder and 28 with subsyndromal symptoms of bipolar disorder
Hajek et al. (2008)	14 affected at-risk 15 unaffected at-risk 19 affected at-risk 21 unaffected at-risk 31 healthy controls Ages 15–30 years old	Voxels in bilateral ventromedial prefrontal cortex	No group differences in N-acetyl aspartate, creatine, choline-containing compounds, myo-inositol levels	At-risk included offspring or siblings Extended sample included at-risk of parents with bipolar disorder type II
Singh et al. (2010)	20 bipolar offspring with bipolar diagnosis 20 bipolar offspring with subthreshold manic symptoms 20 healthy controls Ages 9–18 years old	Voxel in anterior cingulate cortex	Trend for reduced glutamate levels in bipolar offspring with bipolar disorder compared with offspring with subthreshold symptoms and controls No differences in N-acetyl aspartate, creatine, myo-inositol among groups	
Singh et al. (2011)	22 bipolar offspring with nonbipolar mood symptoms 25 healthy controls Ages 9–17 years old	Voxel in cerebellar vermis	Reduced myo-inositol and choline-containing compounds levels in bipolar offspring	
Singh et al. (2013)	64 bipolar offspring (36 with bipolar disorder and 28 with subsyndromal mania) 28 healthy controls Ages 9–18 years old	Voxels in dorsolateral prefrontal cortex	No group differences at baseline or at 5-year follow-up	

BOX 3 FUNCTIONAL MAGNETIC RESONANCE IMAGING (fMRI)

Functional magnetic resonance imaging (fMRI) is an imaging technique that detects changes in regional cerebral perfusion that accompany neuronal activity. In fMRI, the image is obtained by measuring the BOLD (blood oxygenation level dependent) effect, which reflects changes in the magnetic properties of hemoglobin when it becomes saturated or desaturated. The BOLD effect can be used to measure indirectly the increase in neuronal activity when a subject performs a task in the scanner (task-based fMRI) or spontaneous change in BOLD effect at rest (resting-state fMRI), which provides information about regional brain function and physiology (Matthews, Honey, & Bullmore, 2006). The correlations between BOLD signals among different brain regions can also be used to draw inferences about functional connectivity between those areas (van den Heuvel & Hulshoff Pol, 2010).

evaluated, in a cross-sectional design, four groups of adolescents aged between 10 and 20 years old (Welge et al., 2016). In this study, 64 subjects were adolescents with a parent with bipolar disorder. Of those adolescents, 32 were healthy (at-risk group) and 32 had depressive disorder (ultra-high-risk group). The authors were mostly interested in studying the activation patterns of the ventrolateral prefrontal cortex, anterior cingulate cortex, and amygdala. Both at-risk and ultra-high-risk groups exhibited increased activation in the left ventrolateral prefrontal cortex (Brodmann area 44) compared with the controls, which could represent a central biomarker of resilience, as it was observed also among healthy offspring (at-risk group). There was also a decreased activation in Brodmann area 10 in the ultra-high-risk group compared with the offspring who were asymptomatic, suggesting that abnormal activation in these brain regions (Brodmann areas 10 and 44) are relevant to risk progression and in need of further evaluation. A summary of fMRI studies is shown in Table 4.

Using a paradigm to elicit brain networks related to reward processing, Singh, Chang, et al. (2014) investigated regional brain activation and functional connectivity in 20 bipolar offspring and 25 healthy controls, between 8 and 15 years old (Singh, Kelley, et al., 2014). Compared with healthy controls, bipolar offspring had less activation in the pregenual anterior cingulate when anticipating loss, but greater activation in the left orbitofrontal cortex when anticipating rewards. Functional connectivity was also decreased between the pregenual anterior cingulate and the right ventrolateral prefrontal cortex in the anticipation of rewards, and increased between these two regions when anticipating loss, in comparison to the healthy controls. Associations between novelty seeking and increased activation in the amygdala and striatum when anticipating loss, and between impulsivity and increased activation in the insula and striatum when receiving rewards, were also found in the bipolar offspring. Another study using a reward paradigm compared 29 bipolar offspring, 28 offspring of patients with psychiatric morbidity other than bipolar disorder, and 23 healthy controls, all between 7 and 17 years old (Manelis et al., 2016). All participants underwent scanning while performing a number-guessing reward task. Bipolar offspring exhibited a greater activation in the right frontal pole compared with healthy controls. Functional connectivity between the right ventrolateral prefrontal cortex and the bilateral ventral striatum was greater in bipolar offspring than in healthy controls.

Table 4 Summary of findings in functional imaging studies of children and adolescents at risk for bipolar disorder

Task-based imaging

Authors	Sample characteristics	Methods	Results	Comments
Olsavsky et al. (2012)	32 bipolar children 13 healthy bipolar offspring 56 healthy controls Ages 8–18 years old	Emotional face processing task	Greater amygdala activation in both bipolar and at-risk children	
Ladouceur et al. (2013)	15 healthy bipolar offspring 16 healthy controls Ages 8–17 years old	Emotional working memory task	Greater right ventrolateral prefrontal cortex in bipolar offspring. Reduced connectivity between ventrolateral prefrontal cortex and amygdala	
Manelis et al. (2015)	29 bipolar offspring 29 nonbipolar offspring 23 healthy controls Ages 7–17 years old	Emotional face processing task	Greater right amygdala activation during emotional faces in both bipolar and nonbipolar offspring Increased connectivity between right amygdala and anterior cingulate cortex in response to emotional faces	
Welge et al. (2016)	32 first-episode mania patients 32 healthy bipolar offspring 32 bipolar offspring with depression 32 healthy controls Ages 10–20 years old.	Continuous performance task with emotional distracters	Greater activation in left ventrolateral prefrontal cortex (Brodmann area 44) in healthy and depressed bipolar offspring compared with controls Lower activation in right ventrolateral prefrontal cortex (Brodmann area 10) in depressed bipolar offspring and in first-episode mania patients compared with healthy bipolar offspring	
Singh, Chang, et al. (2014)	20 bipolar offspring 25 healthy controls Ages 8–15 years old	Reward processing task	Less activation in pregenual anterior cingulate when anticipating loss, greater activation in left orbitofrontal cortex when anticipating rewards in bipolar offspring Decreased functional connectivity between pregenual anterior cingulate and right ventrolateral prefrontal cortex when anticipating loss and increased when anticipating reward	Association between novelty seeking and increased activation in the amygdala and striatum when anticipating loss and between impulsivity and increased activation in the insula and striatum when receiving rewards were also found in the bipolar offspring

Study	Sample	Analysis	Results
Manelis et al. (2016)	29 bipolar offspring 28 offspring of parents with nonbipolar psychopathology 23 healthy controls Ages 7–17 years old	Reward task (number guessing) reward	Greater activation in right frontal pole in bipolar offspring than in healthy controls Functional connectivity between ventrolateral prefrontal cortex and bilateral striatum was greater in bipolar offspring than healthy controls

Resting state

Study	Sample	Analysis	Results	
Singh, Kelley, et al. (2014)	24 bipolar offspring 25 healthy controls Ages 8–17 years old	Data-driven independent component analyses Hypothesis driven region-of-interest	Increased connectivity in subregion of left executive control network in bipolar offspring Decreased connectivity between left amygdala and pregenual cingulate, between subgenual cingulate and supplemental motor cortex, and between left ventrolateral prefrontal cortex and striatum in bipolar offspring	Decreased connectivity with increased family chaos between the left ventrolateral prefrontal cortex and caudate
Sole-Padulles et al. (2016)	27 offspring of schizophrenia parents 39 bipolar offspring 40 healthy controls Ages 7–19 years old	Independent component analysis	No differences in functional connectivity between bipolar offspring and other groups	Uncorrected group effect of increased connectivity in bipolar offspring compared to the other groups in the right executive control network; however, these results were uncorrected for multiple comparisons
Roberts et al. (2017)	49 bipolar subjects 71 at-risk children 80 healthy controls Ages 16–30 years old	Group analyses Graph theory Machine learning	Decreased connectivity between the left inferior frontal gyrus and left insula and anterior cingulate in at-risk compared with controls	

RESTING-STATE fMRI FINDINGS

Singh, Kelley, et al. (2014) investigated abnormal functional connectivity as a vulnerability marker of bipolar disorder and its association with family chaos in a sample of 24 bipolar offspring and 25 healthy controls, aged 8–17 years old (Singh, Chang, et al., 2014). They used two methodologies: a data-driven independent component analysis (ICA) and a hypothesis-driven region-of-interest (ROI) analysis. In the ICA analysis, they found that bipolar offspring showed increased connectivity in a subregion of the left executive control network (the left ventrolateral prefrontal cortex) compared with the healthy controls. In the ROI analysis, bipolar offspring had decreased connectivity between the left amygdala and pregenual cingulate, between the subgenual cingulate and supplemental motor cortex, and between the left ventrolateral prefrontal cortex and striatum. Moreover, there was decreased connectivity between the left ventrolateral prefrontal cortex and the caudate with increased family chaos. Sole-Padulles et al. (2016) used resting-state fMRI to investigate functional connectivity in a sample of 27 offspring of schizophrenia patients, 39 offspring of bipolar disorder patients, and 40 healthy controls, aged 7–19 years old (Sole-Padulles et al., 2016). They used independent component analysis to evaluate functional connectivity within several resting-state networks and assessed the relationship with gray matter volumes, clinical variables, and cognition. They found no differences in functional connectivity between bipolar offspring and controls, and between bipolar offspring and schizophrenia offspring. There was an uncorrected group effect of increased connectivity in bipolar offspring compared to the other groups in the right executive control network; however, these results were uncorrected for multiple comparisons. Finally, Roberts et al. (2017) used resting-state fMRI to compare 49 subjects with bipolar disorder, 71 at-risk children or siblings of a patient with bipolar disorder type I or II, and 80 healthy controls, between 16 and 30 years old (Roberts et al., 2017). The authors focused on the connectivity between the left inferior frontal gyrus with other brain regions, particularly those related to emotion processing and cognitive control, using between-group analyses, graph theory, and machine learning approaches. They found a decreased connectivity between the left inferior frontal gyrus and left insula and anterior cingulate cortex in the bipolar offspring compared with controls.

SUMMARY OF FUNCTIONAL IMAGING FINDINGS

Results from task-based fMRI studies in youth at risk for bipolar disorder are consistent with prevailing theories that emotion task paradigms elicit brain areas involved with emotion processing and regulation, mainly amygdala (emotion processing) and ventrolateral prefrontal cortex (emotion regulation). The relevance of these findings is that, even in children and adolescents who never had any psychiatric disorders, abnormal brain activation or abnormal functional brain connectivity occurs when processing emotional challenges. Whether that might be an expression of genetic risk and/or to environmental demands is not yet defined. The findings are also consistent with prevailing theories that the disturbance in emotion and attentional processes in bipolar disorder arise from a deficient top-down control of a hypoactive prefrontal cortex over overactive limbic regions (Strakowski, 2012). When studies looked at brain circuits responsible for reward-related behaviors, impulsivity, and inhibitory control, brain areas related to these processes, such as the anterior cingulate, the orbitofrontal cortex, and the striatum, were abnormally activated or had an abnormal connectivity within prefrontal-amygdala-striatum circuits, again suggesting that the familial risk is associated with functional abnormalities in brain areas associated with core features of bipolar disorder, such as hedonism-related behaviors. In addition, some resting-state fMRI findings have revealed abnormal connectivity between areas of the prefrontal-limbic-striatal circuit in bipolar offspring, which is not related to performing a cognitive task during the scan. While one study found abnormal connectivity between the ventrolateral prefrontal cortex and subcortical structures (mainly the amygdala and striatum), and associations of abnormal connectivity with family environment (Singh, Chang, et al., 2014), two other studies found no or only discrete abnormal resting-state connectivity between bipolar offspring and controls (Roberts et al., 2017; Sole-Padulles et al., 2016). The full meaning of these findings is still unknown. Nevertheless, it should be noted that a few positive results have some replication, and draw attention to some specific brain areas that might play a bigger role in predisposition to the disorder, such as the ventrolateral prefrontal cortex and the amygdala and striatum networks, as compared to other brain areas that might be more compromised at other stages of illness. We should again emphasize that all the functional MRI studies were cross-sectional studies and hence there is no evidence to date that any of these abovementioned brain functional abnormalities, as detected by fMRI, is associated with the later development of mood episodes in youth at risk.

CONCLUSION

In this chapter, we reviewed the brain imaging studies of children, adolescents, and young adults at familial risk for bipolar disorder. Such studies are important to investigate the presence of neurobiological abnormalities related to genetic risk to bipolar disorder or to the early development of symptoms in bipolar disorder. The understanding of these neurobiological abnormalities may lead to targeted preventative or early treatment strategies. In this chapter, we also attempted to reconcile findings in at-risk youth with the current prevailing hypothesis of neurophysiology of bipolar disorder. Bipolar disorder is considered to be a dysfunction of emotion and attention processes associated with structural, neurochemical, and functional brain abnormalities within fronto-limbic networks, particularly a dysfunctional prefrontal cortex that seems unable to exert a top-down control of subcortical structures, namely the amygdala and striatum (Strakowski, 2012; Strakowski et al., 2012). In addition, because bipolar disorder often emerges during adolescence, current hypotheses also state that abnormal brain development, particularly in white matter and cortical pruning of prefrontal regions during adolescence, is associated with the pathophysiology of the illness. It has been postulated that this abnormal brain development causes an abnormal connectivity in the anterior emotional control network, impairing the subject's ability to regulate emotion and attention properly, and causing early symptoms of disease as well as enduring and severe mood states (Schneider et al., 2012). The study of at-risk subjects provides an opportunity to examine all of these possibilities.

Structural imaging studies have not found compelling evidence for abnormal gray matter changes or differences in bipolar offspring, either global or localized, particularly the studies that include larger sample sizes, and are likely better powered (Nery et al., 2017; Park et al., 2015; Singh et al., 2008; Sugranyes et al., 2015). Few studies have found increased gray matter volumes in the left parahippocampal gyrus (Ladouceur et al., 2008) and in the inferior frontal gyrus (Hajek et al., 2013), and decreased gray matter volumes or cortical thinning in localized areas of the frontal and temporal gyrus (Hanford, Hall, et al., 2016; Hanford, Sassi, et al., 2016). In contrast, although fewer in number, most structural studies of white matter have found evidence for abnormal white matter tracts that connect frontal to temporal or subcortical brain areas (Frazier et al., 2007; Nery et al., 2017; Roybal et al., 2015), with the exception of one study that was negative (Teixeira et al., 2014). Positive findings, either in gray or white matter, pertain to brain areas of the fronto-limbic circuit, which is consistent with the pathophysiological model of bipolar disorder. The discrepancy regarding the gray matter findings across the studies is still poorly understood. One possible explanation is that gray matter findings in imaging studies are indirect correlates of different neurobiological processes, which may include neurotoxic effects of medication exposure, burden of illness, comorbidities, or developmental or neurodegenerative effects (Nery et al., 2013). Therefore, differences in age distribution (developmental effect), frequency of comorbid conditions (concomitant illness effect), and degree of medication exposure may explain, at least in part, such differences. According to this explanation, differences in gray matter volume, density, or in cortical thickness in adults with established bipolar disorder are more likely to be detected because in those populations, gray matter is exposed to several of these factors for a more substantial period of time. Likewise, if white matter findings are associated with genetic risk, they are more likely to be detected in at-risk youth studies despite of these subjects being medication naïve, or asymptomatic. These speculations provide an intriguing hypothesis that the initial neural changes associated with bipolar disorder *begins* in the white matter tracts that connect prefrontal to subcortical limbic structures, possibly as an effect of genetic predisposition, and subsequently *continues* in gray matter areas associated with mood and attention regulation, as an effect of illness burden.

In contrast, most proton MRS studies have found no neurochemical abnormalities in the prefrontal cortex (dorsolateral prefrontal cortex, ventromedial, or anterior cingulate cortex) of at-risk youth (Gallelli et al., 2005; Hajek et al., 2008; Singh et al., 2010, 2013). Among the imaging techniques reviewed here, proton MRS is the least frequently used in studies of bipolar offspring. It is noteworthy that several brain areas that are very relevant in the pathophysiology of bipolar disorder, such as the hippocampus or striatum, to our knowledge, have not been studied.

Regarding brain function, fMRI studies have provided more compelling evidence for the fronto-limbic dysfunction theory. In several fMRI studies, at-risk youth presented abnormal activation in brain areas of the fronto-limbic network, such as the amygdala and ventrolateral prefrontal cortex, during tasks of emotion processing (Ladouceur et al., 2013; Manelis et al., 2015; Olsavsky et al., 2012; Welge et al., 2016) or reward processing (Manelis et al., 2016; Singh, Kelley, et al., 2014). These findings suggest that a dysfunctional activation of the ventrolateral prefrontal cortex and amygdala are present in at-risk populations prior to the development of mood disorders, and might be a risk marker for bipolar disorder in at-risk youth. Findings from functional connectivity analyses have also corroborated the idea of a dysfunctional connectivity between the prefrontal cortex and amygdala when processing emotional and reward tasks (Ladouceur et al., 2013; Manelis et al., 2015, 2016; Singh, Kelley, et al., 2014), or in resting state (Singh, Chang, et al., 2014).

Further research is necessary to explore the role of brain development on brain structure, neurochemistry, and function in youth at risk for bipolar disorder. Very few of the reviewed studies have explored brain development effects, which is an area in urgent need of further research, given that the first symptoms of bipolar disorder emerge during adolescence, a period during which gray matter undergoes substantial cortical pruning for more efficient synapses (Paus, Keshavan, & Giedd, 2008). For instance, a linear decrease in fractional anisotropy and an increase in radial diffusivity with age in the left corpus callosum and in the right inferior longitudinal fasciculi were found in bipolar offspring, but not in healthy offspring (Versace et al., 2008), supporting the idea that abnormal white matter development—and ultimately an abnormal connectivity between brain areas—occurs in bipolar offspring, and might be associated with the illness onset and progression.

To understand the translation of genetic risk into abnormal brain structure, neurochemistry, or function, and clinical expression of disease in bipolar disorder, an ideal study would have a prospective longitudinal design and a large sample size since only a small minority of all at-risk children and adolescents will eventually develop bipolar disorder. Additionally, such a study would have a healthy comparison sample observed at similar time points, to differentiate abnormal and normal brain development. It is obviously a daunting task to conduct such a study. However, findings from structural neuroimaging studies thus far are largely consistent with the theoretical model of bipolar disorder of an abnormal fronto-limbic network. Longitudinal studies of at-risk populations are likely to be the next frontier to advance the field of prevention and early intervention in bipolar disorder. Such studies are necessary to answer whether all the abovementioned findings are an expression of genetic risk, a resilience marker, or a correlate of very early illness (before it becomes fully developed). Some cross-sectional studies have attempted to address this question by comparing brain changes among bipolar offspring who already developed bipolar disorder, those who are asymptomatic at time of scanning, those who exhibited nonbipolar psychopathology (who, especially in the case of depressive disorders or ADHD, are thought to be on a spectrum of progression from asymptomatic to full bipolar disorder), and healthy comparison youth (Hanford, Hall, et al., 2016; Hanford, Sassi, et al., 2016; Nery et al., 2017; Welge et al., 2016). However, due to their cross-sectional natures, these studies are unable to answer whether specific abnormalities are risk or resilience markers for developing bipolar disorder or are markers of early illness.

Although MRI has great anatomical resolution and provides the best in vivo and low-risk access to the brain, it is unknown to what extent the differences in so-called gray or white matter volumes, cortical thickness, fractional anisotropy, or BOLD signal reflect actual cytoarchitectural or tissue abnormalities (such as change in neuron size or density, neuropil, white matter demyelination, or actual neuronal activation related to brain function in study) (Weinberger & Radulescu, 2016). There is also evidence that other factors may affect the MRI signal, such as exercise, hydration, and brain perfusion. This reflects an inherent limitation of studying the in vivo brain, and the inability to obtain direct measures of in vivo brain tissue. Multimodal imaging modalities (e.g., combining structural and functional, or functional and ^{1}H-MRS) is a recent development of brain imaging that has the potential to extend knowledge about the biological significance of imaging information from what is currently given by unimodal modalities (Calhoun & Sui, 2016; van den Heuvel, Mandl, Kahn, & Hulshoff Pol, 2009).

In summary, most brain imaging studies have found some level of evidence for abnormal structure, regional neurochemical abnormalities, and functional activation patterns in brain areas related to emotion processing and regulation in youth at risk for bipolar disorder. The field of neuroimaging of youth at risk for bipolar disorder is evolving rapidly. We believe that longitudinal designs are the most necessary step to understand the pathway from genetic risk to clinical expression of illness. Efforts in establishing well-characterized cohorts should be a priority in order to unravel early neurobiological abnormalities and create opportunities for primary and secondary prevention of this disabling and severe psychiatric disorder.

REFERENCES

Assaf, Y., & Pasternak, O. (2008). Diffusion tensor imaging (DTI)-based white matter mapping in brain research: a review. *Journal of Molecular Neuroscience, 34*(1), 51–61. https://doi.org/10.1007/s12031-007-0029-0.

Birmaher, B., Axelson, D., Monk, K., Kalas, C., Goldstein, B., Hickey, M. B., et al. (2009). Lifetime psychiatric disorders in school-aged offspring of parents with bipolar disorder: the Pittsburgh bipolar offspring study. *Archives of General Psychiatry, 66*(3), 287–296. https://doi.org/10.1001/archgenpsychiatry.2008.546.

Calhoun, V. D., & Sui, J. (2016). Multimodal fusion of brain imaging data: a key to finding the missing link(s) in complex mental illness. *Biological Psychiatry. Cognitive Neuroscience and Neuroimaging, 1*(3), 230–244. https://doi.org/10.1016/j.bpsc.2015.12.005.

Cecil, K. M. (2013). Proton magnetic resonance spectroscopy: technique for the neuroradiologist. *Neuroimaging Clinics of North America, 23*(3), 381–392. https://doi.org/10.1016/j.nic.2012.10.003.

Cecil, K. M., DelBello, M. P., Sellars, M. C., & Strakowski, S. M. (2003). Proton magnetic resonance spectroscopy of the frontal lobe and cerebellar vermis in children with a mood disorder and a familial risk for bipolar disorders. *Journal of Child and Adolescent Psychopharmacology, 13*(4), 545–555. https://doi.org/10.1089/104454603322724931.

DelBello, M. P., & Geller, B. (2001). Review of studies of child and adolescent offspring of bipolar parents. *Bipolar Disorders, 3*(6), 325–334.

Devlin, J. T., & Poldrack, R. A. (2007). In praise of tedious anatomy. *NeuroImage, 37*(4), 1033–1041. discussion 1050–1038. https://doi.org/10.1016/j.neuroimage.2006.09.055.

Duffy, A., Horrocks, J., Doucette, S., Keown-Stoneman, C., McCloskey, S., & Grof, P. (2014). The developmental trajectory of bipolar disorder. *The British Journal of Psychiatry, 204*(2), 122–128. https://doi.org/10.1192/bjp.bp.113.126706.

Foland-Ross, L. C., Thompson, P. M., Sugar, C. A., Madsen, S. K., Shen, J. K., Penfold, C., et al. (2011). Investigation of cortical thickness abnormalities in lithium-free adults with bipolar I disorder using cortical pattern matching. *The American Journal of Psychiatry, 168*(5), 530–539. https://doi.org/10.1176/appi.ajp.2010.10060896.

Frazier, J. A., Breeze, J. L., Papadimitriou, G., Kennedy, D. N., Hodge, S. M., Moore, C. M., et al. (2007). White matter abnormalities in children with and at risk for bipolar disorder. *Bipolar Disorders, 9*(8), 799–809. https://doi.org/10.1111/j.1399-5618.2007.00482.x.

Gallelli, K. A., Wagner, C. M., Karchemskiy, A., Howe, M., Spielman, D., Reiss, A., et al. (2005). N-acetylaspartate levels in bipolar offspring with and at high-risk for bipolar disorder. *Bipolar Disorders, 7*(6), 589–597. https://doi.org/10.1111/j.1399-5618.2005.00266.x.

Giuliani, N. R., Calhoun, V. D., Pearlson, G. D., Francis, A., & Buchanan, R. W. (2005). Voxel-based morphometry versus region of interest: a comparison of two methods for analyzing gray matter differences in schizophrenia. *Schizophrenia Research, 74*(2–3), 135–147. https://doi.org/10.1016/j.schres.2004.08.019.

Good, C. D., Johnsrude, I. S., Ashburner, J., Henson, R. N., Friston, K. J., & Frackowiak, R. S. (2001). A voxel-based morphometric study of ageing in 465 normal adult human brains. *NeuroImage, 14*(1 Pt 1), 21–36. https://doi.org/10.1006/nimg.2001.0786.

Goodwin, F. K., Jamison, K. R., & Ghaemi, S. N. (2007). *Manic-depressive illness: Bipolar disorders and recurrent depression* (2nd ed.). New York, NY: Oxford University Press.

Gotlib, I. H., Traill, S. K., Montoya, R. L., Joormann, J., & Chang, K. (2005). Attention and memory biases in the offspring of parents with bipolar disorder: indications from a pilot study. *Journal of Child Psychology and Psychiatry, 46*(1), 84–93. https://doi.org/10.1111/j.1469-7610.2004.00333.x.

Gottesman, I. I., & Gould, T. D. (2003). The endophenotype concept in psychiatry: etymology and strategic intentions. *The American Journal of Psychiatry, 160*(4), 636–645. https://doi.org/10.1176/appi.ajp.160.4.636.

Gottesman, I. I., Laursen, T. M., Bertelsen, A., & Mortensen, P. B. (2010). Severe mental disorders in offspring with 2 psychiatrically ill parents. *Archives of General Psychiatry, 67*(3), 252–257. https://doi.org/10.1001/archgenpsychiatry.2010.1.

Hajek, T., Bernier, D., Slaney, C., Propper, L., Schmidt, M., Carrey, N., et al. (2008). A comparison of affected and unaffected relatives of patients with bipolar disorder using proton magnetic resonance spectroscopy. *Journal of Psychiatry & Neuroscience, 33*(6), 531–540.

Hajek, T., Cullis, J., Novak, T., Kopecek, M., Blagdon, R., Propper, L., et al. (2013). Brain structural signature of familial predisposition for bipolar disorder: replicable evidence for involvement of the right inferior frontal gyrus. *Biological Psychiatry, 73*(2), 144–152. https://doi.org/10.1016/j.biopsych.2012.06.015.

Hanford, L. C., Hall, G. B., Minuzzi, L., & Sassi, R. B. (2016). Gray matter volumes in symptomatic and asymptomatic offspring of parents diagnosed with bipolar disorder. *European Child & Adolescent Psychiatry, 25*(9), 959–967. https://doi.org/10.1007/s00787-015-0809-y.

Hanford, L. C., Sassi, R. B., Minuzzi, L., & Hall, G. B. (2016). Cortical thickness in symptomatic and asymptomatic bipolar offspring. *Psychiatry Research, 251*, 26–33. https://doi.org/10.1016/j.pscychresns.2016.04.007.

Hasler, G., Drevets, W. C., Gould, T. D., Gottesman, I. I., & Manji, H. K. (2006). Toward constructing an endophenotype strategy for bipolar disorders. *Biological Psychiatry, 60*(2), 93–105. https://doi.org/10.1016/j.biopsych.2005.11.006.

Kim, E., Garrett, A., Boucher, S., Park, M. H., Howe, M., Sanders, E., et al. (2017). Inhibited temperament and hippocampal volume in offspring of parents with bipolar disorder. *Journal of Child and Adolescent Psychopharmacology, 27*(3), 258–265. https://doi.org/10.1089/cap.2016.0086.

Ladouceur, C. D., Almeida, J. R., Birmaher, B., Axelson, D. A., Nau, S., Kalas, C., et al. (2008). Subcortical gray matter volume abnormalities in healthy bipolar offspring: potential neuroanatomical risk marker for bipolar disorder? *Journal of the American Academy of Child and Adolescent Psychiatry, 47*(5), 532–539. https://doi.org/10.1097/CHI.0b013e318167656e.

Ladouceur, C. D., Diwadkar, V. A., White, R., Bass, J., Birmaher, B., Axelson, D. A., et al. (2013). Fronto-limbic function in unaffected offspring at familial risk for bipolar disorder during an emotional working memory paradigm. *Developmental Cognitive Neuroscience, 5*, 185–196. https://doi.org/10.1016/j.dcn.2013.03.004.

Manelis, A., Ladouceur, C. D., Graur, S., Monk, K., Bonar, L. K., Hickey, M. B., et al. (2015). Altered amygdala-prefrontal response to facial emotion in offspring of parents with bipolar disorder. *Brain, 138*(Pt 9), 2777–2790. https://doi.org/10.1093/brain/awv176.

Manelis, A., Ladouceur, C. D., Graur, S., Monk, K., Bonar, L. K., Hickey, M. B., et al. (2016). Altered functioning of reward circuitry in youth offspring of parents with bipolar disorder. *Psychological Medicine, 46*(1), 197–208. https://doi.org/10.1017/S003329171500166X.

Matthews, P. M., Honey, G. D., & Bullmore, E. T. (2006). Applications of fMRI in translational medicine and clinical practice. *Nature Reviews. Neuroscience, 7*(9), 732–744. https://doi.org/10.1038/nrn1929.

Meyer, S. E., Carlson, G. A., Wiggs, E. A., Martinez, P. E., Ronsaville, D. S., Klimes-Dougan, B., et al. (2004). A prospective study of the association among impaired executive functioning, childhood attentional problems, and the development of bipolar disorder. *Development and Psychopathology, 16*(2), 461–476.

Neil, J. J. (2008). Diffusion imaging concepts for clinicians. *Journal of Magnetic Resonance Imaging, 27*(1), 1–7. https://doi.org/10.1002/jmri.21087.

Nery, F. G., Monkul, E. S., & Lafer, B. (2013). Gray matter abnormalities as brain structural vulnerability factors for bipolar disorder: a review of neuroimaging studies of individuals at high genetic risk for bipolar disorder. *The Australian and New Zealand Journal of Psychiatry, 47*(12), 1124–1135. https://doi.org/10.1177/0004867413496482.

Nery, F. G., Norris, M., Eliassen, J. C., Weber, W. A., Blom, T. J., Welge, J. A., et al. (2017). White matter volumes in youth offspring of bipolar parents. *Journal of Affective Disorders, 209*, 246–253. https://doi.org/10.1016/j.jad.2016.11.023.

Olsavsky, A. K., Brotman, M. A., Rutenberg, J. G., Muhrer, E. J., Deveney, C. M., Fromm, S. J., et al. (2012). Amygdala hyperactivation during face emotion processing in unaffected youth at risk for bipolar disorder. *Journal of the American Academy of Child and Adolescent Psychiatry, 51*(3), 294–303. https://doi.org/10.1016/j.jaac.2011.12.008.

Park, M. H., Garrett, A., Boucher, S., Howe, M., Sanders, E., Kim, E., et al. (2015). Amygdalar volumetric correlates of social anxiety in offspring of parents with bipolar disorder. *Psychiatry Research, 234*(2), 252–258. https://doi.org/10.1016/j.pscychresns.2015.09.018.

Paus, T., Keshavan, M., & Giedd, J. N. (2008). Why do many psychiatric disorders emerge during adolescence? *Nature Reviews. Neuroscience, 9*(12), 947–957. https://doi.org/10.1038/nrn2513.

Phillips, M. L., Ladouceur, C. D., & Drevets, W. C. (2008). A neural model of voluntary and automatic emotion regulation: implications for understanding the pathophysiology and neurodevelopment of bipolar disorder. *Molecular Psychiatry, 13*(9), 829. 833–857. https://doi.org/10.1038/mp.2008.65.

Phillips, M. L., & Swartz, H. A. (2014). A critical appraisal of neuroimaging studies of bipolar disorder: toward a new conceptualization of underlying neural circuitry and a road map for future research. *The American Journal of Psychiatry, 171*(8), 829–843. https://doi.org/10.1176/appi.ajp.2014.13081008.

Roberts, G., Lord, A., Frankland, A., Wright, A., Lau, P., Levy, F., et al. (2017). Functional dysconnection of the inferior frontal Gyrus in young people with bipolar disorder or at genetic high risk. *Biological Psychiatry, 81*(8), 718–727. https://doi.org/10.1016/j.biopsych.2016.08.018.

Roybal, D. J., Barnea-Goraly, N., Kelley, R., Bararpour, L., Howe, M. E., Reiss, A. L., et al. (2015). Widespread white matter tract aberrations in youth with familial risk for bipolar disorder. *Psychiatry Research, 232*(2), 184–192. https://doi.org/10.1016/j.pscychresns.2015.02.007.

Schneider, M. R., DelBello, M. P., McNamara, R. K., Strakowski, S. M., & Adler, C. M. (2012). Neuroprogression in bipolar disorder. *Bipolar Disorders, 14*(4), 356–374. https://doi.org/10.1111/j.1399-5618.2012.01024.x.

Singh, M. K., Chang, K. D., Kelley, R. G., Saggar, M., Reiss, A. L., & Gotlib, I. H. (2014). Early signs of anomalous neural functional connectivity in healthy offspring of parents with bipolar disorder. *Bipolar Disorders, 16*(7), 678–689. https://doi.org/10.1111/bdi.12221.

Singh, M. K., Delbello, M. P., Adler, C. M., Stanford, K. E., & Strakowski, S. M. (2008). Neuroanatomical characterization of child offspring of bipolar parents. *Journal of the American Academy of Child and Adolescent Psychiatry, 47*(5), 526–531. https://doi.org/10.1097/CHI.0b013e318167655a.

Singh, M. K., Jo, B., Adleman, N. E., Howe, M., Bararpour, L., Kelley, R. G., et al. (2013). Prospective neurochemical characterization of child offspring of parents with bipolar disorder. *Psychiatry Research, 214*(2), 153–160. https://doi.org/10.1016/j.pscychresns.2013.05.005.

Singh, M. K., Kelley, R. G., Howe, M. E., Reiss, A. L., Gotlib, I. H., & Chang, K. D. (2014). Reward processing in healthy offspring of parents with bipolar disorder. *JAMA Psychiatry, 71*(10), 1148–1156. https://doi.org/10.1001/jamapsychiatry.2014.1031.

Singh, M., Spielman, D., Adleman, N., Alegria, D., Howe, M., Reiss, A., et al. (2010). Brain glutamatergic characteristics of pediatric offspring of parents with bipolar disorder. *Psychiatry Research, 182*(2), 165–171. https://doi.org/10.1016/j.pscychresns.2010.01.003.

Singh, M. K., Spielman, D., Libby, A., Adams, E., Acquaye, T., Howe, M., et al. (2011). Neurochemical deficits in the cerebellar vermis in child offspring of parents with bipolar disorder. *Bipolar Disorders, 13*(2), 189–197. https://doi.org/10.1111/j.1399-5618.2011.00902.x.

Sole-Padulles, C., Castro-Fornieles, J., de la Serna, E., Romero, S., Calvo, A., Sanchez-Gistau, V., et al. (2016). Altered Cortico-striatal connectivity in offspring of schizophrenia patients relative to offspring of bipolar patients and controls. *PLoS One, 11*(2), e0148045. https://doi.org/10.1371/journal.pone.0148045.

Strakowski, S. M. (2012). In S. M. Strakowski (Ed.), *The bipolar brain: Integrating neuroimaging and genetics* (pp. 253–274). New York: Oxford university press.

Strakowski, S. M., Adler, C. M., Almeida, J., Altshuler, L. L., Blumberg, H. P., Chang, K. D., et al. (2012). The functional neuroanatomy of bipolar disorder: a consensus model. *Bipolar Disorders, 14*(4), 313–325. https://doi.org/10.1111/j.1399-5618.2012.01022.x.

Strakowski, S. M., Delbello, M. P., & Adler, C. M. (2005). The functional neuroanatomy of bipolar disorder: a review of neuroimaging findings. *Molecular Psychiatry, 10*(1), 105–116. https://doi.org/10.1038/sj.mp.4001585.

Sugranyes, G., de la Serna, E., Romero, S., Sanchez-Gistau, V., Calvo, A., Moreno, D., et al. (2015). Gray matter volume decrease distinguishes schizophrenia from bipolar offspring during childhood and adolescence. *Journal of the American Academy of Child and Adolescent Psychiatry, 54*(8), 677–684. e672. https://doi.org/10.1016/j.jaac.2015.05.003.

Teixeira, A. M., Kleinman, A., Zanetti, M., Jackowski, M., Duran, F., Pereira, F., et al. (2014). Preserved white matter in unmedicated pediatric bipolar disorder. *Neuroscience Letters, 579*, 41–45. https://doi.org/10.1016/j.neulet.2014.06.061.

van den Heuvel, M. P., & Hulshoff Pol, H. E. (2010). Exploring the brain network: a review on resting-state fMRI functional connectivity. *European Neuropsychopharmacology, 20*(8), 519–534. https://doi.org/10.1016/j.euroneuro.2010.03.008.

van den Heuvel, M. P., Mandl, R. C., Kahn, R. S., & Hulshoff Pol, H. E. (2009). Functionally linked resting-state networks reflect the underlying structural connectivity architecture of the human brain. *Human Brain Mapping, 30*(10), 3127–3141. https://doi.org/10.1002/hbm.20737.

Versace, A., Almeida, J. R., Hassel, S., Walsh, N. D., Novelli, M., Klein, C. R., et al. (2008). Elevated left and reduced right orbitomedial prefrontal fractional anisotropy in adults with bipolar disorder revealed by tract-based spatial statistics. *Archives of General Psychiatry, 65*(9), 1041–1052. https://doi.org/10.1001/archpsyc.65.9.1041.

Versace, A., Ladouceur, C. D., Romero, S., Birmaher, B., Axelson, D. A., Kupfer, D. J., et al. (2010). Altered development of white matter in youth at high familial risk for bipolar disorder: a diffusion tensor imaging study. *Journal of the American Academy of Child and Adolescent Psychiatry, 49*(12), 1249–1259. e1241. https://doi.org/10.1016/j.jaac.2010.09.007.

Weinberger, D. R., & Radulescu, E. (2016). Finding the elusive psychiatric "lesion" with 21st-century Neuroanatomy: a note of caution. *The American Journal of Psychiatry, 173*(1), 27–33. https://doi.org/10.1176/appi.ajp.2015.15060753.

Welge, J. A., Saliba, L. J., Strawn, J. R., Eliassen, J. C., Patino, L. R., Adler, C. M., et al. (2016). Neurofunctional differences among youth with and at varying risk for developing mania. *Journal of the American Academy of Child and Adolescent Psychiatry, 55*(11), 980–989. https://doi.org/10.1016/j.jaac.2016.08.006.

Yuksel, C., Tegin, C., O'Connor, L., Du, F., Ahat, E., Cohen, B. M., et al. (2015). Phosphorus magnetic resonance spectroscopy studies in schizophrenia. *Journal of Psychiatric Research, 68*, 157–166. https://doi.org/10.1016/j.jpsychires.2015.06.014.

Neurocognitive findings in youth at high risk for bipolar disorder: Potential endophenotypes?

8

Isabelle E. Bauer, Iram F. Kazimi, Cristian P. Zeni, Jair C. Soares
Center of Excellence on Mood Disorders, Department of Psychiatry and Behavioral Sciences, McGovern Medical School, University of Texas Health Science Center at Houston (UTHealth), Houston, TX, United States

CHAPTER OUTLINE

INTRODUCTION

Neurocognitive dysfunction is a fundamental yet frequently ignored or misidentified component of bipolar disorder (BD). While schizophrenia and major depressive disorder have well-established neurocognitive profiles (Bortolato, Miskowiak, Köhler, Vieta, & Carvalho, 2014), the role of cognitive dysfunction as a core endophenotype for BD is still a matter of debate (Bora, Yucel, & Pantelis, 2009; Martínez-Arán, Vieta, Colom, et al., 2004; Martínez-Arán, Vieta, Reinares, et al., 2004; Robinson et al., 2006). In particular, it is still unclear whether deficits are present prior to the onset of the disease, or result from the deleterious effects of repeated mood episodes on the brain (Bauer, Pascoe, Wollenhaupt-Aguiar, Kapczinski, & Soares, 2014; Cardoso, Bauer, Meyer, Kapczinski, & Soares, 2015). This may be due to the heterogeneous nature of cognitive deficit in BD. Research has identified three distinct neurocognitive subgroups, including BD individuals with intact cognitive functioning and superior social cognition, BD individuals with specific cognitive deficits, and BD individuals with global impairment (Burdick et al., 2014). Interestingly, in these studies the group with specific deficits reported the highest number of total mood episodes, and the group with global impairment had a clinical profile intermediate between the intact group and the group with specific deficits. These findings points toward a nonlinear relationship between illness severity and increased severity of cognitive deficits in BD. Whether or not similar cognitive profiles are observed in individuals at high risk for BD is, however, still unknown.

Cognitive psychology has become an important area of research in a number of psychiatric disorders, ranging from severe psychotic illness such as schizophrenia to relatively benign, yet significantly disabling, nonpsychotic illnesses such as somatoform disorder. Cognitive psychology has particularly proven to be helpful in explaining underlying psychopathology and issues related to course, outcome, and treatment strategies for BD. Further, in the last decade cognitive tasks have been used to identify early cognitive markers of vulnerability (Phillips & Kupfer, 2013), and may eventually contribute to prevent the development of full-blown BD in people who are at high risk of the disease.

An increasing number of researchers have used family-based studies (Hajek et al., 2013; Ladouceur et al., 2008; Zalla et al., 2004) to identify markers of vulnerability based on clinical, cognitive, or brain parameters. The concept of *vulnerability* describes the complex interactions between individual genetic susceptibility factors and environmental risk factors, as discussed in previous chapters. An effective vulnerability marker or *endophenotype* should, ideally, be reliable, stable, hereditary, and predict the risk of an individual to develop the specific disease. To satisfy these requirements, a potential marker should meet a set of standard criteria, including (1) increased prevalence in relatives of the affected individual when compared to the general population; (2) association with spectrum disorders in family members and high-risk children; and (3) presence of the marker prior to the manifestation of clinical symptoms (Garver, 1987). From a genetic viewpoint, first-degree relatives (e.g., twins, siblings, and offspring) are an optimal study target as they share up to 50% of their genes and are at high risk (HR) to develop BD (Agrawal & Lynskey, 2008; Gottesman & Gould, 2003). Furthermore, first-degree relatives are more likely than other family members to share a number of environmental, cultural, and behavioral factors that may contribute to the development and maintenance of BD symptoms including cognitive deficits (Valdez, Yoon, Qureshi, Green, & Khoury, 2010).

A better understanding of cognitive functioning in individuals at risk for BD will provide clinicians with new insights into prevention and treatment strategies in HR, and lay the foundation for future clinical research in this rapidly expanding field. This chapter aims to provide an overview of the clinical applications of cognitive performance by examining evidence for cognitive vulnerability markers in youth at high risk (HR) for BD. We will also briefly review current knowledge of the structural and functional neuroanatomy of HR individuals, and then examine, in more detail, the evidence for deficits in a range of cognitive abilities and social cognition in HR individuals.

CLINICAL PROFILE OF INDIVIDUALS AT HIGH RISK FOR BD

The risk to develop BD is approximately 8 times higher in first-degree relatives (e.g., offspring, siblings, twins), and 3.5 times higher in other relatives of patients with BD than those of patients with non-BD major depression (Tsuang, Tohen, & Jones, 2011; Wilde et al., 2014). From a clinical viewpoint, first-degree relatives of BD patients are also approximately 4 times more likely to develop any kind of mood disorders (Chang, Steiner, & Ketter, 2003; Glahn, Bearden, Niendam, & Escamilla, 2004; Lapalme, Hodgins, & LaRoche, 1997), anxiety, and attention deficit hyperactivity disorder (Lapalme et al., 1997; Robinson & Nicol Ferrier, 2006). With regard to comorbid conditions, HR subjects show greater thought disturbances and higher risk for psychiatric disorders other than BD (Narayan, Allen, Cullen, & Klimes-Dougan, 2013). Additionally, a great percentage (10%–34.5%) of HR individuals present with depressive symptoms (Akdemir & Gokler, 2008; Lapalme et al., 1997; Mesman, Nolen, Reichart, Wals, & Hillegers, 2013). Alongside depressive symptoms, there exists an elevated risk of developing substance use disorders (odd ratios above 2) (Duffy, Alda, Hajek, Sherry, & Grof, 2010) and anxiety disorder (odd ratios up to 2.329) (Duffy et al., 2010, 2013). Evidence of the link between HR and attention deficit hyperactivity disorder (Birmaher et al., 2010; Hirshfeld-Becker et al., 2006) and disruptive behavior disorder (Henin et al., 2005) is equally strong.

RECENT NEUROIMAGING FINDINGS
BRAIN STRUCTURE

To date, the search for early neural risk markers in HR individuals has led to mixed results. This is partially due to the lack of studies performing brain assessments at different time points, and the heterogeneity of both clinical presentation and treatment status in HR and BD populations (Abramovic et al., 2016; Chase & Phillips, 2016; López-Jaramillo et al., 2017). Equally relevant are the differences in analyses and neuroimaging methods used, and the relatively arbitrary criteria to define who in an HR group would be more likely to transition to BD. A recent volumetric magnetic resonance imaging study in 115 asymptomatic child and adolescent offspring of BD type I parents and healthy control (HC) offspring (Nery et al., 2017) found decreased white matter volumes in multiple frontal, temporal, and parietal regions in HR compared to HC. Similarly, another study comparing HC, healthy HR offspring, and ultra-HR (UHR) offspring (UHR displayed either hypomanic, depressive, or hyperactive/attentional symptoms) showed gray matter volume reductions in the right orbitofrontal cortex in both offspring groups. Both HR and UHR offspring also displayed abnormalities in the inferior occipital cortex (Lin et al., 2015). Further, deficits in processing speed and visual-spatial memory were observed in UHR but not in HR. The authors argued that, while the abnormalities observed in HR were likely to be inherited, those associated with the UHR offspring indicated clinical deterioration.

White matter abnormalities were primarily found in the bilateral superior longitudinal fasciculus, corpus callosum, and temporal associative tracts of HR compared to HC (Frazier et al., 2007; Gunde et al., 2011). A cross-sectional diffusion tensor imaging study conducted in 20 unmedicated HR and 25 healthy control offspring of healthy parents (HC) (Versace et al., 2010) found that, after correcting for age, there was greater fractional anisotropy and reduced radial diffusion—which suggests intact white matter integrity—in the left corpus callosum and the right inferior longitudinal fasciculus (ILF) of HC compared to HR. By contrast, HR had lower fractional anisotropy and higher radial diffusion—which indicates reduced white matter integrity—in the left corpus callosum. Two studies, comparing BD patients to their biological siblings, and a group of HC, showed that in HR, fractional anisotropy was significantly reduced in the left posterior thalamic radiation, the left sagittal stratum (Sarıçiçek et al., 2016), and the corpus callosum (Sarıçiçek et al., 2016; Sprooten et al., 2016). Interestingly, mean fractional anisotropy values of unaffected siblings were, overall, more subtle than those observed in BD, and were intermediate between HC and BD patients. The question remains as to whether the presence of white matter alterations in HR is the result of altered neurodevelopmental processes or indicates genetic vulnerability to BD (Versace et al., 2010). Additional longitudinal research is therefore required before drawing definitive conclusions.

BRAIN FUNCTION

Overall, HR studies using functional brain imaging techniques during facial emotion processing tasks have consistently shown alterations in the amygdala and inferior frontal gyrus (Brotman et al., 2014; Singh, Kelley, et al., 2014), and reduced functional connectivity between the left amygdala, pregenual and subgenual cingulate, supplementary motor cortex, left ventrolateral prefrontal cortex, and left caudate (Olsavsky et al., 2012). A recent study found that BD and HR subjects displayed common alterations in face-processing regions including the middle-temporal gyrus and dorsolateral prefrontal gyrus. The two groups displayed differential patterns of activation in the fusiform, inferior frontal, superior temporal sulcus, and temporo-parietal junction. These regions are generally involved in face processing, executive function, and social cognition (Wiggins et al., 2017). Other common findings include decreased striatal activity during motor response inhibition tasks (Deveney et al., 2012), and alterations in frontal, cerebellar, vermis, and insula activity during a working memory task (Thermenos et al., 2011). Equally relevant are the reduced activation in the pregenual cingulate during a monetary incentive delay task, and the greater left lateral orbitofrontal cortex activation in response to reward (Singh, Chang, et al., 2014). BD and HR also displayed increased reaction time variability, and reduced relationships between the variation in reaction time and activity in the inferior and middle frontal gyri, precuneus, cingulate cortex, caudate, and postcentral gyrus in both BD and HR (Pagliaccio et al., 2017), along with reduced deactivation in the medial prefrontal cortex when compared to HC (Alonso-Lana et al., 2016).

A key limitation of current fMRI studies is related to variations in spatial normalization and limited reporting of peak activation locations in stereotaxic coordinates across studies. These limitations are typically accounted for by meta-analyses using activation likelihood estimates based on spatial probability distribution centered at given coordinates (Lee, Anumagalla, Talluri, & Pavuluri, 2013). Lee et al.'s (2013) meta-analysis focused on 29 functional magnetic resonance imaging (fMRI) studies including pediatric populations with BD (PBD), HR, and typically developing HC. Across studies, HR showed increased activity in core cognition and emotion-processing brain regions such as the dorsolateral prefrontal cortex, insula, and parietal cortex. When compared to HR and HC, youth with BD displayed a more diverse activation profile with lower activation in the right ventrolateral and dorsolateral prefrontal cortex, and greater activation in the right amygdala, parahippocampal gyrus, medial prefrontal cortex, left ventral striatum, and cerebellum. The altered neural effort in prefrontal regions observed in both groups was viewed as being intrinsic to the disease.

The relevance of fMRI in supporting clinical decision making is highlighted by a recent whole-brain pattern analysis study performed on functional activity recorded during a working memory task (Frangou, Dima, & Jogia, 2017). In this study, patients with BD could be distinguished from HC with an accuracy of 83.5%, and a specificity of 92.3%. High accuracy and specificity values were reached when comparing BD to their healthy relatives and were 81.8% and 90.9%, respectively. These promising findings indicate that the next wave of neuroimaging research in psychiatry should focus on networks susceptible to change prior to the onset of BD, and investigate the neural correlates interventions aimed at preserving neural determinants of resilience in vulnerable individuals.

COGNITIVE FUNCTIONING

Studies in unaffected relatives show that genetic liability explains some, but not all, of the variance in cognitive performance observed in HR individuals (Clark, Kempton, Scarnà, Grasby, & Goodwin, 2005). Indeed, overall first-degree relatives of BD patients present with deficits in verbal learning and memory, verbal fluency, processing speed, and social cognition when compared to control individuals and BD patients (Arts, Jabben, Krabbendam, & Van Os, 2008; Cardenas, Kassem, Brotman, Leibenluft, & McMahon, 2016; Drysdale, Knight, McIntosh, & Blackwood, 2013; Frangou, Haldane, Roddy, & Kumari, 2005). Findings are still equivocal in terms of sustained attention, executive function, and response inhibition (Bauer et al., 2016; Clark et al., 2005; Frías, Palma, Farriols, & Salvador, 2014; Trivedi et al., 2008). A more consistent pattern of results appears when comparing HR individuals specifically to HC, as the HR subjects display specific difficulties in verbal memory, sustained memory, executive functions, and affective and reward processing (Miskowiak et al., 2017). It is, however, confusing that such cognitive deficits are not specific to BD, and resemble those appearing in schizophrenia, schizoaffective disorders, and psychotic BD (Reichenberg et al., 2009). In the next sections we will therefore review relevant literature on intellectual quotient, cognitive abilities (nonemotional or "cold" cognition), and social cognition (emotion-laden cognition or "hot" cognition) in youth at high risk for BD, and assess whether and which cognitive measures constitute credible endophenotypic measures of vulnerability of BD. Given the limited number of studies reporting on cognitive performance in young HR subjects, we also discuss the cognitive performance of adult offspring and siblings of BD patients. An overview of recent cognitive findings in HR (both children and adults) can be found in Table 1.

GENERAL INTELLECTUAL ABILITIES

A few studies have dealt with the question as to whether the intellectual quotient (IQ) of HR individuals follows a different neurodevelopmental trajectory than that of HC. Cardenas et al.'s metaanalysis (2016) of 18 HR studies concluded that intellectual quotient was overall preserved when compared to control participants. Another study including 65 offspring of parents with either BD or schizophrenia showed that IQ was overall comparable between HR groups. However, relative to HC, HR presented an intellectual developmental delay during childhood (Maziade et al., 2011). A large metaanalytical review of the literature on premorbid cognitive functioning (Trotta, Murray, & MacCabe, 2015) retrieved 4 prospective cohort studies that tested cognitive or intellectual functioning prior to illness onset, and 13 retrospective case-control studies that estimated premorbid intellectual functioning based on reading tests conducted after the patient had developed the illness. The authors concluded that, while individuals who developed schizophrenia presented with reduced intellectual function, premorbid function in BD was comparable to that of healthy populations.

Findings from a large student record database of HR adolescents who were later hospitalized for nonpsychotic BD, schizoaffective disorder, and schizophrenia (Reichenberg et al., 2002) showed that adolescents who later developed schizoaffective disorder performed significantly worse than comparison subjects on tasks of nonverbal abstract reasoning and visual-spatial problem solving. They also scored significantly worse than subjects with nonpsychotic BD on some intellectual, reading, and reading comprehension tests. Two studies found discrepancies between verbal IQ (VIQ) and performance IQ (PIQ) in a sample of offspring of BD parents. Full-scale IQ and VIQ did not differ between the control group and HR, but PIQ was significantly lower in HR. There was also a significantly higher incidence of higher VIQ and lower PIQ scores among HR (Decina et al., 1983). Additionally, the HR group performed significantly worse than control subjects on measures of reading, spelling, and arithmetic (McDonough-Ryan et al., 2002).

Notwithstanding the discrepancies, these findings point to altered IQ in HR youth. It is equally important to highlight that intelligence is a heterogeneous concept. To date, studies have used a number of intelligence tests, and possibly assessed different facets of intellectual functioning. Furthermore, academic achievement is only one facet of intelligence. Thus, a proper intellectual assessment should combine a number of estimates such as verbal abilities, mathematics, perceptual organization, and processing speed.

Table 1 Overview of recent cognitive findings in individuals at high-risk for bipolar disorder

	Population	Design	Measures	Outcome
De la Serna et al. (2016)	Children and adolescent BD offspring with disorders other than BD Healthy offspring	Cross-sectional	Wechsler Intelligence Scale for children Verbal and visual memory and learning (Wechsler Memory Scale) Conner's continuous performance test Wisconsin Card Sorting Test	Offspring performed poorly in processing speed, immediate visual recall tasks regardless of psychopathology
Diwadkar et al. (2011)	Pediatric offspring of schizophrenia, offspring of bipolar patients, healthy controls	Cross-sectional	Spatial Working memory task Sustained attention (continuous performance task (CPT)-identical pairs)	Low accuracy rates in BD on CPT compared to HC
Duffy et al. (2009)	Adult unaffected and remitted (medicated for unipolar depression) HR offspring of BD patients, healthy controls	Cross-sectional	Visual backward masking task	Low accuracy rates in BD and HR
Georgiades et al. (2016)	331 pediatric twins/siblings of BD patients	Structural equation modeling	Wechsler Abbreviated Scale of Intelligence California Verbal Learning Test Selected measures of the Cambridge Neuropsychological Test Automated battery	IQ and visual spatial learning have genetic correlations with BD after adjusting for affective symptoms
Gotlib et al. (2005)	Healthy pediatric offspring of BD parents (HR) Healthy controls	Cross-sectional	Emotion Stroop task Self-reference encoding task	HR show attentional bias toward social-threat and manic-irritable words, and better recall of negative words compared to healthy controls

Continued

Table 1 Overview of recent cognitive findings in individuals at high-risk for bipolar disorder—cont'd

	Population	Design	Measures	Outcome
Kulkarni, Jain, Janardhan Reddy, Kumar, and Kandavel (2010)	Unaffected siblings of BD patients Healthy controls	Cross-sectional	Tower of London Rey's auditory verbal learning test (RAVLT) Rey's complex figure test	Siblings showed poor planning and verbal learning/memory compared to HC
McCormack et al. (2016)	Healthy controls Adolescents and young adults with BD parent or sibling diagnosed with BD Patients with BD	Cross-sectional	Wechsler Abbreviated Scale of Intelligence Working memory, attention, language, visuospatial ability from the Repeatable Battery for the Assessment of Neuropsycical Status Executive function and affective processing from the Cambridge Neuropsychological Test Automated Battery Emotion recognition test	Verbal intelligence and affective response inhibition, particularly of negative valence, were impaired in HR compared to HC
Nehra, Grover, Sharma, Sharma, and Sarkar (2014)	Adult unaffected siblings with BD-I HC	Cross-sectional	Wisconsin Card Sorting Test Brief Visuospatial Memory Test-Revised Hopkins Verbal Learning Test Wechsler Adult Intelligence Scale Digit Symbol Test	Siblings performed poorly in verbal learning
Vasconcelos-Moreno et al. (2016)	Adults with BD, unaffected siblings, HC	Cross-sectional	California Verbal Learning Test Stroop Color and Word Test Wisconsin Card Sorting Test	Siblings showed poorer functional performance than HC

"COLD" COGNITION

Child studies

There is a surprisingly small number of cognitive studies in offspring of BD parents, and, to date, no longitudinal study has investigated cognitive development along-side clinical status in this vulnerable population. Three studies in pediatric populations aged less than 16 years old (Costa & McCrae, 1992; McDonough-Ryan et al., 2002; Winters, Stone, Weintraub, & Neale, 1981) provided evidence of impairment in at-risk BD children. In particular, a large sample of HR children with one parent hospitalized for a manic episode (Winters et al., 1981) displayed slow reaction times on a visual search task, and difficulties in word production and communication patterns (Harvey, Weintraub, & Neale, 1982). Interestingly, retrospective intellectual assessments of youth who developed BD found that these children had academic achievement difficulties, specifically in arithmetic (Quackenbush, Kutcher, Robertson, Boulos, & Chaban, 1996), language, social skills, and motor development (Sigurdsson, Fombonne, Sayal, & Checkley, 1999). Further, an increasing body of literature suggests that HR children display cognitive deficits similar to those of children suffering from nonverbal learning disorders. McDonough-Ryan et al.'s study in HR and healthy children aged 8–12 years showed that children of parents with BD encountered difficulties in arithmetic, reading, and spelling. Furthermore, performance intellectual abilities were weaker than verbal skills, thus supporting previous cognitive findings in youth and adults with BD (McDonough-Ryan et al., 2002). While this difference may be simply related to delays in the development of verbal vs. performance skills, it could be speculated that this is related to other factors such as learning disorders, poor attention, and reduced problem-solving skills. Taken together, evidence of cognitive deficits as markers of vulnerability in pediatric HR is still limited and warrants further large-scale investigations adopting cross-sectional and longitudinal designs.

Adult studies

Five of seven studies investigating the neuropsychological function of adult relatives of BD patients reported cognitive deficits (Kéri, Kelemen, Benedek, & Janka, 2001; Kremen et al., 1994; MacQueen et al., 2004; Nicol Ferrier, Chowdhury, Thompson, Watson, & Young, 2004; Pierson, Jouvent, Quintin, Perez-Diaz, & Leboyer, 2000; Sobczak et al., 2002; Zalla et al., 2004). A sustained auditory attention study revealed slow processing speed in HR relative to control subjects (Pierson et al., 2000). This difference remained highly significant when analyses included only relatives with no lifetime history of mental illness. Another study reported significantly impaired performance in long-delay verbal declarative memory in healthy relatives of BD patients. By contrast, measures of executive function, verbal fluency, and visuospatial working memory were intact (Kéri et al., 2001). Partially in line with these findings, relatives of BD I patients showed deficits in planning and measures of delayed verbal declarative memory (Sobczak et al., 2002). Intriguingly, in this study, relatives of patients with BD type II did not show any cognitive deficit. Two studies found increased slowness on a response inhibition task (Zalla et al., 2004) and reduced spatial recognition in relatives of BD patients (Ferrier, Stanton, Kelly, & Scott, 1999). The results remained unchanged after controlling for the presence of psychopathology. In contrast to these studies, Kremen et al. failed to find significant cognitive deficits in female relatives of BD patients (Kremen et al., 1994), and a small pilot study in 7 HR did not find any significant impairments on a visual memory procedure (MacQueen et al., 2004). Taken together, there is emerging evidence that verbal memory and processing speed are markers of vulnerability. Large-scale studies comparing HC, HR, and BD patients are required to confirm this hypothesis.

Twin studies

Twin studies have the advantage of taking into account both genetic and environmental factors, and may be more likely to outline the genetic basis of cognitive deficits in HR. To date, there are no published twin studies in pediatric offspring of parents with BD, or even on pediatric BD. The assessment of cognitive functions in seven adult monozygotic twin pairs discordant for BD showed that the unaffected twins with an affected sibling performed significantly worse than control twins on measures of working memory and attention/mental control (Gourovitch et al., 1999). Both discordant twins performed significantly worse than control twins, but only the affected twin presented short-delay verbal memory deficits. Another larger twin study found no significant deficits between unaffected co-twins and healthy twin pairs in tests of attention, working memory, verbal or nonverbal memory, or verbal learning (Kieseppä et al., 2005). However, after adjusting for gender, the authors did find a significant deficit in verbal learning in the unaffected female co-twins, and once again impairment was shown in delayed verbal recall. Unlike the previous twin study, this study included nonidentical twin pairs. The greater genetic variation between dizygotic twin pairs may explain why this sample of co-twins showed less impairment relative to the healthy comparison group than the monozygotic twins in the earlier study. Based on these findings, the most promising marker for BD in adult HR appears to be verbal memory.

SOCIAL COGNITION

Equally relevant but surprisingly underinvestigated is the role of emotion-laden cognitive functions, as measured by computerized or pen-and-paper tests recording response times and accuracy, as a potential marker of cognitive vulnerability to BD. A review of the literature has shown a consistent pattern of social cognition and affective processing deficits including abnormal reactivity to fear-related emotions, reduced emotional processing, poor ability to recognize social "faux pas," and response inhibition difficulties (Cardenas et al., 2016). A study comparing children of BD parents and children of healthy parents (Gotlib, Traill, Montoya, Joormann, & Chang, 2005) showed that although both groups of children could recall the same number of emotional adjectives on a self-reference task, HR children remembered more negative words (e.g., threat, fear) than HC children. A second study compared children with BD, offspring of BD parents, and healthy controls on tasks of affective processing and sustained attention (Bauer et al., 2015). The findings indicated that BD offspring performed as accurately as healthy volunteers, but were faster in responding to emotional stimuli. This effect was more pronounced among BD offspring who reported mental health symptoms. These results are consistent with the adult HR literature, which shows that relatives of BD patients respond more slowly to depression-like words and make more errors in response to mania-related words (Besnier et al., 2009). They also display greater cognitive interference triggered by negative words in an inhibitory control task (Brand et al., 2012). Taken together, these findings provide preliminary evidence that attentional bias toward affective stimuli may be a vulnerability marker. Further studies should determine whether this bias is independent from current mood state or cognitive beliefs associated with vulnerability to BD.

CONCLUSION

The neural and cognitive deficits observed in HR individuals are likely to reflect the combined effect of genetic vulnerability and other contributing factors such as comorbid conditions, medications, family environment, etc. In general, HR subjects show impairments in fewer domains that BD patients, and are likely to exhibit intermediate performance between patients and HC. Overall, studies in HR highlight the presence of specific deficits in verbal memory, processing speed, and social cognition that are present before the onset of symptoms. Further investigation into cognitive or emotional processing differences between affected individuals and their unaffected relatives is, however, needed to provide insight into psychological risk factors for BD and could enhance the search for specific vulnerability indicators. For instance, Singh et al.'s fMRI study showed that children and adolescent BD offspring have unique patterns of prefrontal and subcortical activation, which may constitute a neural marker of either resilience or vulnerability to BD (Singh, Chang, et al., 2014). Additionally, cognitive impairment in BD patients is often associated with significant psychosocial impairment (Atre-Vaidya et al., 1998; Martínez-Arán et al., 2001; Martínez-Arán, Vieta, Colom, et al., 2004; Martínez-Arán, Vieta, Reinares, et al., 2004; Zarate, Tohen, Land, & Cavanagh, 2000; Zubieta, Huguelet, O'Neil, & Giordani, 2001). Preventing cognitive decline in HR subjects may therefore reduce the risk for clinical, social, and occupational impairment later in life.

A few methodological challenges appear to limit the interpretation and generalizability of cognitive findings in the BD literature, as studies often differ in terms of study design (e.g., practice effects) and sample characteristics (e.g., current mood, severity of affective symptoms and motivation during the testing session (Beblo, Sinnamon, & Baune, 2011; Szmulewicz et al., 2017) (Ferrier et al., 1999). Last but not least, the definition of cognitive functions such as executive abilities is broad, and findings are therefore specific to the task used (Diamond, 2013). For instance, verbal declarative memory and executive function are both broad domains involving several different processes which interact with one another and with other cognitive systems. This may be a further reason for the lack of consistent and informative findings across studies. A more targeted approach focusing on these areas is necessary to clarify the particular nature and specificity of the reported deficits. The fact that some relatives often suffer from other psychiatric disorders (including anxiety and mood and personality disorders, although none had a history of BD) constitutes an issue for identifying a cognitive diagnostic marker of BD, as an ideal vulnerability marker should be able to differentiate patients effectively with BD from HC, HR, and patients with related mood disorders. Future studies may therefore need to focus on selecting common features of mood disorders and their common comorbidities, and investigating their relationship with cognitive dysfunction.

Across studies, the degree of genetic vulnerability to BD varies based on selection criteria and age of the participants. For instance, some studies included first-degree relatives from families with one or two affected members, while others included siblings of patients with sporadic BD. As previously mentioned, the majority of current studies focused on young adult relatives of BD rather than pediatric populations. This may have biased the literature to individuals who developed coping mechanisms to compensate for cognitive weaknesses, or who were more resilient to mood fluctuations.

In sum, neuroimaging findings showed some evidence for structural and functional abnormalities in regions underlying cognitive control and emotion processing in young HR. Cognitive findings in HR are more inconsistent and still preliminary in nature. Poor verbal memory, slow processing speed, and altered social cognition appear to be promising vulnerability markers of BD. The integration of imaging and cognitive assessments using robust methodological approaches (e.g., cluster analyses (Van Rheenen et al., 2016; Wu et al., 2017)) may improve diagnostic accuracy and assist in the development of prevention and treatment strategies for vulnerable individuals.

REFERENCES

Abramovic, L., Boks, M. P., Vreeker, A., Bouter, D. C., Kruiper, C., Verkooijen, S., et al. (2016). The association of antipsychotic medication and lithium with brain measures in patients with bipolar disorder. *European Neuropsychopharmacology*, *26*(11), 1741–1751.

Agrawal, A., & Lynskey, M. T. (2008). Are there genetic influences on addiction: evidence from family, adoption and twin studies. *Addiction*, *103*(7), 1069–1081.

Akdemir, D., & Gokler, B. (2008). Psychopathology in the children of parents with bipolar mood disorder. *Turk Psikiyatri Dergisi*, *19*(2), 133.

Alonso-Lana, S., Valentí, M., Romaguera, A., Sarri, C., Sarró, S., Rodríguez-Martínez, A., et al. (2016). Brain functional changes in first-degree relatives of patients with bipolar disorder: evidence for default mode network dysfunction. *Psychological Medicine*, *46*(12), 2513–2521.

Arts, B., Jabben, N., Krabbendam, L., & Van Os, J. (2008). Meta-analyses of cognitive functioning in euthymic bipolar patients and their first-degree relatives. *Psychological Medicine*, *38*(06), 771–785.

Atre-Vaidya, N., Taylor, M. A., Seidenberg, M., Reed, R., Perrine, A., & Glick-Oberwise, F. (1998). Cognitive deficits, psychopathology, and psychosocial functioning in bipolar mood disorder. *Cognitive and Behavioral Neurology, 11*(3), 120–126.

Bauer, I. E., Frazier, T. W., Meyer, T. D., Youngstrom, E., Zunta-Soares, G. B., & Soares, J. C. (2015). Affective processing in pediatric bipolar disorder and offspring of bipolar parents. *Journal of Child and Adolescent Psychopharmacology, 25*(9), 684–690.

Bauer, I. E., Pascoe, M. C., Wollenhaupt-Aguiar, B., Kapczinski, F., & Soares, J. C. (2014). Inflammatory mediators of cognitive impairment in bipolar disorder. *Journal of Psychiatric Research, 56*, 18–27.

Bauer, I. E., Wu, M.-J., Frazier, T., Mwangi, B., Spiker, D., Zunta-Soares, G. B., et al. (2016). Neurocognitive functioning in individuals with bipolar disorder and their healthy siblings: a preliminary study. *Journal of Affective Disorders, 201*, 51–56.

Beblo, T., Sinnamon, G., & Baune, B. T. (2011). Specifying the neuropsychology of affective disorders: clinical, demographic and neurobiological factors. *Neuropsychology Review, 21*(4), 337–359.

Besnier, N., Richard, F., Zendjidjian, X., Kaladjian, A., Mazzola-Pomietto, P., Adida, M., et al. (2009). Stroop and emotional Stroop interference in unaffected relatives of patients with schizophrenic and bipolar disorders: distinct markers of vulnerability? *The World Journal of Biological Psychiatry, 10*(4 Pt 3), 809–818.

Birmaher, B., Axelson, D., Goldstein, B., Monk, K., Kalas, C., Obreja, M., et al. (2010). Psychiatric disorders in preschool offspring of parents with bipolar disorder: the Pittsburgh bipolar offspring study (BIOS). *American Journal of Psychiatry, 167*(3), 321–330.

Bora, E., Yucel, M., & Pantelis, C. (2009). Cognitive endophenotypes of bipolar disorder: a meta-analysis of neuropsychological deficits in euthymic patients and their first-degree relatives. *Journal of Affective Disorders, 113*(1), 1–20.

Bortolato, B., Miskowiak, K. W., Köhler, C. A., Vieta, E., & Carvalho, A. F. (2014). Cognitive dysfunction in bipolar disorder and schizophrenia: a systematic review of meta-analyses. *Neuropsychiatric Disease and Treatment, 11*, 3111–3125.

Brand, J. G., Goldberg, T. E., Gunawardane, N., Gopin, C. B., Powers, R. L., Malhotra, A. K., et al. (2012). Emotional bias in unaffected siblings of patients with bipolar I disorder. *Journal of Affective Disorders, 136*(3), 1053–1058.

Brotman, M. A., Deveney, C. M., Thomas, L. A., Hinton, K. E., Yi, J. Y., Pine, D. S., et al. (2014). Parametric modulation of neural activity during face emotion processing in unaffected youth at familial risk for bipolar disorder. *Bipolar Disorders, 16*(7), 756–763.

Burdick, K., Russo, M., Frangou, S., Mahon, K., Braga, R., Shanahan, M., et al. (2014). Empirical evidence for discrete neurocognitive subgroups in bipolar disorder: clinical implications. *Psychological Medicine, 44*(14), 3083–3096.

Cardenas, S. A., Kassem, L., Brotman, M. A., Leibenluft, E., & McMahon, F. J. (2016). Neurocognitive functioning in euthymic patients with bipolar disorder and unaffected relatives: a review of the literature. *Neuroscience & Biobehavioral Reviews, 69*, 193–215.

Cardoso, T., Bauer, I. E., Meyer, T. D., Kapczinski, F., & Soares, J. C. (2015). Neuroprogression and cognitive functioning in bipolar disorder: a systematic review. *Current Psychiatry Reports, 17*(9), 75.

Chang, K., Steiner, H., & Ketter, T. (2003). In *Studies of offspring of parents with bipolar disorder. Paper presented at the American journal of medical genetics part C: seminars in medical genetics*.

Chase, H. W., & Phillips, M. L. (2016). Elucidating neural network functional connectivity abnormalities in bipolar disorder: toward a harmonized methodological approach. *Biological Psychiatry: Cognitive Neuroscience and Neuroimaging, 1*(3), 288–298.

Clark, L., Kempton, M. J., Scarnà, A., Grasby, P. M., & Goodwin, G. M. (2005). Sustained attention-deficit confirmed in euthymic bipolar disorder but not in first-degree relatives of bipolar patients or euthymic unipolar depression. *Biological Psychiatry, 57*(2), 183–187.

Costa, P. T., & McCrae, R. R. (1992). Normal personality assessment in clinical practice: the NEO personality inventory. *Psychological Assessment, 4*(1), 5.

de la Serna, E., Vila, M., Sanchez-Gistau, V., Moreno, D., Romero, S., Sugranyes, G., et al. (2016). Neuropsychological characteristics of child and adolescent offspring of patients with bipolar disorder. *Progress in Neuro-Psychopharmacology and Biological Psychiatry, 65*, 54–59.

Decina, P., Kestenbaum, C. J., Farber, S., Kron, L., Gargan, M., Sackeim, H. A., et al. (1983). Clinical and psychological assessment of children of bipolar probands. *The American Journal of Psychiatry, 140*(5), 548–553.

Deveney, C. M., Connolly, M. E., Jenkins, S. E., Kim, P., Fromm, S. J., Brotman, M. A., et al. (2012). Striatal dysfunction during failed motor inhibition in children at risk for bipolar disorder. *Progress in Neuro-Psychopharmacology and Biological Psychiatry, 38*(2), 127–133.

Diamond, A. (2013). Executive functions. *Annual Review of Psychology, 64*, 135–168.

Diwadkar, V. A., Goradia, D., Hosanagar, A., Mermon, D., Montrose, D. M., Birmaher, B., et al. (2011). Working memory and attention deficits in adolescent offspring of schizophrenia or bipolar patients: comparing vulnerability markers. *Progress in Neuro-Psychopharmacology and Biological Psychiatry, 35*(5), 1349–1354.

Drysdale, E., Knight, H. M., McIntosh, A. M., & Blackwood, D. H. (2013). Cognitive endophenotypes in a family with bipolar disorder with a risk locus on chromosome 4. *Bipolar Disorders, 15*(2), 215–222.

Duffy, A., Alda, M., Hajek, T., Sherry, S. B., & Grof, P. (2010). Early stages in the development of bipolar disorder. *Journal of Affective Disorders, 121*(1), 127–135.

Duffy, A., Hajek, T., Alda, M., Grof, P., Milin, R., & Mac Queen, G. (2009). Neurocognitive functioning in the early stages of bipolar disorder: visual backward masking performance in high risk subjects. *European Archives of Psychiatry and Clinical Neuroscience, 259*(5), 263–269.

Duffy, A., Horrocks, J., Doucette, S., Keown-Stoneman, C., McCloskey, S., & Grof, P. (2013). Childhood anxiety: an early predictor of mood disorders in offspring of bipolar parents. *Journal of Affective Disorders, 150*(2), 363–369.

Ferrier, I., Stanton, B. R., Kelly, T., & Scott, J. (1999). Neuropsychological function in euthymic patients with bipolar disorder. *The British Journal of Psychiatry, 175*(3), 246–251.

Frangou, S., Dima, D., & Jogia, J. (2017). Towards person-centered neuroimaging markers for resilience and vulnerability in bipolar disorder. *NeuroImage, 145*, 230–237.

Frangou, S., Haldane, M., Roddy, D., & Kumari, V. (2005). Evidence for deficit in tasks of ventral, but not dorsal, prefrontal executive function as an endophenotypic marker for bipolar disorder. *Biological Psychiatry, 58*(10), 838–839.

Frazier, J. A., Breeze, J. L., Papadimitriou, G., Kennedy, D. N., Hodge, S. M., Moore, C. M., et al. (2007). White matter abnormalities in children with and at risk for bipolar disorder. *Bipolar Disorders, 9*(8), 799–809.

Frías, Á., Palma, C., Farriols, N., & Salvador, A. (2014). Characterizing offspring of bipolar parents: a review of the literature. *Actas espanolas de psiquiatria, 43*(6), 221–234.

Garver, D. L. (1987). Methodological issues facing the interpretation of high-risk studies: biological heterogeneity. *Schizophrenia Bulletin, 13*(3), 525.

Georgiades, A., Rijsdijk, F., Kane, F., Rebollo-Mesa, I., Kalidindi, S., Schulze, K. K., et al. (2016). New insights into the endophenotypic status of cognition in bipolar disorder: genetic modelling study of twins and siblings. *The British Journal of Psychiatry*, https://doi.org/10.1192/bjp.bp.115.167239.

Glahn, D. C., Bearden, C. E., Niendam, T. A., & Escamilla, M. A. (2004). The feasibility of neuropsychological endophenotypes in the search for genes associated with bipolar affective disorder. *Bipolar Disorders*, *6*(3), 171–182.

Gotlib, I. H., Traill, S. K., Montoya, R. L., Joormann, J., & Chang, K. (2005). Attention and memory biases in the offspring of parents with bipolar disorder: indications from a pilot study. *Journal of Child Psychology and Psychiatry*, *46*(1), 84–93.

Gottesman, I. I., & Gould, T. D. (2003). The endophenotype concept in psychiatry: etymology and strategic intentions. *American Journal of Psychiatry*, *160*(4), 636–645.

Gourovitch, M. L., Torrey, E. F., Gold, J. M., Randolph, C., Weinberger, D. R., & Goldberg, T. E. (1999). Neuropsychological performance of monozygotic twins discordant for bipolar disorder. *Biological Psychiatry*, *45*(5), 639–646.

Gunde, E., Novak, T., Kopecek, M., Schmidt, M., Propper, L., Stopkova, P., et al. (2011). White matter hyperintensities in affected and unaffected late teenage and early adulthood offspring of bipolar parents: a two-center high-risk study. *Journal of Psychiatric Research*, *45*(1), 76–82.

Hajek, T., Cullis, J., Novak, T., Kopecek, M., Blagdon, R., Propper, L., et al. (2013). Brain structural signature of familial predisposition for bipolar disorder: replicable evidence for involvement of the right inferior frontal gyrus. *Biological Psychiatry*, *73*(2), 144–152.

Harvey, P. D., Weintraub, S., & Neale, J. M. (1982). Speech competence of children vulnerable to psychopathology. *Journal of Abnormal Child Psychology*, *10*(3), 373–387.

Henin, A., Biederman, J., Mick, E., Sachs, G. S., Hirshfeld-Becker, D. R., Siegel, R. S., et al. (2005). Psychopathology in the offspring of parents with bipolar disorder: a controlled study. *Biological Psychiatry*, *58*(7), 554–561.

Hirshfeld-Becker, D. R., Biederman, J., Henin, A., Faraone, S. V., Dowd, S. T., De Petrillo, L. A., et al. (2006). Psychopathology in the young offspring of parents with bipolar disorder: a controlled pilot study. *Psychiatry Research*, *145*(2), 155–167.

Kéri, S., Kelemen, O., Benedek, G., & Janka, Z. (2001). Different trait markers for schizophrenia and bipolar disorder: a neurocognitive approach. *Psychological Medicine*, *31*(05), 915–922.

Kieseppä, T., Tuulio-Henriksson, A., Haukka, J., Van Erp, T., Glahn, D., Cannon, T. D., et al. (2005). Memory and verbal learning functions in twins with bipolar-I disorder, and the role of information-processing speed. *Psychological Medicine*, *35*(02), 205–215.

Kremen, W. S., Seidman, L. J., Pepple, J. R., Lyons, M. J., Tsuang, M. T., & Faraone, S. V. (1994). Neuropsychological risk indicators for schizophrenia: a review of family studies. *Schizophrenia Bulletin*, *20*(1), 103–119.

Kulkarni, S., Jain, S., Janardhan Reddy, Y., Kumar, K. J., & Kandavel, T. (2010). Impairment of verbal learning and memory and executive function in unaffected siblings of probands with bipolar disorder. *Bipolar Disorders*, *12*(6), 647–656.

Ladouceur, C. D., Almeida, J. R., Birmaher, B., Axelson, D. A., Nau, S., Kalas, C., et al. (2008). Subcortical gray matter volume abnormalities in healthy bipolar offspring: potential neuroanatomical risk marker for bipolar disorder? *Journal of the American Academy of Child & Adolescent Psychiatry*, *47*(5), 532–539.

Lapalme, M., Hodgins, S., & LaRoche, C. (1997). Children of parents with bipolar disorder: a metaanalysis of risk for mental disorders. *The Canadian Journal of Psychiatry*, *42*(6), 623–631.

Lee, M.-S., Anumagalla, P., Talluri, P., & Pavuluri, M. N. (2013). Meta-analyses of developing brain function in high-risk and emerged bipolar disorder. *Frontiers in Psychiatry, 5,* 141.

Lin, K., Xu, G., Wong, N. M., Wu, H., Li, T., Lu, W., et al. (2015). A multi-dimensional and integrative approach to examining the high-risk and ultra-high-risk stages of bipolar disorder. *eBioMedicine, 2*(8), 919–928.

López-Jaramillo, C., Vargas, C., Díaz-Zuluaga, A. M., Palacio, J. D., Castrillón, G., Bearden, C., et al. (2017). Increased hippocampal, thalamus and amygdala volume in long-term lithium-treated bipolar I disorder patients compared with unmedicated patients and healthy subjects. *Bipolar Disorders, 19*(1), 41–49.

MacQueen, G. M., Grof, P., Alda, M., Marriott, M., Young, L. T., & Duffy, A. (2004). A pilot study of visual backward masking performance among affected versus unaffected offspring of parents with bipolar disorder. *Bipolar Disorders, 6*(5), 374–378.

Martínez-Arán, A., Penades, R., Vieta, E., Colom, F., Reinares, M., Benabarre, A., et al. (2001). Executive function in patients with remitted bipolar disorder and schizophrenia and its relationship with functional outcome. *Psychotherapy and Psychosomatics, 71*(1), 39–46.

Martínez-Arán, A., Vieta, E., Colom, F., Torrent, C., Sánchez-Moreno, J., Reinares, M., et al. (2004). Cognitive impairment in euthymic bipolar patients: implications for clinical and functional outcome. *Bipolar Disorders, 6*(3), 224–232.

Martínez-Arán, A., Vieta, E., Reinares, M., Colom, F., Torrent, C., Sánchez-Moreno, J., et al. (2004). Cognitive function across manic or hypomanic, depressed, and euthymic states in bipolar disorder. *American Journal of Psychiatry, 161*(2), 262–270.

Maziade, M., Rouleau, N., Cellard, C., Battaglia, M., Paccalet, T., Moreau, I., et al. (2011). Young offspring at genetic risk of adult psychoses: the form of the trajectory of IQ or memory may orient to the right dysfunction at the right time. *PLoS ONE, 6*(4), e19153.

McCormack, C., Green, M., Rowland, J., Roberts, G., Frankland, A., Hadzi-Pavlovic, D., et al. (2016). Neuropsychological and social cognitive function in young people at genetic risk of bipolar disorder. *Psychological Medicine, 46*(04), 745–758.

McDonough-Ryan, P., DelBello, M., Shear, P. K., Ris, M. D., Soutullo, C., & Strakowski, S. M. (2002). Academic and cognitive abilities in children of parents with bipolar disorder: a test of the nonverbal learning disability model. *Journal of Clinical and Experimental Neuropsychology, 24*(3), 280–285.

Mesman, E., Nolen, W. A., Reichart, C. G., Wals, M., & Hillegers, M. H. (2013). The Dutch bipolar offspring study: 12-year follow-up. *American Journal of Psychiatry, 170*(5), 542–549.

Miskowiak, K. W., Kjærstad, H. L., Meluken, I., Petersen, J. Z., Maciel, B. R., Köhler, C. A., et al. (2017). The search for neuroimaging and cognitive endophenotypes: a critical systematic review of studies involving unaffected first-degree relatives of individuals with bipolar disorder. Neuroscience &amp. *Biobehavioral Reviews, 73,* 1–22.

Narayan, A. J., Allen, T. A., Cullen, K. R., & Klimes-Dougan, B. (2013). Disturbances in reality testing as markers of risk in offspring of parents with bipolar disorder: a systematic review from a developmental psychopathology perspective. *Bipolar Disorders, 15*(7), 723–740.

Nehra, R., Grover, S., Sharma, S., Sharma, A., & Sarkar, S. (2014). Neuro-cognitive functioning in unaffected siblings of patients with bipolar disorder: comparison with bipolar patients and healthy controls. *Indian Journal of Psychiatry, 56*(3), 283.

Nery, F. G., Norris, M., Eliassen, J. C., Weber, W. A., Blom, T. J., Welge, J. A., et al. (2017). White matter volumes in youth offspring of bipolar parents. *Journal of Affective Disorders, 209,* 246–253.

Nicol Ferrier, I., Chowdhury, R., Thompson, J. M., Watson, S., & Young, A. H. (2004). Neurocognitive function in unaffected first-degree relatives of patients with bipolar disorder: a preliminary report. *Bipolar Disorders, 6*(4), 319–322.

Olsavsky, A. K., Brotman, M. A., Rutenberg, J. G., Muhrer, E. J., Deveney, C. M., Fromm, S. J., et al. (2012). Amygdala hyperactivation during face emotion processing in unaffected youth at risk for bipolar disorder. *Journal of the American Academy of Child & Adolescent Psychiatry, 51*(3), 294–303.

Pagliaccio, D., Wiggins, J. L., Adleman, N. E., Harkins, E., Curhan, A., Towbin, K. E., et al. (2017). Behavioral and neural sustained attention deficits in bipolar disorder and familial risk of bipolar disorder. *Biological Psychiatry, 82*(9), 669–678.

Phillips, M. L., & Kupfer, D. J. (2013). Bipolar disorder diagnosis: challenges and future directions. *The Lancet, 381*(9878), 1663–1671.

Pierson, A., Jouvent, R., Quintin, P., Perez-Diaz, F., & Leboyer, M. (2000). Information processing deficits in relatives of manic depressive patients. *Psychological Medicine, 30*(03), 545–555.

Quackenbush, D., Kutcher, S., Robertson, H. A., Boulos, C., & Chaban, P. (1996). Premorbid and postmorbid school functioning in bipolar adolescents: description and suggested academic interventions. *The Canadian Journal of Psychiatry, 41*(1), 16–22.

Reichenberg, A., Harvey, P. D., Bowie, C. R., Mojtabai, R., Rabinowitz, J., Heaton, R. K., et al. (2009). Neuropsychological function and dysfunction in schizophrenia and psychotic affective disorders. *Schizophrenia Bulletin, 35*(5), 1022–1029.

Reichenberg, A., Weiser, M., Rabinowitz, J., Caspi, A., Schmeidler, J., Mark, M., et al. (2002). A population-based cohort study of premorbid intellectual, language, and behavioral functioning in patients with schizophrenia, schizoaffective disorder, and nonpsychotic bipolar disorder. *American Journal of Psychiatry, 159*(12), 2027–2035.

Robinson, L. J., & Nicol Ferrier, I. (2006). Evolution of cognitive impairment in bipolar disorder: a systematic review of cross-sectional evidence. *Bipolar Disorders, 8*(2), 103–116.

Robinson, L. J., Thompson, J. M., Gallagher, P., Goswami, U., Young, A. H., Ferrier, I. N., et al. (2006). A meta-analysis of cognitive deficits in euthymic patients with bipolar disorder. *Journal of Affective Disorders, 93*(1), 105–115.

Sarıçiçek, A., Zorlu, N., Yalın, N., Hıdıroğlu, C., Çavuşoğlu, B., Ceylan, D., et al. (2016). Abnormal white matter integrity as a structural endophenotype for bipolar disorder. *Psychological Medicine, 46*(07), 1547–1558.

Sigurdsson, E., Fombonne, E., Sayal, K., & Checkley, S. (1999). Neurodevelopmental antecedents of early-onset bipolar affective disorder. *The British Journal of Psychiatry, 174*(2), 121–127.

Singh, M. K., Chang, K. D., Kelley, R. G., Saggar, M., Reiss, A. L., & Gotlib, I. H. (2014). Early signs of anomalous neural functional connectivity in healthy offspring of parents with bipolar disorder. *Bipolar Disorders, 16*(7), 678–689.

Singh, M. K., Kelley, R. G., Howe, M. E., Reiss, A. L., Gotlib, I. H., & Chang, K. D. (2014). Reward processing in healthy offspring of parents with bipolar disorder. *JAMA Psychiatry, 71*(10), 1148–1156.

Sobczak, S., Riedel, W., Booij, I., Rot, M. A. H., Deutz, N., & Honig, A. (2002). Cognition following acute tryptophan depletion: difference between first-degree relatives of bipolar disorder patients and matched healthy control volunteers. *Psychological Medicine, 32*(03), 503–515.

Sprooten, E., Barrett, J., McKay, D. R., Knowles, E. E., Mathias, S. R., Winkler, A. M., et al. (2016). A comprehensive tractography study of patients with bipolar disorder and their unaffected siblings. *Human Brain Mapping, 37*(10), 3474–3485.

Szmulewicz, A. G., Valerio, M. P., Smith, J. M., Samamé, C., Martino, D. J., & Strejilevich, S. A. (2017). Neuropsychological profiles of major depressive disorder and bipolar disorder during euthymia. a systematic literature review of comparative studies. *Psychiatry Research, 248*, 127–133.

Thermenos, H. W., Makris, N., Whitfield-Gabrieli, S., Brown, A. B., Giuliano, A. J., Lee, E. H., et al. (2011). A functional MRI study of working memory in adolescents and young adults at genetic risk for bipolar disorder: preliminary findings. *Bipolar Disorders, 13*(3), 272–286.

Trivedi, J. K., Goel, D., Dhyani, M., Sharma, S., Singh, A. P., Sinha, P. K., et al. (2008). Neurocognition in first-degree healthy relatives (siblings) of bipolar affective disorder patients. *Psychiatry and Clinical Neurosciences, 62*(2), 190–196.

Trotta, A., Murray, R., & MacCabe, J. (2015). Do premorbid and post-onset cognitive functioning differ between schizophrenia and bipolar disorder? A systematic review and meta-analysis. *Psychological Medicine, 45*(02), 381–394.

Tsuang, M. T., Tohen, M., & Jones, P. (2011). *Textbook of psychiatric epidemiology.* John Wiley and Sons.

Valdez, R., Yoon, P. W., Qureshi, N., Green, R. F., & Khoury, M. J. (2010). Family history in public health practice: a genomic tool for disease prevention and health promotion. *Annual Review of Public Health, 31*, 69–87.

Van Rheenen, T. E., Bryce, S., Tan, E. J., Neill, E., Gurvich, C., Louise, S., et al. (2016). Does cognitive performance map to categorical diagnoses of schizophrenia, schizoaffective disorder and bipolar disorder? A discriminant functions analysis. *Journal of Affective Disorders, 192*, 109–115.

Vasconcelos-Moreno, M. P., Bücker, J., Bürke, K. P., Czepielewski, L., Santos, B. T., Fijtman, A., et al. (2016). Cognitive performance and psychosocial functioning in patients with bipolar disorder, unaffected siblings, and healthy controls. *Revista Brasileira de Psiquiatria, 38*(4), 275–280.

Versace, A., Ladouceur, C. D., Romero, S., Birmaher, B., Axelson, D. A., Kupfer, D. J., et al. (2010). Altered development of white matter in youth at high familial risk for bipolar disorder: a diffusion tensor imaging study. *Journal of the American Academy of Child & Adolescent Psychiatry, 49*(12). 1249–1259.e1241.

Wiggins, J. L., Brotman, M. A., Adleman, N. E., Kim, P., Wambach, C. G., Reynolds, R. C., et al. (2017). Neural markers in pediatric bipolar disorder and familial risk for bipolar disorder. *Journal of the American Academy of Child & Adolescent Psychiatry, 56*(1), 67–78.

Wilde, A., Chan, H.-N., Rahman, B., Meiser, B., Mitchell, P., Schofield, P., et al. (2014). A meta-analysis of the risk of major affective disorder in relatives of individuals affected by major depressive disorder or bipolar disorder. *Journal of Affective Disorders, 158*, 37–47.

Winters, K. C., Stone, A. A., Weintraub, S., & Neale, J. M. (1981). Cognitive and attentional deficits in children vulnerable to psychopathology. *Journal of Abnormal Child Psychology, 9*(4), 435–453.

Wu, M.-J., Mwangi, B., Bauer, I. E., Passos, I. C., Sanches, M., Zunta-Soares, G. B., et al. (2017). Identification and individualized prediction of clinical phenotypes in bipolar disorders using neurocognitive data, neuroimaging scans and machine learning. *NeuroImage, 145*, 254–264.

Zalla, T., Joyce, C., Szöke, A., Schürhoff, F., Pillon, B., Komano, O., et al. (2004). Executive dysfunctions as potential markers of familial vulnerability to bipolar disorder and schizophrenia. *Psychiatry Research, 121*(3), 207–217.

Zarate, C. A., Tohen, M., Land, M., & Cavanagh, S. (2000). Functional impairment and cognition in bipolar disorder. *Psychiatric Quarterly, 71*(4), 309–329.

Zubieta, J.-K., Huguelet, P., O'Neil, R. L., & Giordani, B. J. (2001). Cognitive function in euthymic bipolar I disorder. *Psychiatry Research, 102*(1), 9–20.

Neuropsychological and social cognitive function in young people at genetic risk of bipolar disorder

Gloria Roberts*,†, **Carina Sinbandhit***,†, **Angela Stuart***,†, **Vivian Leung***,†,
Clare McCormack‡, **Melissa J. Green***,†, **Philip B. Mitchell***,†

School of Psychiatry, University of New South Wales, Randwick, NSW, Australia* *Black Dog
Institute, Prince of Wales Hospital, Randwick, NSW, Australia*† *Division of Behavioral Medicine,
Department of Psychiatry, University Medical Centre, New York, NY, United States*‡

CHAPTER OUTLINE

Bipolar Disorder Vulnerability. https://doi.org/10.1016/B978-0-12-812347-8.00009-9

INTRODUCTION

Identifying early signs of bipolar disorder (BD) is critically important to improve treatment outcome. Having an affected first-degree relative is the strongest determinant of risk for BD (Lau et al., 2017; Lichtenstein et al., 2009); these individuals have a 10-fold higher risk of developing this condition (Merikangas & Yu, 2002). Endophenotypes are intermediate phenotypes that are considered a more promising index of underlying genetic liability than the illness itself. In order for a cognitive marker or any other marker to be accepted as an endophenotype or heritable biomarker, it should show association with the illness, mood state independence, greater prevalence in nonaffected family members than in the general population, heritability, and co-segregation with the illness within families (Gottesman & Gould, 2003). As little is known about the processes that predispose or protect against risk for BD, studying unaffected first-degree relatives could shed light on this vulnerability and hence reveal endophenotypic markers for BD. Cognitive impairments have been well documented in those with established BD (Cullen et al., 2016), whereas findings in relatives of individuals with BD have been inconsistent (Cardenas, Kassem, Brotman, Leibenluft, & McMahon, 2016). Studies that have assessed neurocognition during functional magnetic resonance imaging (fMRI) have helped to improve understanding of abnormalities in neural circuitry associated with genetic risk of developing BD (Lee, Anumagalla, Talluri, & Pavuluri, 2014; Özerdem, Ceylan, & Can, 2016). Discrepant findings in relatives of individuals with BD could be attributed to differences in sample characteristics, such as age or exclusion criteria, where some high-risk or control samples are by definition asymptomatic while others include symptomatic individuals. The peak age of BD onset is before the age of 30 (Merikangas et al., 2011), yet much of the research examining high-risk individuals has included older subjects. Studying unaffected youth provides the valuable opportunity to examine potential heritable biomarkers for the illness without the confounding effects of mood state, psychotropic medications, prior mood episodes, and other non-BD psychopathology.

This chapter summarizes recent neuropsychological findings in young relatives of individuals with BD. Potential confounders of such findings are reported. We used the following criteria for inclusion of studies in this chapter: (1) included neuropsychological data (pen-and-paper tasks, computerized bench tasks, or functional neuroimaging tasks) in relatives of individuals with BD and a comparison control group; (2) included a high-risk sample size of 25 of more (or minimum of 15 in the high-risk group for neuroimaging tasks); (3) included high-risk individuals under 35 years (or a mean age in the high-risk group of less than 35 when an age range was not reported); and (4) were published in peer-reviewed journals in English available between September 2007 and September 2017. Due to the limited number of studies and tasks outside the scanner investigating social cognition in unaffected relatives, exclusion criteria for age and sample size were not applied to this domain of non-MRI tasks. If neuroimaging studies administered a nonsocial cognitive task outside the scanner, these results are not reported unless they met the above inclusion criteria for sample size. Similarly, if social cognition studies also administered a nonsocial cognitive task, these results are not reported unless they met the above age and sample size inclusion criteria. All high-risk cohorts constituted unaffected first-degree relatives with the exception of the Scottish "Bipolar Family Study" (Sprooten et al., 2011; Whalley et al., 2011), where unaffected relatives meet eligibility criteria if they had either a first degree-relative or two second-degree relatives with BD. Table 1 summarizes the demographics and sample characteristics of the selected studies, and Table 2 summarizes the findings detailed below.

NEUROCOGNITION

Neurocognitive function in BD has been studied extensively, and there is substantial evidence of nonsocial cognitive deficits in the euthymic phase of illness (Cullen et al., 2016). Similarly, evidence of neurocognitive deficits in first-degree relatives has been reported in several studies, although results are inconsistent; nearly a decade ago, impaired verbal memory among first-degree relatives (compared to healthy controls) was implicated in a review and separate meta-analysis (Arts, Jabben, Krabbendam, & Van Os, 2008; Balanzá-Martínez et al., 2008). Additionally, the review by Arts et al. (2008) suggested that executive function might be a candidate endophenotype. In another meta-analysis, response inhibition deficits appeared to be the most prominent potential marker, followed by impaired verbal memory, sustained attention, and set shifting (Bora, Yucel, & Pantelis, 2009). In general, these review papers showed that deficits in high-risk individuals were small and intermediate to the BD and control groups. However, no prior meta-analysis or review has limited the focus to young individuals.

In total, 16 studies met our inclusion criteria (10 of these studies also conducted a neuroimaging task that will be discussed later in this chapter). In studies of youth that meet these criteria, nonsocial cognitive findings were inconsistent, with some reporting deficits and others not. Unaffected first-degree relatives were found to have impairments in visuospatial learning and memory (two (de la Serna et al., 2016; Lin et al., 2017) of two studies; see Table 2), verbal learning and memory (two (Deveci et al., 2013; Lin et al., 2017) of four studies), intelligence (none of eight studies), executive functioning (one (Lin et al., 2017) of four studies), attention (none of three studies), working memory (one (Lin et al., 2017) of two studies), and processing speed (one (de la Serna et al., 2016) of one study), compared to young controls. None of the nonemotional cognition studies examined verbal fluency or episodic memory.

Of these 16 studies, in 10 studies that also included a BD comparator group (three of which also had neuroimaging components), both BD patients and unaffected first-degree relatives were found to have impairments in executive functioning (one study; Doyle et al., 2009), working memory (one study; Doyle et al., 2009), and attention (one (Brotman, Rooney, Skup, Pine, & Leibenluft, 2009) of two studies), compared to young controls. Deficits in intelligence (one (Doyle et al., 2009) of seven studies), verbal learning and memory (one study; Doyle et al., 2009), and

Table 1 Sample characteristics of studies using a neurocognitive task and including young unaffected relatives of bipolar disorder patients

First author, year of publication	Bipolar disorder (BD) patients	High-risk relatives (HR)	HR with Axis 1 diagnoses	HR currently using medication	Controls (CON)	CON with Axis 1 diagnoses
Bauer et al. (2015)[E]	BD1: n=5, BD2: n=4, BDNOS: n=10; age (mean)=13.6	FDR: n=16; age (mean)=11.8	ADHD: n=3, DDNOS: n=1, ODD: n=2, GAD: n=1. Diagnostic data were missing for two BD offspring	5HR on medication, but type unspecified	n=23; age (mean)=12.79	NO
Besnier et al. (2009)[E]	N/A	FDR: n=30; age (mean)=41.83	NO	NO	n=60; age (mean)=41.18	NO
Brand et al. (2012)[E]	N/A	FDR: n=20; age (mean)=39.15	NO	NO	n=20; age (mean)=41.90	NO
Brotman, Guyer, et al. (2008)[E]	BD1: n=41, BD2 or BDNOS: n=11, age (mean)=13.30	FDR: n=24; age (mean)=11.45	Total with Axis 1 diagnoses: 38.1%, anxiety: n=5, 23.8%; ADHD: n=2, 9.5%; CD: n=1, 4.8%	NO	n=78; age (mean)=14.43	NO
Brotman, Skup, et al. (2008)[E]	BD1: n=37, age (range)=8.84–18.77, age (mean)=14.16	FDR: n=25; age (range)=7.01–17.56, age (mean)=12.15	Anxiety: n=6; ADHD: n=3	NO	n=36; age (range)=9.47–18.72 age (mean)=14.34	NO
Brotman et al. (2009)[C]	BD1: n=22, BD2 or BDNOS: n=6; age (mean)=14.0	FDR: n=26; age (mean)=12.0	Anxiety: n=5, ADHD: n=3,	NO	n=24; age (mean)=14.3	NO
Brotman et al. (2014)[ME]	BD1: n=15, BD2: n=5; age (range)=8–19, age (mean)=15.6	FDR: n=15; age (range)=8–19, age (mean)=14.5	Anxiety: n=1, ADHD: n=1	NO	n=29; age (mean)=14.9	NO

Study	BD group	FDR/HR group	Comorbid diagnoses	Medication	Control group	Control comorbidity
Calafiore et al. (2017)[S]	BD1: n=28; age (range)=21–65, age (mean)=37.82	FDR: n=27; age (range)=25–62, age (mean)=37.00	NO	NO	n=47; age (range)=21–62, age (mean)=35.02	NO
Chan et al. (2016)[L/ME]	N/A	FDR or two SDR: unaffected HR: n=43; age (mean)=23.8; HR-MDD: n=30; age (mean)=23.4	No Axis 1 diagnoses at baseline. Diagnoses developed after baseline: HR-MDD: n=30, BD1: n=1, BD2: n=1	NO at baseline. Follow-up: antidepressants: n=7. Initially included, but subsequently excluded when examining main findings.	n=54; age (mean)=23.0	NO
de la Serna et al. (2016)[C]	N/A	FDR: n=90; age (mean)=12.52	ADHD: 17.6%, mood disorders: 15.6%, anxiety: 12.2%, disruptive behavior disorders: 3.3%, other: 2.2%	Unspecified	n=107; age (mean)=11.71	ADHD: 7.5%, anxiety: 5.6%, mood disorders: 4.7%, other: 3.7%, disruptive behavior disorders: 1.9%
Deveci et al. (2013)[C]	N/A	FDR: n=30; age (mean)=12.32	NO	NO	n=37; age (mean)=12.48	NO
Doyle et al. (2009)[C]	BD1 and BD2: n=170 (11% BD2); age (range)=7–19, age (mean)=12.2	FDR: n=118; age (range)=7–26, age (mean)=12.8	ADHD: n=30, 25.6%	6% were taking at least one psychotropic medication, but type unspecified	n=79; age (range)=10–18, age (mean)=13.7	ADHD: n=14, 17.7%
Erk et al. (2014)[C/MC]	N/A	FDR: n=59; age (mean)=31.8	NO	NO	n=110; age (mean)=32.7	NO
de Brito Ferreira Fernandes et al. (2016)[E]	BD1: n=23; age (mean)=27	FDR: n=22; age (mean)=32	NO	NO	n=27; age (mean)=27	NO
Hanford et al., 2016[E]	N/A	FDR: n=30; age (range)=8–16, age (mean)=13.4	MDD: n=9, anxiety: n=11, ADHD: n=6, other: n=3	Unspecified	n=20; age (mean)=13.3	NO

Continued

Table 1 Sample characteristics of studies using a neurocognitive task and including young unaffected relatives of bipolar disorder patients—cont'd

First author, year of publication	Bipolar disorder (BD) patients	High-risk relatives (HR)	HR with Axis 1 diagnoses	HR currently using medication	Controls (CON)	CON with Axis 1 diagnoses
Ladouceur et al. (2013)[ME]	N/A	FDR: $n=16$; age (mean) = 14.2	NO	NO	$n=15$; age (mean) = 13.8,	NO
Lelli-Chiesa et al. (2010)[C/ME]	BD1: $n=40$; age (range) = 17–65, (mean) = 44	Unaffected FDR: $n=25$; age (range) = 17–65, age (mean) = 34.9; FDR with non-BD Axis 1 diagnoses: $n=22$, age (range) 16–65, age (mean) = 32.5	MDD: $n=15$, anxiety: $n=7$	NO	$n=50$; age (range) = 17–65, age (mean) = 34.9	NO
Lin et al. (2017)[C]	N/A	FDR: UHR: $n=21$; age (mean) = 14.8; HR: $n=37$, age (mean) = 17.2	UHR: current ADHD symptoms: $n=6$, $n=5$ of these also had subthreshold hypomanic and/or depressive syndromes	NO	$n=48$; age (mean) = 15.6	NO
Linke et al. (2012)[MC]	BD1: $n=19$; age (mean) = 45	FDR: $n=22$; age (mean) = 28	NO	NO	Sample 1 (matched to patients): $n=19$; age (mean) = 45; Sample 2 (matched to HR); $n=22$; age (mean) = 28	NO

Study					
Manelis et al. (2015)[C/ME]	N/A	FDR: n=36; age (mean)=13.81	MDD/DDNOS: n=3, ADHD: n=6, anxiety: n=2, ODD: n=1, phobia: n=2, Tourette's: n=1, ED: n=1	Antipsychotics: n=2, stimulants: n=1, nonstimulants: n=2, benzodiazepine: n=2	HC: n=23; age (mean)=13.74; NBO: n=29; age (mean)=13.83	MDD/DDNOS: n=2, ADHD: n=7, anxiety: n=3, ODD: n=2, phobia: n=2, OCD: n=2
Manelis et al. (2016)[C/MC]	N/A	FDR: n=29; age (mean)=13.81	MDD/DDNOS: n=3, ADHD: n=6, anxiety: n=2, ODD: n=1, phobia: n=2, Tourette's: n=1, ED: n=2	Antipsychotics: n=2, stimulants: n=1, nonstimulants: n=2	HC: n=23; age (mean)=13.74; NBO: n=28; age (mean)=13.93	MDD/DDNOS: n=2, ADHD: n=5, anxiety: n=2, ODD: n=2, phobia: n=2, OCD: n=2
McCormack et al. (2016)[C/E]	BD1: n=27, BD2: n=25; age (range)=16–30, age (mean)=24.6	FDR: n=87; age (range)=16–30; age (mean)=22.0	Affective disorders: n=27, anxiety: n=21, behavioral: n=7, substance: n=7	11 HR participants on psychoactive medications removed from analysis	n=75; age (range)=16–30, age (mean)=30	Non-BD affective disorders: n=13, anxiety: n=5, substance: n=6
Mourão-Miranda et al. (2012)[L/ME]	N/A	FDR: n=16; age (mean)=14.8	NO	NO	n=16; age (mean)=15.3	NO

Continued

Table 1 Sample characteristics of studies using a neurocognitive task and including young unaffected relatives of bipolar disorder patients—cont'd

First author, year of publication	Bipolar disorder (BD) patients	High-risk relatives (HR)	HR with Axis 1 diagnoses	HR currently using medication	Controls (CON)	CON with Axis 1 diagnoses
Papmeyer et al. (2015)[L,C]	N/A	FDR or two SDR: Baseline: unaffected HR: n=92; age (range)=16–25, age (mean)=21.20; HR-MDD: n=19; age (range)=16–25, age (mean)=21.10. Follow-up: unaffected HR: n=63; age (mean)=23.71; HR-MDD: n=20; age (mean)=23.33	No Axis 1 diagnoses at baseline. Diagnoses developed after baseline: HR-MDD: n=19	Antidepressants: n=4	Baseline: n=93; age (range)=16–25, age (mean)=21.01. Follow-up: n=62; age (range)=16–25, age (mean)=22.82	NO
Pompei, Dima, et al. (2011)[C,MC]	BD1: n=39; age (mean)=39.43	Unaffected FDR: n=25; age (mean)=35.0; MDD FDR: n=14; age (mean)=31.2	MDD FDR: n=14	NO	n=48; age (mean)=36.33	NO
Pompei, Jogia, et al., 2011[C,MC]	BD1: n=39; age (mean)=39.43	Unaffected FDR: n=25; age (mean)=35.0; MDD FDR: n=14; age (mean)=31.2	MDD FDR: n=14	NO	n=48; age (mean)=36.33	NO

Reynolds et al. (2014)[S]	N/A	FDR: n=20; age (range)=18–65; age (mean)=34.15	NO	NO	n=20; age (range)=18–65; age (mean)=31.75	NO
Roberts et al., 2013[C/ME]	N/A	FDR: n=47; age (mean)=24.6	Any Axis 1 lifetime diagnosis: n=31, 66%; affective disorder: n=17, 36.2%; anxiety: n=12, 25.5%; behavioral: n=4, 8.5%	HR: SSRI: n=4, methylphenidate: n=1, tricyclic antidepressant: n=1. HC: SSRI: n=1, benzodiazepine: n=1	n=49; age (mean)=23.2	Any Axis 1 lifetime diagnosis: n=13, 26.5%; affective disorder: n=6, 12.2%; anxiety: n=5, 10.2%; current MDD: n=2, 4.1%
Ruocco et al., 2014[E]	BD1 with psychosis: n=248; age (mean)=36.22	FDR: n=286; age (mean)=40.25	NO	Unspecified	n=380; age (mean)=37.71	NO
Santos et al. (2017)[S]	BD1: n=31; age (mean)=41.48	FDR: n=18; age (mean)=50.33	NO	NO	n=31; age (mean)=39.87	NO
Seidel et al. (2012)[E/S]	BD1: n=10, BD2: n=11; age (mean)=46	FDR: n=21; age (mean)=38.43	NO	NO	n=21; age (mean)=41.67	NO
Sepede et al. (2012)[MC]	BD1: n=24; age (mean)=34.8	FDR: n=22; age (mean)=31.5	NO	NO	n=24; age (mean)=32.5	NO
Singh et al. (2014)[MC]	N/A	FDR: n=20; age (mean)=12.7	NO	NO	n=25; age (mean)=11.8	NO

Continued

Table 1 Sample characteristics of studies using a neurocognitive task and including young unaffected relatives of bipolar disorder patients—cont'd

First author, year of publication	Bipolar disorder (BD) patients	High-risk relatives (HR)	HR with Axis 1 diagnoses	HR currently using medication	Controls (CON)	CON with Axis 1 diagnoses
Vierck et al. (2015)[E]	BD1: $n=32$; age (mean)=42.3; BD2: $n=8$; age (mean)=35.3	FDR: $n=24$; age (mean)=33.2	50% HR had a lifetime diagnosis of depression	Antidepressants: $n=6$, atypical antipsychotics, anticonvulsants, and antidepressants for epilepsy: $n=1$, atypical antipsychotics and hypnotics for BPD: $n=1$	$n=40$; age (mean): 36.2	NO
Wang et al. (2015)[S]	N/A	FDR: $n=30$; age (mean)=44.93	NO	NO	$n=44$; age (mean)=46.16	NO
Whalley et al. (2011)[C/MC]	N/A	FDR or two SDR: $n=93$; age (mean)=21.01	NO	NO	$n=70$; age (mean)=20.89	NO
Whalley, Papmeyer, Romaniuk, et al., 2012[C/MC]	N/A	FDR or two SDR: homozygous for risk haplotype: $n=35$; age (mean)=21.75; not homozygous for risk haplotype: $n=46$; age (mean)=21.00	NO	NO	Homozygous for risk haplotype: $n=30$; age (mean)=20.53; not homozygous for risk haplotype: $n=45$; age (mean)=20.99	NO

Whalley, Papmeyer, Sprooten, et al. (2012)[C/MC]	N/A	FDR or two SDR: $n=87$; age (mean)=20.89	NO	NO	$n=71$; age (mean)=20.66	NO
Whalley et al. (2013)L[C/MC]	N/A	FDR or two SDR: unaffected HR: $n=78$; age (mean)=21.12; HR-MDD: $n=20$; age (mean)=20.59	No Axis 1 diagnoses at baseline. Diagnoses developed after baseline: HR-MDD: $n=20$, BD1: $n=1$, BD2: $n=1$	NO at baseline, unspecified for follow-up	$n=58$; age (mean)=20.78	NO
Whalley et al. (2014)L[C/ME]	N/A	FDR or two SDR: Follow-up scan: unaffected HR $n=61$; age (mean)=23.71; HR-MDD $n=20$; age (mean)=22.51 Third scan: unaffected HR $n=39$, HR-MDD $n=31$	No Axis 1 diagnoses at baseline. Diagnoses developed after baseline: Follow-up scan: MDD: $n=20$; Third scan: MDD: $n=31$, BD: $n=1$	NO at baseline. Follow-up: antidepressants: $n=5$. Excluded when examining main findings.	$n=51$; age (mean)=22.80	NO

Continued

Table 1 Sample characteristics of studies using a neurocognitive task and including young unaffected relatives of bipolar disorder patients—cont'd

First author, year of publication	Bipolar disorder (BD) patients	High-risk relatives (HR)	HR with Axis 1 diagnoses	HR currently using medication	Controls (CON)	CON with Axis 1 diagnoses
Whitney et al. (2013)[E,S]	N/A	FDR: $n=24$; age (range) = 9–18, age (mean) = 12.7	BDNOS: $n=5$, current MDD: $n=12$, past MDD: $n=3$, dysthymia: $n=2$, adjustment disorder: $n=1$, ADHD: $n=15$, ODD: $n=5$, GAD: $n=4$, anxiety NOS: $n=2$, phobia: $n=3$	Psychotropic medication: $n=15$, 68%	$n=27$; age (mean) = 13.3	NO

BD, bipolar patients; HR, first-degree relative of patient with bipolar disorder; CON, control participant; BD1, patient with bipolar I disorder; BD2, patient with bipolar 2 disorder; HC, healthy control; NBO, offspring of parents with non-BD psychopathology; BDNOS, patient with bipolar disorder not otherwise specified; FDR, first-degree relative; SDR, second-degree relative; UHR, ultra-high-risk; HR-MDD, high-risk participants that developed MDD during the course of the study; ADHD, attention-deficit hyperactivity disorder; MDD, major depressive disorder; DDNOS, depressive disorder not otherwise specified; ODD, oppositional defiant disorder; GAD, generalized anxiety disorder; CD, conduct disorder; ED, eating disorder; BPD, borderline personality disorder.

[C]: nonsocial cognition; [E]: emotion processing or regulation; [S]: social cognition; [MC]: fMRI nonemotional cognition; [ME]: fMRI emotion processing or regulation; [L]: paper that examines longitudinal data.

Table 2 Summary of neurocognitive tasks in studies using a neurocognitive task and including a young unaffected relative of bipolar disorder patients

Domain	Neurocognitive test	Deficit BD<CON	No Deficit BD=CON	Deficit HR<CON	No Deficit HR=CON	Deficit BD<HR	No Deficit BD=HR
Intelligence	GAI				de la Serna et al. (2016)		
	Kent E-G-Y Test		Deveci et al. (2013)		Deveci et al. (2013)		Deveci et al. (2013)
	NART		Whalley et al. (2013)[mL], Whalley et al. (2014)[mL]		Whalley et al. (2011)[m], Whalley, Papmeyer, Romaniuk, et al. (2012)[m], Whalley, Papmeyer, Sprooten, et al. (2012)[m], Whalley et al. (2013)[mL], Whalley et al. (2014)[mL]		Whalley et al. (2011)[mL], Whalley et al. (2013)[mL], Whalley et al. (2014)[mL]
	FSIQ	Doyle et al. (2009)			Doyle et al. (2009)	Doyle et al. (2009)	
	MWTB				Erk et al. (2014)[m]		
	WAIS-R IQ		Lelli-Chiesa et al. (2010)[m] Pompei, Dima, et al. (2011)[m] Pompei, Jogia, et al. (2011)[m]		Lelli-Chiesa et al. (2010)[m] Pompei, Dima, et al. (2011)[m] Pompei, Jogia, et al. (2011)[m]		Lelli-Chiesa et al. (2010)[m] Pompei, Dima, et al. (2011)[m] Pompei, Jogia, et al. (2011)[m]
	WASI IQ		Brotman et al. (2009) McCormack et al. (2016)		Brotman et al. (2009) Manelis et al. (2015)[m] Manelis et al. (2016)[m] McCormack et al. (2016) Roberts et al. (2013)[m]		Brotman et al. (2009) McCormack et al. (2016)
	WASI Matrix Reasoning		McCormack et al. (2016)		McCormack et al. (2016)		McCormack et al. (2016)
	WASI Vocab		McCormack et al. (2016)	McCormack et al. (2016)			McCormack et al. (2016)
Total		1	8	1	16	1	8

Continued

Table 2 Summary of neurocognitive tasks in studies using a neurocognitive task and including a young unaffected relative of bipolar disorder patients—cont'd

Domain	Neurocognitive test	Deficit BD < CON	No Deficit BD = CON	Deficit HR < CON	No Deficit HR = CON	Deficit BD < HR	No Deficit BD = HR
Executive Functioning	IED			Papmeyer et al. (2015)[L]			
	Monetary Incentive Delay Task[M]				Singh et al. (2014)[m]		
	Reversal Learning Task[M]	Linke et al. (2012)[m]		Linke et al. (2012)[m]			
	Reward Task[M]						
	SCWT[M]	Doyle et al. (2009)	Pompei, Dima, et al. (2011)[m] Pompei, Jogia, et al. (2011)[m]		Manelis et al. (2016)[m] de la Serna et al. (2016) Doyle et al. (2009) Pompei, Dima, et al. (2011)[m] Pompei, Jogia, et al. (2011)[m]	Doyle et al. (2009)	Pompei, Dima, et al. (2011)[m] Pompei, Jogia, et al. (2011)[m]
	WCST—Categories				Deveci et al. (2013)		
	Tower of London			Lin et al. (2017)			
	WCST—Perseverative Errors	Doyle et al. (2009)		Doyle et al. (2009)		Doyle et al. (2009)	
	WCST				de la Serna et al. (2016)		
Total		2	2	4	7	1	2
Attention	CPT				Lin et al. (2017)		
	CPT—Target Detection[M]	Sepede et al. (2012)[m]		Sepede et al. (2012)[m]			
	CPT—II				de la Serna et al. (2016)		
	Flanker CPT	Brotman et al. (2009)		Brotman et al. (2009)			Brotman et al. (2009)

	McCormack et al. (2016)		de la Serna et al. (2016)	McCormack et al. (2016) / Deveci et al. (2013)		McCormack et al. (2016)
RBANS						
TMT	3	0	2	4	0	2
Total						
Processing Speed						
PSI (From the WISC-IV)	Doyle et al. (2009)			Doyle et al. (2009)	Doyle et al. (2009)	
WAIS III Digit Symbol Coding	Doyle et al. (2009)				Doyle et al. (2009)	
WISC III/WAIS-III Symbol Search			de la Serna et al. (2016)			
WISC-IV PSI			de la Serna et al. (2016)			
Total	1	0	1	1	1	0
Working Memory						
Seidman CPT Interference	Doyle et al. (2009)		Doyle et al. (2009)		Doyle et al. (2009)	
Seidman CPT Memory	Doyle et al. (2009)		Doyle et al. (2009)		Doyle et al. (2009)	
Seidman CPT Vigilance	Doyle et al. (2009)			Doyle et al. (2009)	Doyle et al. (2009)	
WAIS-IV Digit Span				Papmeyer et al. (2015)[L]		
WISC-III/WAIS-III Arithmetic	Doyle et al. (2009)			Doyle et al. (2009)	Doyle et al. (2009)	
WISC-IV WMI			Lin et al. (2017)	de la Serna et al. (2016)		
WMS-III Spatial-Span						
Total	1	0	2	2	1	0
Episodic Memory						
Episodic Memory Task[M]				Erk et al. (2014)[m]		
Total	0	0	0	1	0	0

Continued

Table 2 Summary of neurocognitive tasks in studies using a neurocognitive task and including a young unaffected relative of bipolar disorder patients—cont'd

Domain	Neurocognitive test	Deficit BD < CON	No Deficit BD = CON	Deficit HR < CON	No Deficit HR = CON	Deficit BD < HR	No Deficit BD = HR
Verbal Learning and Memory	CVLT	Doyle et al. (2009)		Papmeyer et al. (2015)[L]	Doyle et al. (2009)	Doyle et al. (2009)	
	HVLT-R			Lin et al. (2017)	Erk et al. (2014)[m]		
	VLMT						
	RAVLT			Deveci et al. (2013)			
	TOMAL				de la Serna et al. (2016)		
Total		1	0	3	3	1	0
Visual-Spatial Learning and Memory	Brief Visuospatial Memory Test–Revised			Lin et al. (2017)			
	RCF						
	WMS-III			de la Serna et al. (2016)	de la Serna et al. (2016)		
Total		0	0	2	1	0	0
Verbal Fluency	Hayling Sentence Completion Test[M]				Whalley et al. (2011)[m] Whalley, Papmeyer, Romaniuk, et al. (2012)[m] Whalley, Papmeyer, Sprooten, et al. (2012)[m] Whalley et al. (2013)[mL]		
Total		0	0	0	4	0	0
Social Cognition and Emotional processing	Affectiveness Responsiveness	Seidel et al. (2012)		McCormack et al. (2016)	Seidel et al. (2012)	Seidel et al. (2012)	
	Affective Go/No Go Task[M]	Bauer et al. (2015)	McCormack et al. (2016)	Roberts et al. (2013)[m]	Bauer et al. (2015) Brand et al. (2012)	Bauer et al. (2015)	McCormack et al. (2016)
Total		0	0	0	4	0	0

Task				
DANVA			Whitney et al. (2013)	Brotman, Guyer, et al. (2008)
Dynamic Faces Task[M]	Brotman, Guyer, et al. (2008)	Brotman, Guyer, et al. (2008), Hanford et al. (2016)	Manelis et al. (2015)[m]	
Dynamic Faces—Post-Scan Emotional Labelling Task			Manelis et al. (2015)[m]	
Emotional Expression Multimorph Task	Brotman, Skup, et al. (2008)	Brotman, Skup, et al. (2008)		Brotman, Skup, et al. (2008)
Emotion Recognition Subtest (Fear)	de Brito Ferreira Fernandes et al. (2016)		de Brito Ferreira Fernandes et al. (2016)	de Brito Ferreira Fernandes et al. (2016)
Emotion Recognition Task	Seidel et al. (2012)	Seidel et al. (2012)	Seidel et al. (2012)	Seidel et al. (2012)
Emotional Face Gender-Labelling Task[M]			Mourão-Miranda et al. (2015)[mL]	
Emotional Perspective Taking Task			Seidel et al. (2012)	Seidel et al. (2012)
ER-40	Ruocco et al. (2014) Vierck et al. (2015)	Ruocco et al. (2014) Vierck et al. (2015)		
Facial Expression Recognition Task	Vierck et al. (2015)			Vierck et al. (2015)

Continued

Table 2 Summary of neurocognitive tasks in studies using a neurocognitive task and including a young unaffected relative of bipolar disorder patients—cont'd

Domain	Neurocognitive test	Deficit BD < CON	No Deficit BD = CON	Deficit HR < CON	No Deficit HR = CON	Deficit BD < HR	No Deficit BD = HR
	Emotional Memory Task[M]				Whalley et al. (2014)[mL]		
	Emotional Memory—Post-Scan Task				Whalley et al. (2014)[mL]		
	Emotional Stroop Task				Besnier et al. (2009)		
	Implicit Memory Task[m]				Chan et al. (2016)[mL]		
	Post-Scan Ekman Faces Task				Chan et al. (2016)[mL]		
	IRI	Seidel et al. (2012)			Seidel et al. (2012)		Seidel et al. (2012)
	Happé Stories Task			Reynolds et al. (2014)			
	MSCEIT		Calafiore et al. (2017)		Calafiore et al. (2017)		Calafiore et al. (2017)
	N-Back Task[M]				Ladouceur et al. (2013)[m]		
	N-Back—Post-Scan Emotion Labelling Task				Ladouceur et al. (2013)[m]		
	Parametric Faces Paradigm[M]	Brotman et al. (2014)[m]		Brotman et al. (2014)[m]			
	Picture Sequencing Task				Reynolds et al. (2014)		

Test						
RVP						
Sad Facial Affect Discrimination Task[M]				Bauer et al. (2015)		
The Eyes Task	Lelli-Chiesa et al. (2010)[m]	Bauer et al. (2015)	Santos et al. (2017)	Lelli-Chiesa et al. (2010)[m]; Reynolds et al. (2014)	Lelli-Chiesa et al. (2010)[m]	Bauer et al. (2015)
Total	9	2	9	14	2	6
ToM						
False Belief Commands				Wang et al. (2015)		
False Belief Probes				Wang et al. (2015)		
Hinting Task		Santos et al. (2017)		Santos et al. (2017)		Santos et al. (2017)
MASC	Santos et al. (2017)		Santos et al. (2017)			Santos et al. (2017)
NEPSY						
Reality Commands				Whitney et al. (2013); Wang et al. (2015)		
Reality Probes				Wang et al. (2015)		
Second-Order False Belief	Santos et al. (2017)	Santos et al. (2017)		Santos et al. (2017)		Santos et al. (2017)
Total	1	1	1	3	0	1

Total, total number of studies that reported a finding; For studies that report both the presence and absence of a deficit between two groups using different neurocognitive tests, only the deficit finding will be counted in the total; BD, patients with bipolar; HR, bipolar relatives at genetic high-risk of developing bipolar; CON, control subjects; GAI, General Ability Index; NART, National Adult Reading Test; FSIQ, Full-Scale Intelligence Quotient; MWTB, Mehrfachwahl-Wortschatz-Intelligenztest; WASI-R IQ, Wechsler Adult Intelligence Scale—Revised Intelligence Quotient; WASI, Wechsler Abbreviated Scale of Intelligence; WTAR, Wechsler Test of Adult Reading; IED, Intra-Extra Dimensional Set Shift; SCWT, Stroop Color Word Test; WCST, Wisconsin Card Sorting Test; CPT, Continuous Performance Test; RBANS, Repeatable Battery for Assessment of Neuropsychological Status; TMT, Trail Making Test; PSI, Processing Speed Index; WAIS, Wechsler Adult Intelligence Scale; WISC, Wechsler Intelligence Scale for Children; WMS III, Wechsler Memory Scale III; WMI, Working Memory Index... CVLT, California Verbal Learning Test; HVLT-R, Hopkins Verbal Learning Test-Revised; VLMT, Verbal Learning and Memory Task; RAVLT, Rey Auditory Verbal Learning Test; TOMAL, Test of Memory and Learning; RCF, Rey Complex Figure; DANVA, Diagnostic Analysis of Nonverbal Accuracy; ER-40, Penn Emotional Recognition Task; IRI, Interpersonal Reactivity Index; MSCEIT, Mayer-Salovey-Caruso Emotional Intelligence Test; RVP, Rapid Visual Processing. MASC, Movie for the Assessment of Social Cognition; NEPSY, A Development NEuroPSYchological Assessment. [m], neuroimaging study; [L], paper that examines longitudinal data.

attention (two (Brotman et al., 2009; McCormack et al., 2016) of two studies) were found between BD patients and controls. In these studies that found a difference between BD patients and controls, although the differences between unaffected first-degree relatives of BD participants and controls were not significant in these domains, the performance of unaffected first-degree relatives was intermediate to that of BD patients and controls on all task measures, except for Digit Symbol/Coding of the Wechsler Intelligence Scale for Children-Third Edition (WISC-III)/Wechsler Adult Intelligence Scale-Third Edition (WAIS-III) and the Color Naming measure of the Stroop Color Word Task (SCWT; Doyle et al., 2009) where high-risk subjects either performed worse than BD and controls, or quite similarly. Furthermore, Doyle et al. (2009) found that BD patients showed impairments in intelligence, executive functioning, working memory, episodic memory, and verbal learning and memory compared to siblings of those with BD. However, this was the only study that found a deficit in intelligence in BD patients compared to unaffected first-degree relatives. No deficits were found in attention when comparing BD patients to unaffected first-degree relatives (two studies: Brotman et al., 2009; McCormack et al., 2016). None of the studies that included a BD comparator group investigated visuospatial learning and memory, or verbal fluency domains.

Table 2 provides a more comprehensive list of results including details of the specific psychometric instruments used. For intelligence testing, where studies used the same tests, the same results were found. Specifically, three studies used the National Adult Reading Test (NART; Whalley et al., 2011; Whalley et al., 2014; Whalley et al., 2013), three the WAIS-R IQ (Lelli-Chiesa et al., 2010; Pompei, Dima, Rubia, Kumari, & Frangou, 2011; Pompei, Jogia, et al., 2011), and three the WASI-IQ (Brotman et al., 2014; Manelis et al., 2015; Roberts et al., 2013), with no deficit found between the BD, high-risk, and control groups. Again, for executive functioning, three studies used the SCWT (de la Serna et al., 2016; Deveci et al., 2013; Doyle et al., 2009); all reported no differences in task performance between high-risk participants and controls. All the other nonemotional cognitive tasks reported in Table 2 were only used in one study each.

Studies that only investigated differences between unaffected first-degree relatives and controls consistently reported impairments in visuospatial learning and memory, and processing speed. Notably, high-risk individuals were not found to be impaired on these domains in the meta-analyses and reviews that included a broader age range of unaffected relatives. This suggests that these behavioral deficits may be specific to a younger population. In support of a possible candidate endophenotype in youth, studies that included a BD group showed deficits in processed speed in both BD patients and unaffected first-degree compared to controls. Visual spatial learning and memory, however, were not examined in any of the studies that included a BD comparator group. In one study that included a BD group, executive functioning and working memory also showed deficits in executive functioning, and working memory in both BD patients and unaffected compared to controls. This deficit in executive functioning is in accordance with a meta-analysis that included a broader age range (Arts et al., 2008), but needs to be interpreted with caution given that this interpretation is based on just one study. Overall, intelligence seems to be spared in high-risk individuals. Further, results on the remaining domains (attention, episodic memory, and verbal learning and memory) are also mixed, with some studies reporting deficits in unaffected first-degree relatives, while others find these domains to be intact.

SOCIAL COGNITION

Social cognition refers to a complex set of processes that enable adaptive social interaction, such as emotion perception, theory of mind (mentalizing), and other processes that facilitate effective interpersonal relations (Lewis, 2012). In a meta-analysis of 20 social cognition studies in euthymic patients with BD, significant performance decrements of small magnitude were found for recognition of facial affect expressions, whereas impairments of moderate magnitude were noted across theory of mind (ToM) tasks (Samame, Martino, & Strejilevich, 2015). However, research on social cognition in unaffected relatives at genetic risk of BD is scant and provides mixed results, with only some evidence for behavioral deficits. Of the seven studies that have investigated social cognition in unaffected relatives of BD and controls (see Table 2), only two (Hanford, Sassi, & Hall, 2016; Reynolds, Van Rheenen, & Rossell, 2014) have shown behavioral deficits in the high-risk group. Of the nine studies that included a BD group, five (Brotman, Guyer, et al., 2008; Brotman, Skup, et al., 2008; Ruocco et al., 2014; Santos et al., 2017; Vierck, Porter, & Joyce, 2015) reported behavioral deficits in both the BD and high-risk group compared to controls, while two (Bauer et al., 2015; Seidel et al., 2012) found behavioral deficits in the BD group compared to the high-risk and control groups. Investigations have focused mainly on three central processes within this construct: emotion processing, emotion regulation, and ToM. We will now look at the individual studies in more detail.

EMOTION PROCESSING

Emotion processing, a central aspect of social cognition, encompasses the capability to identify and discriminate "basic emotions," which are thought to be innate and have universally recognizable facial expressions (Phillips, Drevets, Rauch, & Lane, 2003). In the research literature, facial expressions are the most commonly utilized stimuli. Most facial emotion recognition tasks require subjects to name the emotions displayed in pictures of posed facial expressions, including happiness, sadness, anger, disgust, fear, and surprise. Nine studies have investigated facial emotion recognition as an endophenotype candidate for BD. In two studies, children and adolescents with a first-degree relative with BD made more errors identifying facial emotions (Brotman, Guyer, et al., 2008) and required significantly more intense emotional information to identify and correctly label facial emotions compared to healthy child and adolescent controls (Brotman, Skup, et al., 2008). The group effects in these studies did not differ based on the facial emotion expressed. However, in a more recent study, children and adolescents at high risk were more likely to make errors in identifying sad and angry faces compared to healthy controls (Hanford et al., 2016). In another youth sample that included a BD group (McCormack et al., 2016), BD patients, but not high-risk participants, demonstrated higher accuracy for recognition of disgusted facial expressions compared to controls. Whitney et al. (2013) also did not report a facial emotion AGN recognition deficit between young high-risk individuals and healthy controls.

One study that investigated three components of empathy (emotion recognition, perspective taking, and affective taking) in adult BD patients, their first-degree relatives, and healthy controls reported the control group to be more accurate in recognizing facial emotions than both relatives and patients (Seidel et al., 2012). Compared to a healthy control group, BD patients, but not their unaffected first-degree relatives, showed differences in affective responsiveness reflecting difficulties in identifying emotions experienced in a given situation. Also using an adult sample, Vierck et al. (2015) found that both BD patients and their relatives had slower emotion recognition than healthy controls in the absence of accuracy differences; furthermore, there was no evidence of emotion-specific differences. In another recently reported adult sample, first-degree relatives did not differ significantly from healthy controls on emotion recognition (de Brito Ferreira Fernandes et al., 2016). However, compared with healthy controls, euthymic BD patients gave significantly fewer correct responses for fearful faces, and took significantly longer to recognize happy faces, but not when compared with first-degree relatives (de Brito Ferreira Fernandes et al., 2016). A large sample of psychotic BD patients aged 15–65 ($n=248$) and their first-degree relatives ($n=286$) showed deficits in recognition of angry and neutral faces compared to controls (Ruocco et al., 2014). Deficits increased progressively from psychotic BD to schizoaffective disorder to schizophrenia (Ruocco et al., 2014).

Three of the above studies used the Diagnostic Analysis of Nonverbal Accuracy scale (DANVA; Brotman, Guyer, et al., 2008; Hanford et al., 2016; Whitney et al., 2013), and of these, two reported a deficit (Brotman, Guyer, et al., 2008; Hanford et al., 2016). All other facial emotion recognition tasks were only used once. Taken together, some level of facial emotion recognition impairment is evident in high-risk individuals. However, in the studies that found differences in emotion recognition of a particular emotion, the specific emotion that showed group differences was inconsistent between studies.

EMOTION INHIBITION

In an adult sample, unaffected first-degree relatives of BD patients responded more slowly to depression-like words, and made more errors in response to manic-related words during the emotional Stroop Color Word Test (SCWT) compared to healthy controls (Besnier et al., 2009). Among the tasks used to assess this domain, the Affective Go/No Go (AGN) task was the only measure used by more than one group of researchers. Child offspring of BD patients displayed intact performance accuracy, but quicker response times than healthy controls in an AGN task (Bauer et al., 2015). In the same study, individuals with BD responded faster to correct trials and committed an elevated number of commission errors across all affective conditions relative to healthy controls (Bauer et al., 2015). Contrary to this finding, in a young sample, BD participants showed no deficits compared to controls and high-risk individuals in an AGN task (McCormack et al., 2016). However, participants with BD-I did show deficits in the AGN task compared to controls, and high-risk participants made more errors of commission than controls during all trials of the AGN task when the stimuli were negative or neutral (McCormack et al., 2016). In line with this result, adults at high genetic risk of developing BD showed a response bias to negatively valenced stimuli compared to healthy controls during an AGN task (Brand et al., 2012). Overall, results are somewhat inconclusive, but the findings suggest that individuals at high genetic risk appear to have a response bias to negatively valenced stimuli compared to controls.

THEORY OF MIND

Theory of mind (ToM), also referred to as mentalizing, is the cognitive ability to attribute mental states (such as beliefs, desires, and intentions) to others, as separate to the self (Bora et al., 2009). Several types of measures, with varying levels of complexity, have been used to assess this construct. A recent meta-analysis of ToM studies in BD patients reported small-to-medium effect sizes for tasks measuring cognitive ToM and small effect sizes for tasks measuring affective ToM in favor of controls versus euthymic BD patients (Samame et al., 2015).

There has been little research addressing this domain of cognition in individuals at high genetic risk of BD. In a symptomatic high-risk sample of children and adolescents (reporting past or current moderate mood symptoms), Whitney et al. (2013) reported no significant group differences in ToM as measured by the Developmental Neuropsychological Assessment (NEPSY), a test that assesses multiple ToM constructs. The authors suggested that their results may have been due to either psychopathology in the high-risk group or the healthy control population having higher error rates than those previously reported in children and adolescents of the same age (Whitney et al., 2013). These symptomatic offspring at risk of BD did, however, have significant impairment in social reciprocity, including impairments in social awareness, social cognition, and social motivation (Whitney et al., 2013). In another young sample of BD patients, high-risk offspring, and controls, no group differences were evident in the interpretation of conversational remarks meant literally (i.e., sincere remarks and lies) or nonliterally (i.e., sarcasm), as well as in the ability to make judgments about the thoughts, intentions, and feelings of speakers during The Awareness of Social Inference Test (TASIT; McCormack et al., 2016).

Adult relatives at high risk of developing BD performed significantly worse on verbal (Happé Stories Task), but not visual (Picture Sequencing Task) or higher-order (Reading the Eyes in the Mind Task) ToM tasks compared to healthy controls

(Reynolds et al., 2014). Consistent with the verbal domain, Wang, Roberts, Liang, Shi, and Wang (2015) found no difference in performance between individuals at high risk of developing BD and healthy controls on a revised verbal computerized referential task requiring ToM, in a sample of relatively older adults. In a recent study of adults, no behavioral differences were evident between BD patients, unaffected first-degree relatives, and healthy controls in the Mayer-Salovey-Caruso Emotional Intelligence Test (MSCEIT; Calafiore, Rossell, & Van Rheenen, 2017). Additionally, in an older sample that included a BD group, Santos et al. (2017) administered first and second-order false-belief tasks, the hinting task, and the Movie for the Assessment of Social Cognition (MASC). Compared to healthy controls, relatives at high genetic risk and BD patients had worse performance on the MASC test—a complex test which uses different channels (auditory, verbal, and emotional). No group differences were evident on the remaining tests that just used one channel (Santos et al., 2017). Taken together, the few available studies in first-degree relatives to date have shown a mixture of subtle deficits and no deficits; however, the diversity of tasks restricts direct comparisons between tasks. Therefore, simple ToM tasks such as false-belief stories might not be adequately sensitive to detect subtle ToM deficits.

Overall, the studies reported here provide some evidence that social cognition impairments may precede the onset of BD in high-risk individuals. The evidence is more compelling for facial emotional recognition. ToM showed subtle effects with complex tasks, but results are inconclusive because of diversity of tasks employed.

NEUROCOGNITION DURING FUNCTIONAL MRI

Neuroimaging studies in youth at high risk of BD have examined components of nonsocial cognition with a focus on executive functioning, attention, and verbal fluency. As seen in Table 2, of the six studies that compared unaffected relatives at high risk of BD and healthy controls, no cognitive behavioral deficits have been reported on measures of executive functioning (two studies: Manelis et al., 2016; Singh et al., 2014), episodic memory (one study: Erk et al., 2014), and verbal fluency (three studies: Whalley, Papmeyer, Romaniuk, et al., 2012; Whalley et al., 2013; Whalley et al., 2011). In the four studies that included a BD comparator group, BD patients and unaffected offspring were impaired on measures of executive functioning (one of three studies: Linke et al., 2012) and attention (one study: Sepede et al., 2012). Findings from studies using emotion processing tasks conducted during MRI are varied, as seen in Table 2, with some suggesting a behavioral deficit while others show comparable performance between groups. Of the three studies that compared unaffected high-risk relatives and controls during emotion processing and regulation, there is evidence to suggest an emotional processing and regulation deficit (one of two studies: Roberts et al., 2013). Of the two studies that included a BD group, there is also limited evidence to suggest a behavioral deficit in both the BD and high-risk group compared to controls (one of two studies: Brotman et al., 2014) or in the BD group compared to the high-risk and control groups (one of two studies: Lelli-Chiesa et al., 2010). Given that these studies and tasks have been designed mostly for neuro-imaging comparisons, performance differences are not necessarily expected. In fact, implicit tasks are chosen for some neuroimaging experiments to avoid any neural differences being confounded by performance differences.

We will next look at the behavioral and neural results of these papers in more detail.

NONEMOTIONAL NEUROIMAGING TASKS

Three studies have investigated reward processing in relatives of individuals with BD. One study targeted anticipation and feedback of reward and loss during a monetary incentive task in children at high risk of BD and healthy controls (Singh et al., 2014). In the absence of behavioral deficits, high-risk children showed decreased activation in the pregenual cingulate cortex during loss anticipation, and increased activation in the left IFG when receiving feedback of reward. Increased connectivity between the pregenual cingulate cortex and right ventrolateral frontal gyrus during anticipation of loss, and weaker connectivity during reward processing, were also evident in the high-risk group. Another study conducted by Manelis et al. (2016) also reported no behavioral differences between young unaffected relatives and controls. Moreover, they found significantly greater frontal pole activation and greater negative functional connectivity between the bilateral ventral striatum and the right ventrolateral prefrontal cortex in high-risk subjects compared to controls. In adults, Linke et al. (2012) used a probabilistic reversal-learning task to investigate defects in the reward system. No difference in reaction time was observed between the groups. However, high-risk subjects won less money (in terms of reward), relative to healthy controls. BD patients showed greater activation in the medial and right lateral orbito-frontal cortex (OFC), the right amygdala, the dorsal amygdala-anterior cingulate cortex (ACC), and the putamen during rule reversal, whereas reward elicited a greater activation only in the medial OFC compared to healthy controls. High-risk relatives also revealed higher levels of activity than healthy controls in the medial OFC when presented with reward or rule reversal, as well as during punishment. In addition to these findings, high-risk relatives had greater activation of the right amygdala during reward compared to BD patients and healthy controls. Lastly, BD patients showed greater activation in the right lateral OFC and the putamen during rule reversal compared to their unaffected relatives.

Erk et al. (2014) found no performance differences in relatives of BD patients compared with healthy controls in an episodic memory task. During this task, diminished activation was demonstrated in the bilateral hippocampus and the subgenual part of the pregenual ACC in relatives of BD patients compared to healthy controls.

In a test of attention using the Continuous Performance Task (CPT), high-risk relatives and BD patients had lower accuracy during the correct target condition compared to healthy controls (Sepede et al., 2012). During this condition, BD patients had significantly lower right insula activation compared to healthy controls and high-risk subjects. During the most difficult condition, high-risk subjects showed increased activity in the bilateral inferior parietal lobule and the left insula. Finally, during the incorrect target condition, in both BD and high-risk subjects, there was increased activity of the posterior part of the middle cingulate cortex and bilaterally in the insula relative to healthy controls. The SCWT showed no behavioral differences between BD patients, first-degree relatives of BD patients, and healthy controls, for both accuracy and reaction times (Pompei, Jogia, et al., 2011). This study showed that patients and relatives displayed significantly lower activation levels in the superior and inferior parietal lobules compared to healthy controls. BD patients also showed significantly less activity in the right ventrolateral prefrontal cortex (VLPFC) and in the head of the left caudate nucleus compared to both controls and their relatives. In the same sample, findings from a subsequent psychophysiological interaction (PPI) analysis suggest that the VLPFC and subcortical regions are involved in a dynamic interplay during this task (Pompei, Dima, et al., 2011).

The Hayling Sentence Completion Test has been used as a measure of verbal production in BD risk groups on more than one occasion by the same group. No significant differences in performance were seen between first- or second-degree relatives of BD patients relative to healthy controls. When the difficulty of the task increased, the high-risk group showed significantly higher activation in the left amygdala (Whalley et al., 2011). Successive studies found have reported differential effects of the DGKH gene (Whalley, Papmeyer, Romaniuk, et al., 2012) and polygenic risk scores (Whalley, Papmeyer, Sprooten, et al., 2012) in healthy controls compared to the high-risk group during this task.

EMOTION PROCESSING AND REGULATION NEUROIMAGING TASKS

While Ladouceur et al. (2013) found no significant differences in accuracy or reaction times, they did find greater right ventrolateral prefrontal cortex (VLPFC) activation in response to positive emotional distracters and reduced VLPFC modulation of the amygdala to both the positive and negative emotional distracters in an unaffected high-risk group relative to healthy controls during an affective working memory task. In a dynamic faces task that measured emotional processing and regulation, no behavioral differences were evident between offspring of parents with bipolar disorder, offspring of nonbipolar parents, and controls (Manelis et al., 2015). This study showed that abnormal functional connectivity patterns in the amygdala-left VLPFC and amygdala-anterior cingulate cortex (ACC) distinguished offspring of parents with bipolar disorder from those of nonbipolar parents and healthy controls.

During an AGN task, Roberts et al. (2013) found a lack of recruitment of the inferior frontal gyrus (IFG) during inhibition of fearful faces in the high-risk group compared to a control group. No performance differences were evident for accuracy, but high-risk subjects showed *increased* accuracy for fearful face stimuli compared to controls (Roberts et al., 2013). Brotman et al. (2014) administered a parametric face paradigm with both explicit and implicit components. Behaviorally, both subjects with BD and high-risk youth rated faces as less hostile relative to healthy controls. In response to increasing anger expressed on the face, subjects with BD and high-risk youth showed decreased modulation in the amygdala and IFG relative to healthy controls. Amygdala dysfunction was present across both implicit and explicit rating conditions, but IFG modulation deficits were specific to the explicit condition. With increasing happiness, high-risk youth showed aberrant modulation in the IFG.

Results from a sad affect facial discrimination task revealed behavioral differences for BD patients only, and not for unaffected relatives compared to healthy controls (Lelli-Chiesa et al., 2010). Signal changes in the amygdala and ventromedial prefrontal cortex (VMPFC) were associated with COMT Val158Met polymorphism, but showed no association with the group. None of the emotion processing regulation tasks were directly comparable.

Overall, behavioral performance was intact whereas neuronal deficits were evident, suggesting that brain imaging may be a more sensitive tool for investigating differences that relate to neurocognitive alterations. Differences in brain activation along with subtle differences in neurocognitive performance may therefore represent a heritable endophenotype of bipolar disorder. As these MRI tasks were designed for the scanner, behaviorally they were not directly comparable to tasks that were completed outside the scanner. Although neuroimaging tasks were also not directly comparable on a neuronal level, regions involved in emotional processing and emotional regulation, including the IFG, ACC, insula, VLPFC, OFC, limbic regions, and particularly the amygdala, were functionally altered in unaffected first-degree relatives compared to controls. In studies that included a comparison BD group, overall brain signal in the high-risk group was intermediate to the BD and control group. These same regions also show abnormalities in older subjects at genetic risk of BD (Piguet, Fodoulian, Aubry, Vuilleumier, & Houenou, 2015). Similar regions showed alterations during both nonemotional cognitive and emotional processing, while altered limbic activation including connectivity was mainly related to emotional processing. Alterations manifested as both hypoactivation and hyperactivation, with reports of hyperactivation being more abundant. Hyperactivation is generally interpreted as compensation, whereas hypoactivation is interpreted as an inherent difficulty in engaging a particular brain region or a decreased effort, with the exception of the default network which is more active in the absence of a task (Grady, 1998).

AXIS 1 DIAGNOSES AND OTHER SUBGROUP AUXILIARY ANALYSES

While there are advantages to studying individuals that have not yet passed the peak age of BD onset, the samples contain a mix of those who are resilient (i.e., who will not go on to develop BD) and those who are truly at risk (i.e., will go on to develop BD). When using a cross-sectional approach in high-risk samples, it is not possible to make this distinction and thus it is difficult to conclude if cognitive deficits represent true familial risk or resilience. It is known that high-risk individuals with an Axis 1 diagnosis of recurrent major depressive disorder (MDD) or anxiety disorders are at greatest risk for developing manic or hypomanic episodes (Duffy, Alda, Hajek, Sherry, & Grof, 2010; Johnson, Cohen, & Brook, 2000). Thus, studying subgroups of high-risk participants with and without non-BD psychopathology may help disambiguate differences associated with risk vs. resilience. In instances where high-risk participants do have non-BD psychopathology, studies that define controls on the basis of family history and include non-BD psychopathology allow us to see if group differences are due to underlying genetic risk rather than comorbid non-BD psychopathology. In contrast, studies that recruit a "supernormal" healthy control group without Axis I diagnoses are not equipped to make this differentiation. For the cross-sectional studies reported in Tables 1 and 2, within the high-risk and control groups, we will next look at inclusion and exclusion criteria for Axis 1 diagnoses and, when applicable, report subgroup analyses.

NONEMOTIONAL NEUROCOGNITION
HIGH-RISK GROUPS

In total, there were six studies that examined differences in nonsocial cognition in participants at high genetic risk of developing bipolar compared to controls. Five studies included high-risk individuals with Axis 1 diagnoses (Brotman et al., 2009; de la Serna et al., 2016; Doyle et al., 2009; Lin et al., 2017; McCormack et al., 2016), of which four conducted further analysis. After removing high-risk participants with lifetime diagnoses of attention-deficit hyperactivity disorder, Brotman et al. (2009) found that the intrasubject variability in reaction times remained significantly greater for high-risk participants compared to healthy controls. This pattern was also found after removing high-risk participants with anxiety disorders. Similarly, both de la Serna et al. (2016) and Doyle et al. (2009) found that the presence of psychopathology in high-risk participants did not influence their findings. Subgroup analysis by McCormack et al. (2016) focused on the history of psychosis in the bipolar proband of their high-risk participants. Specifically, they found that there were no differences in the cognitive function of high-risk participants whose proband had a history of psychosis compared to those whose proband did not.

One study excluded high-risk individuals with Axis 1 diagnoses (Deveci et al., 2013). However, this study conducted further analysis focusing on participants who were over 11 years of age and found that the deficits present when comparing high-risk participants to healthy controls in the Rey Auditory Verbal Learning and Memory Test, Auditory Consonant Trigram Test, and the Trail Making Test (Part A) remained significant. Additionally, they also found that high-risk participants over 11 years old performed worse on the Controlled Word Association Test and the Digit Span Forward compared to healthy controls over 11 years old, whereas these deficits were not present when participants of all ages were included in the analysis.

The studies that had neuroimaging in addition to a nonsocial cognition component outside the scanner are included in the neuroimaging section detailed below.

CONTROL GROUPS

Three studies examining nonsocial neurocognition in unaffected first-degree relatives compared to healthy controls excluded individuals with Axis 1 diagnoses from their control group (Brotman et al., 2009; Deveci et al., 2013; Lin et al., 2017). Of the remaining studies that included control individuals with Axis 1 diagnoses, two conducted further analysis (de la Serna et al., 2016; Doyle et al., 2009). Both these studies found that the presence of psychopathology in control participants did not influence their findings.

SOCIAL COGNITION
HIGH-RISK GROUPS

Sixteen reports have examined participants at high genetic risk of developing BD compared with controls on social cognition. Of these, nine excluded high-risk individuals who had an Axis 1 diagnosis. For the remaining seven studies, an Axis 1 diagnosis was not an exclusion factor for first-degree relatives, and five of these studies conducted further analysis based on this.

Three of these studies subdivided their high-risk participants into those with an Axis 1 diagnosis and those without (Bauer et al., 2015; Brotman, Guyer, et al., 2008; Hanford et al., 2016), and found that high-risk participants with an Axis 1 diagnosis showed faster reaction times on the AGN task compared to healthy controls for all stimuli, as well as for positive and negative stimuli specifically. However, they did not find any differences in reaction times when comparing high-risk participants with and without Axis 1 diagnoses, or when comparing the high-risk participants without Axis 1 diagnoses and healthy controls. In terms of error rates, there were no significant differences in the number of error rates between the high-risk groups, or between either of the high-risk groups compared to healthy controls.

Contrary to the above finding, Brotman, Guyer, et al. (2008) found that high-risk participants without an Axis 1 diagnosis made more errors in the facial expressions subtests of the DANVA scale compared to healthy controls. Similarly, when comparing high-risk participants who presented with symptoms of an Axis 1 disorder and high-risk participants with no symptoms, Hanford et al. (2016) found that symptomatic high-risk participants made more errors when labeling sad faces on the DANVA than healthy controls. Additionally, both symptomatic and asymptomatic high-risk participants made greater numbers of errors labeling angry faces compared to healthy controls. Further, both symptomatic and asymptomatic high-risk participants made more errors labeling low-intensity faces, and asymptomatic high-risk participants made more errors on high-intensity faces compared to healthy controls. However, there was no significant correlation between symptom severity and total errors.

Brotman, Guyer, et al. (2008) conducted analyses to determine whether characteristics of the bipolar relatives of high-risk participants influence performance on an emotion identification task. Specifically, this study found no differences in the error rates of high-risk participants who had a bipolar sibling compared to those with a bipolar parent. Lastly, Brotman, Guyer, et al. (2008), Brotman, Skup, et al. (2008) repeated their analyses after excluding those with Axis 1 diagnoses, after which the findings of both studies remained the same. Similarly, Whitney et al. (2013) found no main effect of diagnostic category or medication exposure in their analysis.

CONTROLS

Of the 16 studies that examined social and/emotional processing, 15 stated that the presence of a current or lifetime Axis 1 diagnosis was an exclusion criterion for their control group. The one study that included Axis 1 diagnoses (McCormack et al., 2016) did not conduct any further analysis on controls with vs. without Axis 1 diagnoses.

NEUROIMAGING STUDIES INVOLVING NONSOCIAL COGNITION
HIGH-RISK GROUPS

In total, 10 studies involved participants completing a nonemotional cognitive neuroimaging task. Of these, seven excluded high-risk individuals with Axis 1 diagnoses (Erk et al., 2014; Linke et al., 2012; Sepede et al., 2012; Singh et al., 2014; Whalley, Papmeyer, Romaniuk, et al., 2012; Whalley, Papmeyer, Sprooten, et al., 2012; Whalley et al., 2011). The remaining three studies conducted further analyses after subdividing high-risk participants into those who had a lifetime diagnosis of major depression compared to those who did not. During the SCWT, Pompei, Jogia, et al. (2011) found that high-risk participants with MDD and BD patients had hypoactivation in fronto-striatal regions compared to healthy controls and high-risk participants with no Axis 1 diagnoses. Using data from this study, Pompei, Dima, et al. (2011) conducted a PPI analysis that revealed that BD patients and high-risk participants with MDD had frontal-basal ganglia abnormalities compared to healthy controls and high-risk participants with no Axis 1 diagnoses. Increased connectivity was evident between dorsal and ventral prefrontal regions in high-risk participants with no Axis 1 diagnoses relative to the other groups. Lastly, Manelis et al. (2016) conducted their analyses again after excluding participants with psychopathology and those taking psychotropic medication, and their findings remained unchanged.

CONTROLS

For 9 out of 10 studies, current or lifetime diagnosis of an Axis 1 disorder were exclusion criteria for the healthy control groups. The final study found that their results remained unchanged when repeating analyses after removing participants with psychopathology and those taking psychotropic medication (Manelis et al., 2016).

NEUROIMAGING STUDIES INVOLVING EMOTIONAL PROCESSING

HIGH-RISK GROUPS

Five studies have examined emotional processing tasks while participants underwent an MRI. Of these, one excluded high-risk individuals that had Axis 1 diagnoses (Ladouceur et al., 2013). The four remaining studies whose high-risk group included those with Axis 1 diagnoses conducted further analyses. For instance, Lelli-Chiesa et al. (2010) found that in the right VLPFC, the Met158 allele was associated with greater activation and the Val158 allele was associated with lower activation in all family members with affective morbidity compared with relatives without a psychiatric diagnosis and healthy controls. The remaining three studies (Brotman et al., 2014; Manelis et al., 2015; Roberts et al., 2013) conducted their analyses again after excluding those with psychopathology and those currently taking psychotropic medication. The findings of all these studies remained the same.

CONTROLS

The majority (three out of five) of studies had control groups that comprised only those without an Axis 1 diagnosis. The remaining two studies (Manelis et al., 2015; Roberts et al., 2013) excluded participants who were currently depressed and currently taking psychotropic medication, and found that their results remained significant.

The presence of Axis 1 diagnoses in high-risk participants influenced behavioral scores during an AGN task (Bauer et al., 2015) and in the identification of facial emotional expressions (Brotman, Guyer, et al., 2008; Hanford et al., 2016). The presence of Axis 1 diagnoses in the high-risk group did not influence performance of nonemotional cognitive tasks. However, during the SCWT, brain signal differences were evident between high-risk subgroups in the absence of behavioral differences (Pompei, Dima, et al., 2011), thus making fMRI a promising tool in identifying subtle differences that may differ between those who are resilient vs. those who will go on to develop BD in high-risk samples. The use of operationalized criteria for defining risk and resilience in future studies would assist in improving our understanding of the complex changes observed in first-degree relatives.

Overall, the majority of control groups comprised only those without an Axis 1 diagnosis. The few studies that investigated non-BD psychopathology in control participants reported that the presence of Axis 1 diagnoses did not influence their significant findings. Results in all the studies that excluded high-risk participants with Axis 1 diagnoses need to be interpreted with caution, as the study design does not allow the distinction of differences that are attributed to underlying genetic risk from differences that are associated with comorbid non-BD psychopathology. In these studies that recruited a "supernormal" healthy control group without Axis I diagnoses, it is possible that differences between groups may have been amplified.

LONGITUDINAL STUDIES

Up to this point in the chapter, we have only reported cross-sectional findings. As cross-sectional studies of young high-risk cohorts come with limitations, longitudinal studies are needed to disentangle whether neurocognitive abnormalities reflect the early effects of familial vulnerability to mood disorders or rather if they emerge at illness onset, as well as whether abnormalities can predict future mood disorder onset. Some of the aforementioned high-risk cohorts are enrolled in longitudinal studies. In particular, the McIntosh/Whalley group ("Bipolar Family Study") has clinically examined individuals at familial risk of BD 2 years apart on three occasions (with a repeated neuropsychological battery after a 2-year follow-up interval); the Phillips group ("Bipolar Offspring Study") conducts clinical assessments 2 years apart as part of an ongoing study; and also in an ongoing study, the Mitchell group ("Bipolar Kids and Sibs Study") conducts annual clinical assessments (with a repeated neuropsychological and neuroimaging battery after a 2-year follow-up interval).

In young patients and offspring, Mourão-Miranda et al. (2012) employed an emotional face task with both implicit and explicit emotion recognition. No behavioral differences were evident, but Gaussian Process Classifiers (GPC) found that the morphed happy faces task differentiated the high-risk group from healthy controls. In contrast, the task using morphed fearful faces could not discriminate between the groups. Follow-up lasting 12–45 months revealed that GPC predictive probabilities were significantly more accurate for relatives who developed a mood or anxiety disorder compared to unaffected relatives who did not develop any psychiatric disorder. These results from the Phillips group suggest that a combination of neuroimaging and machine learning may help to predict those at risk of developing psychopathology versus those who are resilient.

Four longitudinal reports have been published from the Scottish "Bipolar Family Study." Whalley et al. (2013) reported that 20 previously unaffected relatives developed MDD, while one developed BD-I and another BD-II over a period of 2 years. Consistent with baseline reports in the earlier study, after 2 years Whalley et al. continued to find no significant behavioral difference across all groups during the Hayling Sentence Completion Test. However, when the difficulty of the task increased, relatives affected by MDD showed increased activation in the insula bilaterally and the inferior parietal cortex compared to healthy controls and high-risk subjects who remained well. During an emotional memory task that was undertaken at the second assessment, the high-risk participants who had developed MDD ($n=20$) demonstrated increased thalamic activation compared to high-risk subjects who were well ($n=61$). No behavioral differences were evident. Healthy high-risk individuals who became unwell after the second scanning assessment ($n=11$) showed increased activation of thalamus, insula, and anterior cingulate compared to those who remained well (Whalley et al., 2015). Implicit and explicit facial emotion processing tasks were also administered within the same scanning session that took place at the second assessment (Chan et al., 2016). This study found functional abnormalities of the ACC together with emotion recognition deficits in the explicit task, in high-risk individuals affected by MDD

($n=30$ (22 developed MDD between T1 and T2; 8 between T2 and T3)) compared to both controls and high-risk subjects who remained well. Papmeyer et al. (2015) used a neuropsychological test-battery that investigated differences and longitudinal changes in the domains of attentional processing, working memory, verbal learning and memory, and cognitive flexibility. This battery was administered at baseline and after 2 years in the McIntosh sample of unaffected young adults at high familial risk of mood disorders and healthy controls. Reduced long delay verbal memory and extradimensional set-shifting performance across both time points were found in the high-risk group that did not develop psychopathology relative to healthy controls. The high-risk group that went on to develop MDD ($n=20$, who did not have an MDD diagnosis at baseline) displayed decreased extradimensional set-shifting abilities across both time points as compared to the healthy control group only. No significant performance differences were evident between the two high-risk groups (Papmeyer et al., 2015). Multimodal longitudinal analyses are underway in the Mitchell sample (the largest cohort that has included genetics, neuropsychological testing, and neuroimaging).

In summary, these studies suggest the subtle neurocognitive abnormalities across time may constitute familial trait markers for vulnerability to BD. Future research in these cohorts will further enhance our understanding of developmental trajectories.

CONCLUSION

The neurocognitive profile in young unaffected relatives of BD patients is still not clarified. Even though some deficits are evident, overall, nonsocial neurocognitive functioning seems mainly to be spared. Impairments in visuospatial learning and memory consistency and processing speed appear as the most suitable possible neurocognitive candidate endophenotypes in youth. There is more compelling evidence that tasks involving an emotional component and high-level ToM tasks may represent potential trait markers for BD. When functional brain differences have been revealed in first-degree relatives, they have generally been evident in the absence of behavioral differences. Brain imaging may therefore be a more sensitive tool to investigate subtle differences that relate to neurocognitive alterations. In partial fulfillment of criteria for an endophenotype, some deficits were found in those at high genetic risk for BD, as well as in those with BD when compared with controls. Future studies should use larger samples, longitudinal designs, and a comprehensive neuropsychological approach that covers complex tests (e.g., higher-level ToM tasks) and incorporates multimodal techniques (e.g., neuroimaging in addition to behavioral performance scores) that are more sensitive to subtle cognitive differences.

REFERENCES

Arts, B., Jabben, N., Krabbendam, L., & Van Os, J. (2008). Meta-analyses of cognitive functioning in euthymic bipolar patients and their first-degree relatives. *Psychological Medicine*, *38*(6), 771–785.

Balanzá-Martínez, V., Rubio, C., Selva-Vera, G., Martinez-Aran, A., Sanchez-Moreno, J., Salazar-Fraile, J., et al. (2008). Neurocognitive endophenotypes (endophenocognitypes) from studies of relatives of bipolar disorder subjects: a systematic review. *Neuroscience & Biobehavioral Reviews, 32*(8), 1426–1438.

Bauer, I. E., Frazier, T. W., Meyer, T. D., Youngstrom, E., Zunta-Soares, G. B., & Soares, J. C. (2015). Affective processing in pediatric bipolar disorder and offspring of bipolar parents. *Journal of Child and Adolescent Psychopharmacology, 25*(9), 684–690.

Besnier, N., Richard, F., Zendjidjian, X., Kaladjian, A., Mazzola-Pomietto, P., Adida, M., et al. (2009). Stroop and emotional Stroop interference in unaffected relatives of patients with schizophrenic and bipolar disorders: distinct markers of vulnerability? *The World Journal of Biological Psychiatry, 10*(4 Pt 3), 809–818.

Bora, E., Yucel, M., & Pantelis, C. (2009). Cognitive endophenotypes of bipolar disorder: a meta-analysis of neuropsychological deficits in euthymic patients and their first-degree relatives. *Journal of Affective Disorders, 113*(1), 1–20.

Brand, J. G., Goldberg, T. E., Gunawardane, N., Gopin, C. B., Powers, R. L., Malhotra, A. K., et al. (2012). Emotional bias in unaffected siblings of patients with bipolar I disorder. *Journal of Affective Disorders, 136*(3), 1053–1058.

Brotman, M. A., Deveney, C. M., Thomas, L. A., Hinton, K. E., Yi, J. Y., Pine, D. S., et al. (2014). Parametric modulation of neural activity during face emotion processing in unaffected youth at familial risk for bipolar disorder. *Bipolar Disorders, 16*(7), 756–763.

Brotman, M. A., Guyer, A. E., Lawson, E. S., Horsey, S. E., Rich, B. A., Dickstein, D. P., et al. (2008). Facial emotion labeling deficits in children and adolescents at risk for bipolar disorder. *American Journal of Psychiatry, 165*(3), 385–389.

Brotman, M. A., Rooney, M. H., Skup, M., Pine, D. S., & Leibenluft, E. (2009). Increased intra-subject variability in response time in youths with bipolar disorder and at-risk family members. *Journal of the American Academy of Child & Adolescent Psychiatry, 48*(6), 628–635.

Brotman, M. A., Skup, M., Rich, B. A., Blair, K. S., Pine, D. S., Blair, J. R., et al. (2008). Risk for bipolar disorder is associated with face-processing deficits across emotions. *Journal of the American Academy of Child & Adolescent Psychiatry, 47*(12), 1455–1461.

Calafiore, D., Rossell, S. L., & Van Rheenen, T. E. (2017). Cognitive abilities in first-degree relatives of individuals with bipolar disorder. *Journal of Affective Disorders.*

Cardenas, S. A., Kassem, L., Brotman, M. A., Leibenluft, E., & McMahon, F. J. (2016). Neurocognitive functioning in euthymic patients with bipolar disorder and unaffected relatives: a review of the literature. *Neuroscience & Biobehavioral Reviews, 69*, 193–215.

Chan, S. W. Y., Sussmann, J. E., Romaniuk, L., Stewart, T., Lawrie, S. M., Hall, J., et al. (2016). Deactivation in anterior cingulate cortex during facial processing in young individuals with high familial risk and early development of depression: fMRI findings from the Scottish bipolar family study. *Journal of Child Psychology and Psychiatry, 57*(11), 1277–1286.

Cullen, B., Ward, J., Graham, N. A., Deary, I. J., Pell, J. P., Smith, D. J., et al. (2016). Prevalence and correlates of cognitive impairment in euthymic adults with bipolar disorder: a systematic review. *Journal of Affective Disorders, 205*, 165–181.

de Brito Ferreira Fernandes, F., Gigante, A. D., Berutti, M., Amaral, J. A., de Almeida, K. M., de Almeida Rocca, C. C., et al. (2016). Facial emotion recognition in euthymic patients with bipolar disorder and their unaffected first-degree relatives. *Comprehensive Psychiatry, 68*, 18–23.

de la Serna, E., Vila, M., Sanchez-Gistau, V., Moreno, D., Romero, S., Sugranyes, G., et al. (2016). Neuropsychological characteristics of child and adolescent offspring of patients with bipolar disorder. *Progress in Neuro-Psychopharmacology & Biological Psychiatry, 65*, 54–59.

Deveci, E., Ozan, E., Kirpinar, I., Oral, M., Dalogu, A. G., Aydin, N., et al. (2013). Neurocognitive functioning in young high-risk offspring having a parent with bipolar I disorder. *Turkish Journal of Medical Sciences, 43*(1), 110–117.

Doyle, A. E., Wozniak, J., Wilens, T. E., Henin, A., Seidman, L. J., Petty, C., et al. (2009). Neurocognitive impairment in unaffected siblings of youth with bipolar disorder. *Psychological Medicine, 39*(8), 1253–1263.

Duffy, A., Alda, M., Hajek, T., Sherry, S. B., & Grof, P. (2010). Early stages in the development of bipolar disorder. *Journal of Affective Disorders, 121*(1-2), 127–135.

Erk, S., Meyer-Lindenberg, A., Schmierer, P., Mohnke, S., Grimm, O., Garbusow, M., et al. (2014). Hippocampal and frontolimbic function as intermediate phenotype for psychosis: evidence from healthy relatives and a common risk variant in CACNA1C. *Biological Psychiatry, 76*(6), 466–475.

Gottesman, I. I., & Gould, T. D. (2003). The endophenotype concept in psychiatry: etymology and strategic intentions. *American Journal of Psychiatry, 160*(4), 636–645.

Grady, C. L. (1998). Brain imaging and age-related changes in cognition. *Experimental Gerontology, 33*(7), 661–673.

Hanford, L. C., Sassi, R. B., & Hall, G. B. (2016). Accuracy of emotion labeling in children of parents diagnosed with bipolar disorder. *Journal of Affective Disorders, 194*, 226–233.

Johnson, J. G., Cohen, P., & Brook, J. (2000). Associations between bipolar disorder and other psychiatric disorders during adolescence and early adulthood: a community-based longitudinal investigation. *American Journal of Psychiatry, 157*(10), 1679–1681.

Ladouceur, C. D., Diwadkar, V. A., White, R., Bass, J., Birmaher, B., Axelson, D. A., et al. (2013). Fronto-limbic function in unaffected offspring at familial risk for bipolar disorder during an emotional working memory paradigm. *Developmental Cognitive Neuroscience, 5*, 185–196.

Lau, P., Hawes, D. J., Hunt, C., Frankland, A., Roberts, G., & Mitchell, P. B. (2017). Prevalence of psychopathology in bipolar high-risk offspring and siblings: a meta-analysis. *European Child & Adolescent Psychiatry*.

Lee, M.-S., Anumagalla, P., Talluri, P., & Pavuluri, M. N. (2014). Meta-analyses of developing brain function in high-risk and emerged bipolar disorder. *Frontiers in Psychiatry, 5*.

Lelli-Chiesa, G., Kempton, M. J., Jogia, J., Tatarelli, R., Girardi, P., Powell, J., et al. (2010). The impact of the Val158Met catechol-O-methyltransferase genotype on neural correlates of sad facial affect processing in patients with bipolar disorder and their relatives. *Psychological Medicine, 41*(4), 779–788.

Lewis, M. (2012). *Social cognition and the acquisition of self*. Springer Science and Business Media.

Lichtenstein, P., Yip, B. H., Björk, C., Pawitan, Y., Cannon, T. D., Sullivan, P. F., et al. (2009). Common genetic determinants of schizophrenia and bipolar disorder in Swedish families: a population-based study. *The Lancet, 373*(9659), 234–239.

Lin, K., Lu, R., Chen, K., Li, T., Lu, W., Kong, J., et al. (2017). Differences in cognitive deficits in individuals with subthreshold syndromes with and without family history of bipolar disorder. *Journal of Psychiatric Research, 91*, 177–183.

Linke, J., King, A. V., Rietschel, M., Strohmaier, J., Hennerici, M., Gass, A., et al. (2012). Increased medial orbitofrontal and amygdala activation: evidence for a systems-level endophenotype of bipolar I disorder. *American Journal of Psychiatry, 169*(3), 316–325.

Manelis, A., Ladouceur, C. D., Graur, S., Monk, K., Bonar, L. K., Hickey, M. B., et al. (2015). Altered amygdala-prefrontal response to facial emotion in offspring of parents with bipolar disorder. *Brain, 138*(9), 2777–2790.

Manelis, A., Ladouceur, C. D., Graur, S., Monk, K., Bonar, L. K., Hickey, M. B., et al. (2016). Altered functioning of reward circuitry in youth offspring of parents with bipolar disorder. *Psychological Medicine, 46*(1), 197–208.

McCormack, C., Green, M. J., Rowland, J. E., Roberts, G., Frankland, A., Hadzi-Pavlovic, D., et al. (2016). Neuropsychological and social cognitive function in young people at genetic risk of bipolar disorder. *Psychological Medicine, 46*(4), 745–758.

Merikangas, K., Jin, R., He, J.-P., Kessler, R. C., Lee, S., Sampson, N. A., et al. (2011). Prevalence and correlates of bipolar spectrum disorder in the world mental health survey initiative. *Archives of General Psychiatry, 68*(3), 241–251.

Merikangas, K., & Yu, K. (2002). Genetic epidemiology of bipolar disorder. *Clinical Neuroscience Research, 2*(3), 127–141.

Mourão-Miranda, J., Oliveira, L., Ladouceur, C. D., Marquand, A., Brammer, M., Birmaher, B., et al. (2012). Pattern recognition and functional neuroimaging help to discriminate healthy adolescents at risk for mood disorders from low risk adolescents. *PLoS ONE, 7*(2), e29482.

Özerdem, A., Ceylan, D., & Can, G. (2016). Neurobiology of risk for bipolar disorder. *Current Treatment Options in Psychiatry, 3*(4), 315–329.

Papmeyer, M., Sussmann, J. E., Hall, J., McKirdy, J., Peel, A., Macdonald, A., et al. (2015). Neurocognition in individuals at high familial risk of mood disorders with or without subsequent onset of depression. *Psychological Medicine, 45*(15), 3317–3327.

Phillips, M. L., Drevets, W. C., Rauch, S. L., & Lane, R. (2003). Neurobiology of emotion perception I: the neural basis of normal emotion perception. *Biological Psychiatry, 54*(5), 504–514.

Piguet, C., Fodoulian, L., Aubry, J.-M., Vuilleumier, P., & Houenou, J. (2015). Bipolar disorder: functional neuroimaging markers in relatives. *Neuroscience & Biobehavioral Reviews, 57*, 284–296.

Pompei, F., Dima, D., Rubia, K., Kumari, V., & Frangou, S. (2011). Dissociable functional connectivity changes during the Stroop task relating to risk, resilience and disease expression in bipolar disorder. *NeuroImage, 57*(2), 576–582.

Pompei, F., Jogia, J., Tatarelli, R., Girardi, P., Rubia, K., Kumari, V., et al. (2011). Familial and disease specific abnormalities in the neural correlates of the Stroop task in bipolar disorder. *NeuroImage, 56*(3), 1677–1684.

Reynolds, M. T., Van Rheenen, T. E., & Rossell, S. L. (2014). Theory of mind in first degree relatives of individuals with bipolar disorder. *Psychiatry Research, 219*(2), 400–402.

Roberts, G., Green, M. J., Breakspear, M., McCormack, C., Frankland, A., Wright, A., et al. (2013). Reduced inferior frontal gyrus activation during response inhibition to emotional stimuli in youth at high risk of bipolar disorder. *Biological Psychiatry, 74*(1), 55–61.

Ruocco, A. C., Reilly, J. L., Rubin, L. H., Daros, A. R., Gershon, E. S., Tamminga, C. A., et al. (2014). Emotion recognition deficits in schizophrenia-spectrum disorders and psychotic bipolar disorder: Findings from the bipolar-schizophrenia network on intermediate phenotypes (B-SNIP) study. *Schizophrenia Research, 158*(1), 105–112.

Samame, C., Martino, D. J., & Strejilevich, S. A. (2015). An individual task meta-analysis of social cognition in euthymic bipolar disorders. *Journal of Affective Disorders, 173*, 146–153.

Santos, J. M., Pousa, E., Soto, E., Comes, A., Roura, P., Arrufat, F. X., et al. (2017). Theory of mind in euthymic bipolar patients and first-degree relatives. *The Journal of Nervous and Mental Disease, 205*(3), 207–212.

Seidel, E.-M., Habel, U., Finkelmeyer, A., Hasmann, A., Dobmeier, M., & Derntl, B. (2012). Risk or resilience? Empathic abilities in patients with bipolar disorders and their first-degree relatives. *Journal of Psychiatric Research, 46*(3), 382–388.

Sepede, G., De Berardis, D., Campanella, D., Perrucci, M. G., Ferretti, A., Serroni, N., et al. (2012). Impaired sustained attention in euthymic bipolar disorder patients and non-affected relatives: an fMRI study. *Bipolar Disorders*, *14*(7), 764–779.

Singh, M. K., Kelley, R. G., Howe, M. E., Reiss, A. L., Gotlib, I. H., & Chang, K. D. (2014). Reward processing in healthy offspring of parents with bipolar disorder. *JAMA Psychiatry*, *71*(10), 1148–1156.

Sprooten, E., Sussmann, J. E., Clugston, A., Peel, A., McKirdy, J., Moorhead, T. W. J., et al. (2011). White matter integrity in individuals at high genetic risk of bipolar disorder. *Biological Psychiatry*, *70*(4), 350–356.

Vierck, E., Porter, R. J., & Joyce, P. R. (2015). Facial recognition deficits as a potential endophenotype in bipolar disorder. *Psychiatry Research*, *230*(1), 102–107.

Wang, Y.-G., Roberts, D. L., Liang, Y., Shi, J.-F., & Wang, K. (2015). Theory-of-mind understanding and theory-of-mind use in unaffected first-degree relatives of schizophrenia and bipolar disorder. *Psychiatry Research*, *230*(2), 735–737.

Whalley, H. C., Papmeyer, M., Romaniuk, L., Johnstone, E. C., Hall, J., Lawrie, S. M., et al. (2012). Effect of variation in diacylglycerol kinase eta (DGKH) gene on brain function in a cohort at familial risk of bipolar disorder. *Neuropsychopharmacology*, *37*(4), 919–928.

Whalley, H., Papmeyer, M., Sprooten, E., Romaniuk, L., Blackwood, D., Glahn, D., et al. (2012). The influence of polygenic risk for bipolar disorder on neural activation assessed using fMRI. *Translational Psychiatry*, *2*(7), e130.

Whalley, H. C., Sussmann, J. E., Chakirova, G., Mukerjee, P., Peel, A., McKirdy, J., et al. (2011). The neural basis of familial risk and temperamental variation in individuals at high risk of bipolar disorder. *Biological Psychiatry*, *70*(4), 343–349.

Whalley, H. C., Sussmann, J. E., Romaniuk, L., Stewart, T., Kielty, S., Lawrie, S. M., et al. (2015). Dysfunction of emotional brain systems in individuals at high risk of mood disorder with depression and predictive features prior to illness. *Psychological Medicine*, *45*(6), 1207–1218.

Whalley, H. C., Sussmann, J. E., Romaniuk, L., Stewart, T., Papmeyer, M., Sprooten, E., et al. (2013). Prediction of depression in individuals at high familial risk of mood disorders using functional magnetic resonance imaging. *PLoS ONE*, *8*(3). e57357.

Whitney, J., Howe, M., Shoemaker, V., Li, S., Sanders, E. M., Dijamco, C., et al. (2013). Socioemotional processing and functioning of youth at high risk for bipolar disorder. *Journal of Affective Disorders*, *148*(1), 112–117.

Cognitive and neural basis of hypomania: Perspectives for early detection of bipolar disorder

10

Giuseppe Delvecchio*, Alessandro Pigoni*, A.C. Altamura*, Paolo Brambilla*,†

Department of Neurosciences and Mental Health, Fondazione IRCCS Ca' Granda Ospedale Maggiore Policlinico, University of Milan, Milan, Italy* Translational Psychiatry Program, Department of Psychiatry and Behavioral Sciences, University of Texas Health Sciences Center at Houston, Houston, TX, United States†

CHAPTER OUTLINE

INTRODUCTION

Mood disorders can be classified along a spectrum defined by the extent and severity of mood alteration, from unipolar to bipolar depression. Individuals with unipolar disorder are characterized by the presence of only depressive episodes, whereas those with bipolar disorder (BD) alternate manic/hypomanic episodes to depressive episodes, alone or in combination. Specifically, BD is a recurrent chronic disorder characterized by fluctuations from highs and lows of mood, thinking, and activity, which may vary in terms of severity, duration, and frequency (Grande, Berk, Birmaher, & Vieta, 2016). Moreover, it can be classified as BD type I (BD I), which is diagnosed after one full-blown manic episode, and as BD type II (BD II), which is characterized by depressive and hypomanic episodes, the latter being less severe manifestations of psychomotor activation compared to mania (DSM-5, American Psychiatric Association, 2013).

Although in the last 2 decades researchers have started to drive their attention toward the study of hypomania (Merikangas & Lamers, 2012), the idea that BD II is a less severe and destabilizing form of BD I, with lower morbidity and a better prognosis, is still present, hence the identification of BD II as "soft bipolar disorder" (Benazzi, 2007). However, BD II can still have a serious impact on the way an individual would normally live her/his life. Indeed, this disorder can be associated with cognitive limitations, which are often underrated, and may heavily influence the functionality and the quality of life (Altamura et al., 2016; Vinberg, Mikkelsen, Kirkegaard, Christensen, & Kessing, 2017).

Moreover, BD II is often difficult to diagnose accurately as it can be confused with recurrent unipolar disorders (Bond, Noronha, Kauer-Sant'Anna, Lam, & Yatham, 2008) or personality disorders (Renaud, Corbalan, & Beaulieu, 2012). As previously mentioned, the core feature of BD II is hypomania, which can be described as persistent elevated mood, increased activity, overthinking, and, not rarely, increased functionality. Such fluctuations are common in life and might be experienced in nonclinical populations, particularly when facing stressful events and often regarded as favorable (Brand et al., 2015). Compared to healthy subjects, patients experiencing hypomania do not differ substantially in quality of life or distress, but they often show socially positive and advantageous facets with a higher likelihood to get married and to earn more money (Brand et al., 2015). The "bright" side of hypomania is negatively associated with peer problems, especially because hypomanic symptoms are often considered and perceived to be favorable for social interactions (Angst et al., 2005). However, as first described by Hantouche, Angst, and Akiskal (2003), besides this "sunny" side, hypomania could also present with a "dark" side, associated with socially negative aspects, peer problems, disinhibited behaviors, and irritability. Therefore, although often presenting with "positive" features, hypomania also brings "negative" aspects characterized by increased anxiety, irritability and perceived stress (Brand, Gerber, Pühse, & Holsboer-Trachsler, 2011). Nonetheless, although the "dark" aspects of hypomania were first described in 2003 by Hantouche and colleagues, they are still often underreported (Brand et al., 2011).

Furthermore, the experience of hypomania is linked to a proneness to develop further mood disorders even in nonclinical populations, although subjects tend to underestimate it and regard it as a positive state (Seal, Mansell, & Mannion, 2008; Fletcher, Parker, Paterson, and Synnott, 2013). From a clinical point of view, although according to DSM-5 the symptoms associated to manic or hypomanic phases can overlap (American Psychiatric Association, 2013; Benazzi, 2007), they also differ to some extent. Indeed, while overactivity, reduced sleep, and elevated, expansive, or irritated mood are present in both phases, psychosis and hospitalization are present only in mania, while everyday functioning is only mildly impaired in hypomania. Moreover, while both can lead to mental overactivity, mania is associated with flight of ideas (speedy, disconnected thinking) plus rapid, loquacious, and nonstop talking, whereas in hypomania the severity of thinking disorder can vary between creative thinking, crowded or racing thoughts, and, only rarely, flight of ideas. Similarly, evaluation of actions and decision making are also different: hypomania can include moderate risk-taking behaviors, while mania often leads to severe risk-taking behaviors (Fletcher, Parker, Paterson, et al., 2013).

Given all these factors, hypomania may be regarded as part of a spectrum of mood swings ranging from depression to mania (Benazzi, 2007). Individuals with BD often report hypomanic episodes early in their life, before the clear clinical onset. Therefore, assessing previous incidents of hypomania requires attention, given the fact that underrecognition of hypomanic features is common. Patients often do not recognize their mood accurately and therefore they might perceive hypomania as favorable and desirable (Lee et al., 2016).

Interestingly, hypomania has also been found to be associated with selective neurocognitive and neurobiological deficits (Aminoff et al., 2013; Brooks 3rd, Bearden, Hoblyn, Woodard, & Ketter, 2010; Martino, Igoa, Marengo, Scápola, & Strejilevich, 2011; Romero et al., 2016). Indeed, from a biological prospective, the majority of magnetic resonance imaging (MRI) studies identified specific gray matter alterations in frontal and prefrontal regions as well as white matter abnormalities in traits connecting prefrontal regions and in corpus callosum. Additionally, functional and metabolic investigations reported that hypomania is associated with brain anomalies in areas accounting for emotion regulation and reward, such as the amygdala and insula. Finally, from a cognitive point of view, selective deficits have been observed in several abilities, including executive functions, sustained attention, and verbal memory. These features may negative influence clinical outcome and global functioning in both clinical and nonclinical populations. Therefore, the study of both neurocognitive and morphological characteristics of hypomania may greatly help in further developing preventive strategies and early interventions, along with a better knowledge of mood disorders. This chapter aimed at providing an overview of cognitive and morpho-functional correlates of hypomania, summarizing and debating the current available literature.

COGNITIVE BASIS OF HYPOMANIA

Although several behavioral studies have tried to explore the cognitive profile associated with hypomania, the results are far from conclusive (Brooks 3rd et al., 2010; Martino et al., 2011; Pålsson et al., 2013; Sparding et al., 2015), mainly due to the small samples employed and to the evaluation of a wide range of cognitive functions. For these reasons, a clear picture of the neuropsychological deficits linked to hypomania is still difficult to see. Nonetheless, we will try to delineate an overview of the studies exploring cognition in hypomania in both nonclinical and clinical populations, as reported in full detail in Table 1.

NONCLINICAL POPULATIONS

Hypomania is a psychological condition that can be experienced not only by individuals affected by mood disorders but also by nonclinical, healthy people, who can express mild hypomanic features often overlapping with hyperthymic temperament (Brand et al., 2015; Harada et al., 2013). Therefore, hypomania might be regarded

Table 1 Studies investigating cognitive deficits and the clinical profile of hypomania in clinical and nonclinical populations

Authors	Sample (M/F)	Diagnosis	Psychopathological measures	Questionnaires	Main results
Dempsey et al. (2011)	353 (76/277)	Nonclinical population	HPS CESD ISS	HIQ IDQ RPA RRS	Hypomania associated with positive cognitive styles (appraisal) and increased rumination
Kirkland et al. (2015)	989	Nonclinical population	ASRM BDI HPS PANAS	BFAS SHS	Hypomania risk predicted by extraversion neuroticism measures. Hypomania associated with increased risk of developing BD and manic episodes at follow up
Brand et al. (2011)	862 (223/639)	Nonclinical population	HCL-32 Von Zerssen's Depression Scale	ISI FEPS SAQ QoL SWL PSS SVF CEI IPC	"Dark side" hypomania associated with more depressive symptoms, sleep disturbances, perceived stress, and negative coping strategies "Bright side" associated with lower stress scores, higher levels of exploration, self-efficacy, and physical activity
Mason et al. (2012)	32	Nonclinical population	HPS DAS	TCIP	Hypomania associated with more immediate choices than the control group Hypomania evidenced greater differentiation between delayed and immediate outcomes in early attention-sensitive and later reward-sensitive components
Lardi Robyn et al. (2012)	79 (40/39)	Nonclinical population	MDQ PANAS	SDM SDFP	Hypomanic symptoms associated with enhanced retrieval of memories describing positive relationships and to reduce future projections Hypomanic individuals described more recent events and produced self-defining memories

Study	N	Population			Findings
Seal et al. (2008)	12	Nonclinical population	MDQ DASS ASRM WASA		Hypomania considered as a positive experience causing no problems
Dodd et al. (2011)	68	Nonclinical population	GBI	HAPPI ISAT	Hypomania associated with positive appraisals of internal states and with more extreme ratings in the appraisal task
Hosang et al. (2017)	1440 (610/830)	Nonclinical pediatric population	HCL-16	SMFQ; Child Anxiety Sensitivity Index; SPEQ; PLISK-Q; SDQ; BMSLSS; PSQI	Hypomania described as active-elated and irritable/risk-taking Hypomanic symptoms were significantly correlated with psychotic-like experiences, internalizing and externalizing problems, and reduced life satisfaction relative to the active-elated dimension Adolescents at "high risk" for bipolar disorder reported more psychopathology relative to the comparison group
Stange et al. (2013)	105 (41/64)	65 BD I 40 BD II		ASQ	Extreme attributions predicted a higher likelihood of transition of phase
Malhi, Ivanovski, Hadzi-Pavlovic, et al. (2007)	25	25 BD I 25 HC	HAM-D MADRS YMRS	TMT COWAT RAVLT SDMT	Hypomania impaired in recognition, recall and at generating semantic words compared to HC
Dickstein et al. (2016)	11 (7/4)	81 BD I 11 BD II 28 BD-NOS 55 HC Pediatric population	K-SADS (DRS) K-SADS (MRS)	CANTAB	BD children and adolescents showed impairments in sustained attention and information processing for emotionally valenced words

Continued

Table 1 Studies investigating cognitive deficits and the clinical profile of hypomania in clinical and nonclinical populations—cont'd

Authors	Sample (M/F)	Diagnosis	Psychopathological measures	Questionnaires	Main results
Fletcher et al. (2013a)	13 (6/7)	13 BD II	MINI	MDQ ISS	Hypomania described as an enjoyable state, preferable to depression, but characterized by interpersonal difficulties The impact of chronicity referred to shifts in coping strategies over time, moving from maladaptive to adaptive behavioral responses
Lex et al. (2011)	15 (11/4)	15 hypomanic BD I 26 remitted BD I 21 HC	BDI BRMES BRMAS	DAS ASQ EST	Hypomanic patients were significantly slower for all word conditions and they showed learning deficits
Scott and Pope (2003)	77 (32/45)	16 MDD; 26 remitted BD; 38 depressed BD; 13 hypomanic BD	ISS Well-being ISS Activation score	SEQ DAS SAS	Hypomanic patients reported higher levels of dysfunctional beliefs than individuals in remission, but lower than depressed subjects Hypomanic subjects recorded the highest levels of negative as well as positive self-esteem
Fletcher et al. (2013b)	417	94 BD I 114 BD II 109 MDD 100 HC	MINI ISS	BC RPA RSQ CIPM CERQ	BD I and BD II were more likely to ruminate about positive affect and engage in risk taking BD II were less likely to seek support when faced with stress and to engage in strategies to downregulate hypomania

Study	N (sex)	Sample	Clinical measures	Cognitive tests	Results
Sparding et al. (2015)	44 (20/24)	64 BD I 44 BD II 86 HC	YMRS MADRS GAF MINI AUDIT DUDIT CGI	The Claeson-Dahl Verbal Learning and Retention Test D-KEFS CPT RCFT	Both BD I and BD II were cognitively impaired compared to HC No differences between BD I and BD II
Schenkel et al. (2012)	79	27 BD I 19 BD II 33 HC Pediatric population	YMRS CDRS-R WASI WASH-U-KSADS	TMT Digit Span (WMS) CVLT-C CPT PCET VOLT SST COWAT SSP	BD I performed worse than HC in attention, executive function, working memory, visual memory, and verbal learning and memory BD I performed worse compared to BD II on all cognitive domains, with the exception of working memory BD II performed worse on verbal learning and memory compared to HC
Schenkel et al. (2014)	50 (34/16)	17 BD I 8 BD II 25 HC Pediatric population	YMRS CDRS-R WASI KSADS-PL SCQ-LF	MET CEPTT IPR	BD I reported greater peer difficulties compared to HC and performed more poorly on the MET and the cognitive condition of the CEPTT, but did not differ significantly on the emotional condition No significant group differences between BD II and HC
Pálsson et al. (2013)	196 (87/109)	67 BD I 43 BD II 86 HC	YMRS MADRS CGI	D-KEFS; Claeson-Dahl learning and memory test RCFT	Both BD I and BD II were cognitively impaired compared to HC No significant group differences between BD I and BD II

Continued

Table 1 Studies investigating cognitive deficits and the clinical profile of hypomania in clinical and nonclinical populations—cont'd

Authors	Sample (M/F)	Diagnosis	Psychopathological measures	Questionnaires	Main results
Romero et al. (2016)	92 (33/59)	46 BD II 46 HC	HAM-D YMRS	TEMPS-A RAVLT ROCF SCWT WCST Digit Span TMT Inverse Digit Span FAS WAIS-III Vocabulary Subtest	Correlations between hyperthymic BD and verbal memory, attention and verbal fluency
Bourne et al. (2015)	279 (89/190)	183 BD I 96 BD II	QIDS ASRM	RAVLT	BD I patients were significantly more impaired compared to BD II patients on all five verbal learning and memory tasks
Cotrena, Branco, Shansis, and Fonseca (2016)	203 (73/130)	37 BD I 35 BD II 45 MDD 89 HC	YMRS HAM-D	TMT SCWT WCST	Patients with BD I displayed more widespread cognitive impairment than other groups Decision-making ability and attentional control were able to distinguish between patients with BD I and BD II
Brooks 3rd et al. (2010)	27 (24/3)	10 BD I 6 BD II 11 HC	YMRS MADRS	CPT	BD I and BD II committed significant more omission and commission errors than HC

Study	Sample	Clinical measures	Cognitive measures	Findings
Aminoff et al. (2013)	199 (82/117)	128 BD I 71 BD II	YMRS PANSS IDS-C PAS	CVLT-II; N-back DSST D-KEFS	Both groups associated with impairments in verbal memory, with elevated presenting polarity explaining more of the variance in this cognitive domain BD I and history of psychosis specifically related to impairment in semantic fluency
Torrent et al. (2006)	106 (43/63)	38 BD I 33 BD II 35 HC	YMRS HAM-D GAF	WCST SCWT FAS Digit Span TMT CVLT	Both BD I and BD II patients showed significant deficits in most cognitive tasks. BD II had an intermediate level of performance between BD I and HC group in verbal memory and executive functions
Martino et al. (2011)	126 (70/56)	48 BD I 39 BD II 39 HC	YMRS HAM-D GAF	Forward Digit Span TMT Memory Battery of Signoret WCST	BD II patients performed worse n psychomotor speed, verbal memory, and executive functioning compared to HC No differences between BD I and II BD II cognitive impairments have a negative influence on functional outcome
Fletcher et al. (2013b)	93 (34/59)	93 BD II	QIDS-SR: 14.4		Risk-taking behaviors associated with hypomania, leading to interpersonal conflict, substantial financial burden, and feelings of guilt, shame, and remorse Less than one-fifth of participants agreed that hypomania should be treated because of the associated risks
Brooks 3rd et al. (2009)	27 (24/3)	10 BD I 6 BD II 11 HC	YMRS MADRS	CVLC	Patients with bipolar disorder, relative to healthy comparison subjects, had significantly poorer delayed free verbal recall

Continued

Table 1 Studies investigating cognitive deficits and the clinical profile of hypomania in clinical and nonclinical populations—cont'd

Authors	Sample (M/F)	Diagnosis	Psychopathological measures	Questionnaires	Main results
Li et al. (2012)	51 (17/34)	17 BD I 17 BD II 17 HC	YMRS HAM-D	Tests for Attentional Performance Go/No-Go Word List Test Face Memory Test WCST	Executive function was significantly worse in BD I patients than in BD II

BD, Bipolar Disorder; BD-NOS, Bipolar Disorder Non-Other Specified; HC, Healthy Controls; YMRS, Young Mania Rating Scale; MADRS, Montgomery-Asberg Depression Rating Scale; HAM-D, Hamilton Rating Scale for Depression; MINI, Mini-International Neuropsychiatric Interview; MDQ, Mood Disorder Questionnaire; HPS, Hypomanic Personality Scale; ISI, Insomnia Severity Index; FEPS, Dysfunctional sleep-related thoughts; QoL, Quality of life; PSS, Perceived Stress Scale; CEI, Curiosity and Exploration Inventory; IPC, Internal, Powerful others and Chance Scale; TMT, Trail Making Test; COWAIT, Controlled Oral Word Association Test; RAVLT, Rey Auditory Verbal Learning Test; SDMT, Symbol Digit Modalities Test; CANTAB, Cambridge Neuropsychological Automated Testing Battery; BDI, Beck Depression Inventory; BRMAS, Bech Rafaelsen Melancholia Scale; the Bech Rafaelsen Mania Scale; ASQ, Attributional Style Questionnaire; EST, Emotional Stroop Test; SEQ, Rosenberg Self-Esteem Questionnaire; DAS, Dysfunctional Attitudes Scale; SAS, Sociotropy Autonomy Scale; GBI, General Behavior Inventory; CT, semistructured interview for cyclothymic temperament; D-KEFS, Delis-Kaplan Executive Function System; CPT, Continuous Performance Test; RCFT, Rey Complex Figure Test; CVLT, California Verbal Learning Test; CVLC-C, California Verbal Learning Test-Child Version; CDRS, Child Depression Rating Scale; MET, Mind in the Eyes Task; CEPTT, Cognitive and Emotional Perspective Taking Task; IPR, Index of Peer Relations; FAS, Verbal Fluency Task; WAIS, Wechsler Adult Intelligence Scale; TEMPS-A, Temperament Evaluation of Memphis, Pisa, Paris and San Diego auto questionnaire version; QIDS, Quick Inventory of Depressive Symptomatology; ASRM, Altman Self-Rating Mania scale; WCST, Wisconsin Card Sorting Test; SCWT, Stroop Color-Word Interference test; HCL-32, Hypomania Check List-32; PCET, Conditional Exclusion Test; VOLT, Visual Object Learning Test; SSP, Spatial Span Test; ROCF, Rey-Osterrieth Complex Figure; CPT, Continuous Performance Test; DSST, Digit Symbol Substitution Test; RPA, Responses to Positive Affect questionnaire; CIPM, Coping Inventory for Prodromes of Mania; CERQ, Cognitive Emotion Regulation Questionnaire; HAPPI, Hypomanic Attitudes and Positive Predictions Inventory; ISAT, Internal State Appraisals Task; HIQ, Hypomania Interpretations Questionnaire; Responses to Positive Affect Scale; RRS, Ruminative Responses Scale; PANAS, Positive and Negative Affect Schedule; BFAS, Big Five Aspects Scale; SHS, Subjective Happiness Scale; CESD, Center for Epidemiological Studies Depression Scale; IDQ, Interpretations of Depression Questionnaire; ISS, Internal States Scale; TCIP, Two Choice Impulsivity Paradigm; SAQ, Somatosensory Amplification Questionnaire; SVF, Coping with stress questionnaire; SWL, Satisfaction with life; SDM, Self-defining memory; SDFP, self-defining future projection tasks; K-SADS (DRS), Schedule for Affective Disorders and Schizophrenia-depression rating scale; K-SADS (MRS), Schedule for Affective Disorders and Schizophrenia—Manic rating scale; DASS, Depression, Anxiety, and Stress scale; WASA, Work and Social Adjustment scale; BC, Brief Cope; RSQ, Response Styles Questionnaire; AUDIT, Alcohol Use Disorders Identification Test; DUDIT, Drug Use Disorders Identification Test; CGI, Clinical Global Impression; WASI, Wechsler Abbreviated Scale of Intelligence; CDRS, Children's depression rating scale; WASH-U-KSADS, Washington University St. Louis Kiddie Schedule for Affective Disorders and Schizophrenia; WMS, Weschler Memory Scale; SST, Set Shifting Test; KSADS-PL, Schedule for Affective Disorders and Schizophrenia-Present and Lifetime Version; SCQ-LF, Social Communication Questionnaire-Lifetime Version; PANSS, Positive and Negative Symptom Scale; IDS-C, Depressive Symptoms-Clinician rated; PAS, Premorbid Adjustment Scale; GAF, Global Assessment of Functioning; HCL-16, Hypomania Checklist-16; SMFQ, Short version of the Moods and Feelings questionnaire; SPEQ, Psychotic Experiences Questionnaire; PLIKS-Q, Psychotic-Like Symptoms Questionnaire; SDQ, Strengths and Difficulties Questionnaire; PSQI, Pittsburgh Sleep Quality Index; BMSLSS, Brief Multidimensional Students Life Satisfaction Scale; CASI, Child Anxiety Sensitivity Index; QIDS-SR, Quick Inventory of Depressive Symptomatology-Self Report.

as a mental state, yet elusive, between heath and disease. Indeed, many studies have been conducted on nonclinical populations regarding hyperthymic temperament and mild hypomanic features (Dempsey, Gooding, & Jones, 2011; Kirkland, Gruber, & Cunningham, 2015; Fletcher, Parker, Paterson, et al., 2013). However, most of these studies did not assess cognitive functions directly, but only explored functional and temperamental aspects through specific questionnaires.

From a functional point of view, nonclinical populations generally present a double-sided cognitive style, reflecting the observed sides of hypomania, namely "sunny" and "dark" hypomania. The majority of studies reported a positive appraisal of internal states and behaviors along with positive cognitive styles, extraversion and assertiveness, and positive relationships in hypomanic individuals (Dempsey et al., 2011; Dodd, Mansell, Morrison, & Tai, 2011; Kirkland et al., 2015). On the other hand, many researchers also reported a negative side of hypomania, suggesting that a subsample of subjects with hypomania experienced maladaptive behavioral responses, higher irritability, and more risk-taking behaviors including excessive alcohol or drug use, dangerous driving, and endangering sexual activities (Dempsey et al., 2011; Fletcher, Parker, Paterson, et al., 2013; Kirkland et al., 2015).

Two large studies in hypomanic nonclinical populations reported the presence of positive cognitive styles, described as a positive state of mind towards external events, and increased extraversion together with an increase rate of rumination, irritability, and more depressive symptoms (Dempsey et al., 2011; Kirkland et al., 2015). In addition, Kirkland et al. (2015) evaluated, in healthy undergraduate students and community participants ($n = 989$), the risk of developing mood disorders by employing two clinical scales: the Altman Self-Rating Mania Scale (ASRM) and the Beck Depression Inventory (BDI). Participants who scored above the high-risk threshold on the ASRM had an increased risk for the development of manic episodes at a 13-year follow up. Similarly, Brand et al. (2011) found that almost 20% of 800 healthy undergraduate students were currently in a hypomanic state, some with "bright" ($n = 94$) and others with "dark" ($n = 69$) hypomania. Compared to nonhypomanic participants and the "bright side" group, "dark side" hypomanic participants reported more depressive symptoms, sleep disturbances, somatic complaints, perceived stress, negative coping strategies, and lower self-efficacy. In contrast, "bright side" hypomanic participants had lower stress scores, more positive self-instructions, and higher levels of exploration, self-efficacy, and physical activity. Therefore, the authors reported that besides the positive feelings and increased self-esteem associated with hypomania, their sample of hypomanic individuals also showed difficulties and disturbances in everyday life.

With regards to cognitive alterations, Mason, O'Sullivan, Blackburn, Bentall, and El-Deredy (2012) showed that nonclinical subjects experiencing hypomania, as assessed through the Hypomanic Personality Scale, tended to have impulsive decision making and risk-taking behaviors with respect to the control group. These results highlight how an evident positive behavior may be associated with cognitive alterations in nonclinical populations. Additionally, a very recent study highlighted that risk-taking behaviors characterized a pediatric population experiencing hypomania

(Hosang, Cardno, Freeman, & Ronald, 2017), where hypomanic features correlated not only with behavioral alterations and drug use, but also with psychotic-like experiences, internalizing and externalizing problems, and reduced life satisfaction. Finally, Lardi Robyn, Ghisletta, and Van der Linden (2012) showed that individuals with hypomanic features are more prone to retrieve positive memory and show an impairment in the ability to produce solid plans for the future, which is consistent with studies conducted in BD (Au et al., 2013). Therefore, these data are of fundamental importance to characterize a nonclinical population with an increased risk of developing mood disorders.

CLINICAL POPULATIONS

Mild cognitive alterations in BD II have been detected by several studies, and some did not find any differences with BD I by showing abnormalities in sustained attention and executive functions in both groups (Brooks 3rd et al., 2010; Martino et al., 2011; Pålsson et al., 2013; Sparding et al., 2015). Moreover, it has been reported that alterations in verbal memory and verbal fluency seem to characterize BD II patients (Aminoff et al., 2013; Romero et al., 2016).

However, it is important to highlight that a general consensus has been reached on the evidence suggesting that BD II presents less severe cognitive impairments than BD I, with particular regard to verbal learning and memory (Bourne et al., 2015; Schenkel, West, Jacobs, Sweeney, & Pavuluri, 2012; Torrent et al., 2006).

Interestingly, hypomanic patients are impaired in both recognition and recall of words, and performed poorly when generating (Malhi, Ivanovski, Hadzi-Pavlovic, et al., 2007) or learning (Lex, Hautzinger, & Meyer, 2011) words from specific semantic (emotional-valued) areas. Moreover, patients experiencing hypomania were more likely to give extreme judgments (both positively and negatively valenced) (Stange et al., 2013), and to report positive and negative self-esteem (Scott & Pope, 2003). Also, Fletcher et al. (2013b), reported that both BD I and BD II patients were more prone to engage in risk-taking behaviors and to report dysfunctional coping strategies, with BD II patients being less likely to seek support. Similarly, an independent study from the same group reported in BD II patients increased interpersonal conflicts, substantial financial burden, and feelings of guilt, shame, and remorse, in consequence to risky behaviors (Fletcher et al., 2013a).

Deficits in verbal memory, sustained attention, and information processing of emotionally valenced words have also been reported in children and adolescents with BD II in two independent studies (Dickstein et al., 2016; Schenkel et al., 2012), suggesting that they may represent core cognitive markers of BD II. Interestingly, these results were replicated by a recent study (Schenkel, Chamberlain, & Towne, 2014).

All together, these findings support the evidence that hypomania is characterized by selective cognitive alterations, such as deficits in executive functions, sustained attention, and verbal memory in both clinical and nonclinical populations, potentially constituting an intermediate phenotype between healthy cognition and BD I.

STRUCTURAL BASIS OF HYPOMANIA (TABLE 2)

In the last 3 decades, MRI techniques proved to be a promising tool for the identification of the neural underpinnings of psychiatric features, including hypomania. However, the available MRI studies comparing BD I with BD II in small samples are often unbalanced toward BD I. Also, to the best of our knowledge, no MRI studies have tried to explore hypomania in nonclinical populations.

Despite the common belief regarding BD II as a soft, nonserious disease, most of the MRI studies on this disorder found cortical and subcortical alterations when compared to healthy controls. As mentioned above, although widespread gray matter reductions have been observed in BD I compared to BD II, white matter deficits seem to reveal a multifaceted situation. Specifically, studies assessing white matter integrity in BD II with diffusion imaging techniques reported altered white matter architecture in the corpus callosum, ventral and medial prefrontal cortex, anterior cingulate, and insula in BD II (Ambrosi et al., 2016, 2013; Yasuno et al., 2016; Ha et al., 2011). Interestingly, altered white matter in prefrontal regions in BD II was associated with impairments in executive functions and with manic symptoms (Liu et al., 2010).

With regards to gray matter alterations, most of the studies found differences between the two subtypes of BD, with the exception of two reports (Gutiérrez-Galve et al., 2012; Maller et al., 2015). Prefrontal alterations seem to characterize BD II patients who, in turn, show a relative integrity of temporal regions, which are in contrast often found to be altered in BD I (Abé et al., 2016; Ambrosi et al., 2013; Ha et al., 2011; Ha, Ha, Kim, & Choi, 2009). This difference may account for some clinical features of the two BD subtypes. Indeed, psychosis, one of the key symptoms characterizing BD I and not BD II, has been proposed to rely on temporal alterations and therefore it is not surprising to see the presence of widespread gray matter alterations in BD I (Mathew et al., 2014). In line with this evidence, a recent study showed that both BD I and BD II patients display lower gray matter volumes in frontal and ventromedial prefrontal regions when compared to healthy controls (Abé et al., 2016; Ha et al., 2009, 2011). Additionally, decreased gray matter volumes in the middle frontal and superior temporal gyri (Ambrosi et al., 2013), as well as in the thalamus and basal ganglia, have been found in BD II patients compared to healthy controls (Haznedar et al., 2005).

In general, frontal and prefrontal cortices are designated to modulate superior functions, such as planning, attention, and executive functions (Kozicky et al., 2013; McKenna, Sutherland, Legenkaya, & Eyler, 2014). Therefore, alterations in these structures observed in both BD I and BD II patients, although to a larger extent in BD I, might sustain the cognitive deficits associated with BD II and hypomania: attention, memory, and executive functions (Kozicky et al., 2013).

In summary, the available MRI evidence reported in BD II patients shows frontal and prefrontal gray matter deficits as well as abnormalities in white matter traits connecting prefrontal regions and corpus callosum, which may ultimately subtend the cognitive alterations often described in BD II and hypomania. Nonetheless, the

Table 2 Structural magnetic resonance studies in hypomania

Authors	Sample (M/F)	Diagnosis	Psychopathological measures	Cerebral tissues	MRI acquisition	Results
Abé et al. (2016)	225 (93/132)	81 BD I 59 BD II 85 HC	ADE	GM	1.5T Whole-brain study	Both BD I and BDII associated with lower volume, thickness, and surface area in frontal regions BDI but not BD II had abnormally low cortical volume and thickness in temporal and medial prefrontal areas
Tighe et al. (2012)	83 (36/47)	26 psychotic BD I 7 nonpsychotic BD I 12 BD II 7 unaffected family members 31 HC	K-SADS-PL	WMH	1.5T	Positive linear trend by familiarity and type of affectedness
Maller, Thaveenthiran, Thomson, McQueen, and Fitzgerald (2014)	62 (35/37)	16 BD I 15 BD I 31 HC	HAM-D	GM	1.5T Whole-brain study	Reduced regional brain volume and thickness among BD subjects, but also between BD I when compared to BD II White matter integrity differed between BD I and BD II
Ambrosi et al. (2013)	41 (11/30)	20 BD II 21 HC	HAM-D YMRS	GM WM	1.5T Whole-brain study	Decreased GM in the right middle frontal gyrus and in the right superior temporal gurus of BD II Widespread FA reduction in BD II in all major WM tracts, including cortico-cortical association tracts

Ambrosi et al. (2016)	100 (52/48)	25 BD I 25 BD II 50 HC	YMRS HAM-D	WM	3T	BD II showed lower FA but no significant ADC differences in the right inferior longitudinal fasciculus, left internal capsule, cortico-spinal tract, and cerebellum
Yasuno et al. (2016)	38 (18/20)	16 BD II; 22 MDD	YMRS MADRS	WM GM	3T Whole-brain study	Alteration in FA in the corpus callosum and interhemispheric connectivity of the ventral prefrontal cortex and insula cortex
Maller et al. (2015)	71 (31/40)	17 depressed BD I 18 depressed BD II 36 depressed HC	HAM-D	GM	1.5T Whole-brain study	No volumetric differences between BD I and BD II subjects Significantly greater ratio of total brain volume to ICV among females with BD II compared to females with BD I
Pompili et al. (2008)	99 (57/42)	40 BD I 21 BD II 38 MDD		WMH	1.5T	Subanalysis of the association between suicide attempts and WMH revealed a concordant association for BD I and MDD, but not BD II
Haznedar et al. (2005)	69	11 BD I 6 BD II 16 cyclothymic 36 HC	YMRS CARS-M	WM	1.5T Whole-brain study	BD I and BD II showed volumetric changes in the thalamus and basal ganglia compared to HC BD II presented the greatest anomaly, with a larger right thalamus and a smaller left thalamus
Kieseppä et al. (2014)	65 (29/36)	13 BP I 15 BP II 16 MDD 21 HC	YMRS BDI	WMH	1.5T	BD I group predicted higher deep WMH grade, which predicted worse performance on the Visual Span Forward test
Gutiérrez-Galve et al. (2012)	36	25 BD I 11 BD II		GM	1.5T ROI study	Premorbid IQ associated with frontal cortical volume Cortical parameters were not different in BD I and BD II patients

Continued

Table 2 Structural magnetic resonance studies in hypomania—cont'd

Authors	Sample (M/F)	Diagnosis	Psychopathological measures	Cerebral tissues	MRI acquisition	Results
Ha et al. (2009)	69 (24/45)	23 BD I 23 BD II 23 HC	HAM-D	GM	1.5T Whole-brain study	Compared to HC, both BD II and BD I presented reduced GM in the ventromedial prefrontal regions BD I but not BD II had widespread GM reductions in the bilateral frontal, temporal, parietal, and parahippocampal cortices
Ha et al. (2011)	46 (10/36)	12 BD I 12 BD II 22 HC	YMRS HAM-D	GM WM	1.5T Whole-brain study	Compared to HC, both BD I and BD II showed decreased FA in the corpus callosum, cingulate, and right prefrontal regions, and increased ADC in the medial frontal, anterior cingulate, insula, and temporal regions BD I had an FA decrease in the right temporal WM and an ADC increase in the frontal, temporal, parietal, and thalamic regions, compared to BD II
Liu et al. (2010)	48 (17/31)	14 BD I 13 BD II 21 HC	YMRS HAM-D MADRS	WM	1.5T Whole-brain study	BD I and BD II presented alterations in FA in the thalamus, anterior cingulate, and inferior frontal areas BD II showed more alterations in the temporal and inferior prefrontal regions. These alterations correlated with executive functions and YMRS

BD, Bipolar Disorder; BDI, Beck Depression Inventory; HC, Healthy controls; YMRS, Young Mania Rating Scale; MADRS, Montgomery-Asberg Depression Rating Scale; HAM-D, Hamilton Rating Scale for Depression; WMH, White Matter Hyperintensities; FA, Fractional anisotropy; ADC, apparent diffusion coefficient; K-SADS-PL, Schedule for Affective Disorders and Schizophrenia—Lifetime version; ADE, Affective Disorder Evaluation; GM, gray matter; WM, white matter; CARS-M, Clinician-Administered Rating Scale for Mania; ICV, Intra-Cranial Volume.

relatively limited number of studies focusing specifically on BD II and hypomania as well as the small sample sizes employed limit the consistency and the generalization of these results. Therefore, future structural and diffusion MRI studies coupled with cognitive investigations enrolling large samples of hypomanic patients are needed to confirm and validate these data.

Fig. 1 depicts the neural network consistently found altered in hypomania.

FUNCTIONAL BASIS OF HYPOMANIA (TABLE 3)

Functional MRI (fMRI) provides in vivo dynamic measures reflecting brain regional activation of an individual at a specific time point. Although fMRI studies in BD have led to heterogeneous findings, there is a general consensus on the presence of an overactivity of subcortical structures, including the amygdala, hippocampus, and basal ganglia, coupled with a reduction in activity in prefrontal regions in this disorder (Townsend & Altshuler, 2012). Interestingly, recent meta-analyses suggest that this pattern of activation/deactivation can be seen in both rest and active conditions (Chen, Suckling, Lennox, Ooi, & Bullmore, 2011; Wegbreit et al., 2014).

To the best of our knowledge, no fMRI studies investigating hypomania in non-clinical populations have been published so far. Therefore, the available data allowed us to discuss only the evidence reported by studies comparing BD I and/or BD II and healthy controls. As in the structural studies, fMRI studies employed small samples and, often, different tasks, limiting the generalizability of the findings.

Most of the fMRI studies in BD II investigated emotional response. Overall, these studies reported alterations in regions implicated in generation and modulation of affect, including prefrontal cortices, posterior cingulate, caudate, thalamus, and insula (Hulvershorn et al., 2012; Malhi et al., 2004; Rey et al., 2014). Furthermore, Hulvershorn et al. (2012) showed that the hypomanic group had decreased left lateral orbitofrontal cortex activation compared to both euthymic and healthy controls, increased dorsal anterior cingulate cortex activation compared with the depressed subjects, and increased dorsolateral prefrontal cortical activation compared with all other groups. Corroborating these findings, Caseras et al. (2015) showed that BD II patients had increased activation in the amygdala, nucleus accumbens, and dorsolateral prefrontal cortex, as well as increased inverse functional connectivity between the prefrontal cortex and amygdala in comparison to healthy subjects.

Furthermore, other networks have been proven to be altered in BD II patients, especially those related to reward response. Indeed, two independent fMRI studies (Caseras, Lawrence, Murphy, Wise, & Phillips, 2013; O'Sullivan, Szczepanowski, El-Deredy, Mason, & Bentall, 2011) reported that the ventral striatal activity in response to reward was significantly increased in BD II compared to both BD I and healthy controls. Moreover, this higher activation correlated with value and prediction errors, which is consistent with the hypothesis that hypomania presents enhanced perception of the value of goals that may lead to reward (O'Sullivan et al., 2011). O'Sullivan et al. (2011) also showed a stronger value-related medial temporal

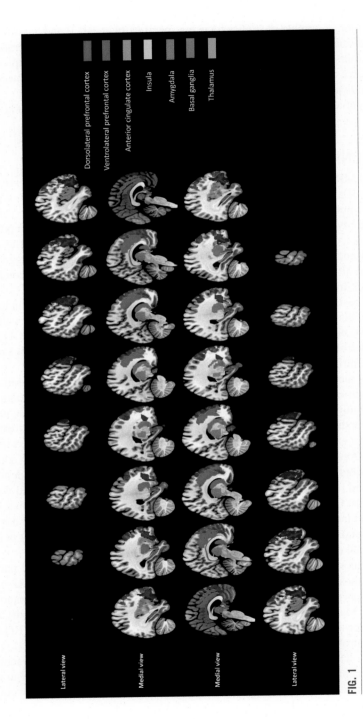

FIG. 1

Brain regions consistently found altered in hypomania.

Table 3 Functional magnetic resonance studies in hypomania

Authors	Sample (M/F)	Diagnosis	Psychopathological measures	Tasks	MRI acquisition	Results
Malhi et al. (2004)	20 (0/20)	10 BD I/BD II 10 HC	HAM-D MADRS YMRS	video-based emotional activation task	1.5T	Increased activation in BD in the caudate and thalamus compared to HC
Rey et al. (2014)	24 (16/8)	12 BD 12 HC	YMRS MADRS	Emotional conflict task	3T	Deactivation in several key areas in response to emotion-conflict trials, including the posterior cingulate cortex during hypomania
Hulvershorn et al. (2012)	105 (49/66)	30 depressed BD 30 hypomanic BD 15 euthymic BD 30 HC	YMRS HAM-D	Negative facial emotion matching task	3T	Euthymic and hypomanic BD had increased activation in the insula The hypomanic group exhibited decreased left lateral orbitofrontal cortex activation and increased dorsal anterior cingulate and dorsolateral prefrontal activation

Continued

Table 3 Functional magnetic resonance studies in hypomania—cont'd

Authors	Sample (M/F)	Diagnosis	Psychopathological measures	Tasks	MRI acquisition	Results
Caseras et al. (2015)	55 (19/36)	16 BD I 19 BD II 20 HC	YMRS HAM-D	Emotion regulation paradigm	1.5T	BD I showed slower reaction times to targets, and increased blood oxygenation level-dependent (BOLD) responses in the amygdala, accumbens, and dorsolateral prefrontal cortex, but not increased inverse functional connectivity BD II showed increased BOLD responses in the amygdala, accumbens, and dorsolateral prefrontal cortex, and increased inverse functional connectivity between the prefrontal cortex and amygdala
O'Sullivan et al. (2011)	24 (not specified)	12 BD II 12 HC	HPS DAS	Reward processing task	3T	Higher striatal activation in BD II Value-related medial temporal activation was stronger in BD II Increased insula activation in response to expected, but absent, reward in BD II
Caseras et al. (2013)	52 (19/33)	17 BD I 15 BD II 20 HC	YMRS HAM-D HCL-32	Reward-processing task	3T	Significantly greater ventral striatal activity in BD II compared with the other groups Greater putamen volume in BD II, with left putamen correlating positively with left ventral striatal activity to reward anticipation

Caligiuri et al. (2006)	10 (4/6)	10 BD assessed in different phases	SMS	SRT	1.5 T	Hypomanic states exhibited increased caudate activity bilaterally and the globus pallidus of the left hemisphere compared to depressed
Spielberg et al. (2016)	90 (not specified)	30 hypomanic BD 30 depressed BD 30 HC	YMRS HAM-D	–	3T Rs-fMRI	BD evidenced hyperconnectivity in a network involving right amygdala Hypomanic symptoms were associated with hyperconnectivity in an overlapping network
Yang et al. (2016)	61 (29/32)	19 MDD positive for Hypomania 18 negative for Hypomania 24 HC	HCL-32	–	3T Rs-fMRI	Hypomania showed significant increase of regional homogeneity in the right medial superior frontal cortex, left inferior parietal cortex, and middle/inferior temporal cortex, and decrease of regional homogeneity in the left postcentral cortex and cerebellum
Altinay et al. (2016)	90 (36/54)	30 depressed BD 30 hypomanic BD 30 HC	MINI YMRS HAM-D	–	3T Rs-fMRI	Hypomania showed unique increased connectivity between left dorsal caudate and midbrain regions as well as increased connectivity between ventral striatum inferior and thalamus

Continued

Table 3 Functional magnetic resonance studies in hypomania—cont'd

Authors	Sample (M/F)	Diagnosis	Psychopathological measures	Tasks	MRI acquisition	Results
Malhi, Ivanovski, Wen, et al. (2007)	18 (4/14)	9 hypomanic BD 9 HC	YMRS HAM-D MADRS	–	1.5T MRI proton spectroscopy	Compared to HC, hypomanic patients had lower mean metabolite levels across the anterior cingulate and frontal cortex regions as well as a smaller difference in mean metabolite levels between the basal ganglia and frontal cortex
Dager et al. (2004)	58 (26/32)	17 BD II 15 BD I 26 HC	HAM-D YMRS	–	1.5T PEPSI	Patients with BD exhibited elevated GM lactate and gamma-aminobutyric acid levels. In BD II, elevated Glx levels and trend elevated lactate levels predicted a statistically significant model for GM and WM chemicals
Brooks 3rd et al. (2010)	27 (24/3)	10 BD I 6 BD II 11 HC	YMRS MADRS	CPT	(18) F-FDG-PET	In BD patients (both type I and type II), commission errors were strongly related to hypometabolism in inferior frontal gyrus and hypermetabolism in paralimbic regions, whereas omission errors were strongly related to hypometabolism in dorsolateral prefrontal and hypermetabolism in limbic/paralimbic regions

Brooks 3rd et al. (2009)	27 (24/3)	10 BD I 6 BD II 11 HC	YMRS MADRS	CVLT-II	(18) F-FDG-PET	In BD patients (both type I and type II) recall deficits were associated with hypometabolism in dorsolateral prefrontal cortex and hypermetabolism in paralimbic regions (hippocampus, parahippocampal gyrus, and superior temporal gyrus)
Liu et al. (2010)	51 (17/34)	17 BD I 17 BD II 17 HC	YMRS HAM-D	Tests for Attentional Performance; Go/No-Go task Word List Test and Face Memory Test; WCST	(18) F-FDG PET	BD I compared to BD II had significantly lower glucose uptake in the bilateral anterior cingulium, insula, striatum, and part of the prefrontal cortex, and higher glucose uptake in the left parahippocampus

BD, Bipolar Disorder; HC, Healthy Controls; YMRS, Young Mania Rating Scale; MADRS, Montgomery-Asberg Depression Rating Scale; HAM-D, Hamilton Rating Scale for Depression; (18) F-FDG PET, fluorodeoxyglucose positron emission tomography; PEPSI, proton echo-planar spectroscopic imaging; DAS, Dysfunctional Attitudes scale; HPS, Hypomanic Personality Scale; ReHo, Regional homogeneity; SRT, simple reaction time; SMS, Simplified Mood Scale; CVLT-II, California Verbal Learning Test; Rs-fMRI, Resting State-Functional Magnetic Resonance Imaging; MINI, Mini International Neuropsychiatric Interview.

activation in subjects with hypomania, suggesting that, in these individuals, positive memories were easier to remember and recall. With regard to subcortical regions, increased activation in both the insula and putamen has also been found in BD II during reward tasks (Caseras et al., 2013; O'Sullivan et al., 2011).

Interestingly, research on hypomania also focused on resting-state fMRI (rs-fMRI), a promising technique supposed to explore the baseline status of brain activity. Although performed on small samples, the available rs-fMRI studies on hypomania showed increased activation in the medial superior frontal cortex, inferior parietal cortex, and middle/inferior temporal cortex (Yang et al., 2016). Spielberg et al. (2016) showed that hypomanic patients had hyperconnectivity in a specific network involving the right amygdala. Additionally, Altinay, Hulvershorn, Karne, Beall, and Anand (2016) found that hypomanic patients also showed increased connectivity between the left dorsal caudate and midbrain regions as well as increased connectivity between the ventral striatum and thalamus.

Furthermore, some studies explored brain metabolism in subjects with hypomania by means of the fluorodeoxyglucose (18-F-FDG) positron emission tomography (PET), but their results are far from conclusive. Indeed, while some PET studies reported cortical abnormalities (Brooks 3rd et al., 2010, 2009), others failed to detect differences (Li et al., 2012) in hypomanic patients compared to healthy controls. Of interest, two consecutive studies from Brooks 3rd et al. (2010, 2009) reported prefrontal hypometabolism (inferior frontal gyrus, dorsolateral prefrontal cortex) and paralimbic hypermetabolism (hippocampus, parahippocampal gyrus, and superior temporal gyrus), in association with deficits in attention and memory proficiency.

Finally, two studies explored metabolite and neurotransmitter distribution throughout the brain in both BD I and BD II (Dager et al., 2004; Malhi, Ivanovski, Wen, et al., 2007), although they were limited by the assessment of different metabolites and by small samples. First, Dager et al. (2004) suggested a shift in energy state from oxidative phosphorylation towards glycolysis, as predicted by GMr lactate and Glx elevations in medication-free BD II patients. Second, Malhi, Ivanovski, Wen, et al. (2007) found lower mean metabolite levels across the anterior cingulate and frontal cortex regions, and a smaller difference in mean metabolite levels between the basal ganglia and frontal cortex in patients experiencing hypomania compared to euthymic BD patients and healthy controls.

In summary, these findings suggest that hypomania is characterized by functional and metabolic alterations in selective brain regions important for emotion regulation and reward response, such as the prefrontal cortex, amygdala, and insula. An abnormal activation within these networks may support some core deficits of hypomania, especially in attribution of emotional value and in response to reward stimuli. Similarly, rs-fMRI investigations reported alterations in networks involving the amygdala and prefrontal cortices, which may further sustain the presence of hypersalience, reflecting the altered attribution of value and self-esteem in hypomania. However, as already stated, the generalizability of these findings is limited by small sample size and the use of different tasks, therefore future larger longitudinal functional MRI studies are warranted to further understand brain metabolism in the presence of hypomania in BD patients even when a full-blown episode is recovered.

CONCLUSION

Hypomanic features are present in both clinical and nonclinical populations, and characterize BD type II in clinical populations. Therefore, the importance of assessing hypomania with multimodal approaches in healthy populations is crucial to further understand the pathophysiology of BD. With regard to BD, hypomania is considered to constitute an intermediate-severity phenotype between healthy controls and BD I, characterized by less severe manifestations.

Several studies suggest that hypomania has negative consequences in terms of quality of life, social interactions, and cognition, although it is often coupled with an experience of global well functioning. Indeed, hypomanic subjects often engage in risky behaviors, including spending significant amounts of money, excessive alcohol or drug use, dangerous driving, and endangering sexual activities, ultimately causing interpersonal conflict, substantial financial burden, and feelings of guilt, shame, and remorse.

From a cognitive point of view, selective alterations have been found during hypomania, including deficits in executive functions and verbal and learning memory. These cognitive alterations seem to represent a core feature of hypomania, being also detected in children and adolescents.

Therefore, the investigation of hypomania might be of crucial importance for addressing future therapeutic and rehabilitation programs. Adolescence, in particular, is a critical period for neurodevelopment where subjects start to learn specific coping strategies and cognitive patterns. The presence of hypomanic features during adolescence might therefore determine a higher vulnerability to hexogen insults (such as substance and alcohol abuse), ultimately leading to the expression of a full-blown hypomanic or manic episode (Do & Mezuk, 2013). This evidence should lead to an increased emphasis on psychoeducation in the management of nonclinical hypomanic adolescents to prevent misleading behaviors which may augment the risk of developing a mood episode.

Finally, neuroimaging studies reported that hypomania, as investigated in BD II, is often associated with selective gray matter alterations in neural networks sustaining emotion regulation and reward, including prefrontal and limbic regions, in both clinical and nonclinical populations (Fig. 1). These findings are not completely surprising, given the fact that structural abnormalities in these areas have often been reported in BD I and manic patients, who often had more severe and extensive gray matter reductions in these areas compared to BD II patients (Abé et al., 2016; Ha et al., 2009, 2011). These studies provide deeper insight into the neural underpinnings of hypomania, helping to understand the neural and cognitive basis of this mood state, which transversally affects both psychiatric and healthy populations. This moves us closer to the identification of the pathophysiological processes associated with hypomania, ultimately allowing for a better understanding of clinical psychopathology.

In conclusion, hypomania is characterized by a number of behavioral modifications, ultimately causing unpleasant experiences such as irritability, lack of attention, engaging in risky situations, and peer problems. Further investigation of the neural and cognitive basis of hypomania will help to better understand the roots of mood swings as well as the transition mechanisms leading from clinical high risk to the full expression of BD.

REFERENCES

Abé, C., Ekman, C. J., Sellgren, C., Petrovic, P., Ingvar, M., & Landén, M. (2016). Cortical thickness, volume and surface area in patients with bipolar disorder types I and II. *Journal of Psychiatry Neuroscience, 41*, 240–250.

Altamura, A. C., Delvecchio, G., Marotta, G., Oldani, L., Pigoni, A., Ciappolino, V., et al. (2016). Structural and metabolic differentiation between bipolar disorder with psychosis and substance-induced psychosis: an integrated MRI/PET study. *European Psychiatry, 41*, 85–94.

Altinay, M. I., Hulvershorn, L. A., Karne, H., Beall, E. B., & Anand, A. (2016). Differential resting-state functional connectivity of striatal subregions in bipolar depression and hypomania. *Brain Connectivity, 6*, 255–265.

Ambrosi, E., Chiapponi, C., Sani, G., Manfredi, G., Piras, F., Caltagirone, C., et al. (2016). White matter microstructural characteristics in Bipolar I and Bipolar II Disorder: a diffusion tensor imaging study. *Journal of Affective Disorder, 189*, 176–183.

Ambrosi, E., Rossi-Espagnet, M. C., Kotzalidis, G. D., Comparelli, A., Del Casale, A., Carducci, F., et al. (2013). Structural brain alterations in bipolar disorder II: a combined voxel-based morphometry (VBM) and diffusion tensor imaging (DTI) study. *Journal of Affective Disorder, 150*, 610–615.

American Psychiatric Association. (2013). *Diagnostic and statistical manual of mental disorders, fifth edition (DSM 5)*. Arlington, VA: American Psychiatric Publishing.

Aminoff, S. R., Hellvin, T., Lagerberg, T. V., Berg, A. O., Andreassen, O. A., & Melle, I. (2013). Neurocognitive features in subgroups of bipolar disorder. *Bipolar Disorder, 15*, 272–283.

Angst, J., Adolfsson, R., Benazzi, F., Gamma, A., Hantouche, E., Meyer, T. D., et al. (2005). The HCL-32: towards a self-assessment tool for hypomanic symptoms in outpatients. *Journal Affective Disorder, 88*, 217–233.

Au, R. W., Ungvari, G. S., Lee, E., Man, D., Shum, D. H., Xiang, Y. T., et al. (2013). Prospective memory impairment and its implications for community living skills in bipolar disorder. *Bipolar Disorder, 15*, 885–892.

Benazzi, F. (2007). Bipolar disorder--focus on bipolar II disorder and mixed depression. *Lancet, 369*, 935–945.

Bond, D. J., Noronha, M. M., Kauer-Sant'Anna, M., Lam, R. W., & Yatham, L. N. (2008). Antidepressant-associated mood elevations in bipolar II disorder compared with bipolar I disorder and major depressive disorder: a systematic review and meta-analysis. *Journal of Clinical Psychiatry, 69*, 1589–1601.

Bourne, C., Bilderbeck, A., Drennan, R., Atkinson, L., Price, J., Geddes, J. R., et al. (2015). Verbal learning impairment in euthymic bipolar disorder: BDI v BDII. *Journal Affective Disorder, 182*, 95–100.

Brand, S., Foell, S., Bajoghli, H., Keshavarzi, Z., Kalak, N., Gerber, M., et al. (2015). Tell me, how bright your hypomania is, and I tell you, if you are happily in love! among young adults in love, bright side hypomania is related to reduced depression and anxiety, and better sleep quality. *International Journal of Psychiatry in Clinical Practice*, *19*, 24–31.

Brand, S., Gerber, M., Pühse, U., & Holsboer-Trachsler, E. (2011). "Bright side" and "dark side" hypomania are associated with differences in psychological functioning, sleep and physical activity in a non-clinical sample of young adults. *Journal Affective Disorder*, *131*, 68–78.

Brooks, J. O., 3rd, Bearden, C. E., Hoblyn, J. C., Woodard, S. A., & Ketter, T. A. (2010). Prefrontal and paralimbic metabolic dysregulation related to sustained attention in euthymic older adults with bipolar disorder. *Bipolar Disorder*, *12*, 866–874.

Brooks, J. O., 3rd, Rosen, A. C., Hoblyn, J. C., Woodard, S. A., Krasnykh, O., & Ketter, T. A. (2009). Resting prefrontal hypometabolism and paralimbic hypermetabolism related to verbal recall deficits in euthymic older adults with bipolar disorder. *American Journal of Geriatric Psychiatry*, *17*, 1022–1029.

Caligiuri, M. P., Brown, G. G., Meloy, M. J., Eberson, S., Niculescu, A. B., & Lohr, J. B. (2006). Striatopallidal regulation of affect in bipolar disorder. *Journal Affective Disorder*, *91*, 235–242.

Caseras, X., Lawrence, N. S., Murphy, K., Wise, R. G., & Phillips, M. L. (2013). Ventral striatum activity in response to reward: differences between bipolar I and II disorders. *American Journal of Psychiatry*, *170*, 533–541.

Caseras, X., Murphy, K., Lawrence, N. S., Fuentes-Claramonte, P., Watts, J., Jones, D. K., et al. (2015). Emotion regulation deficits in euthymic bipolar I versus bipolar II disorder: a functional and diffusion-tensor imaging study. *Bipolar Disorder*, *17*, 461–470.

Chen, C. H., Suckling, J., Lennox, B. R., Ooi, C., & Bullmore, E. T. (2011). A quantitative meta-analysis of fMRI studies in bipolar disorder. *Bipolar Disorder*, *13*, 1–15.

Cotrena, C., Branco, L. D., Shansis, F. M., & Fonseca, R. P. (2016). Executive function impairments in depression and bipolar disorder: association with functional impairment and quality of life. *Journal of Affective Disorder*, *190*, 744–753.

Dager, S. R., Friedman, S. D., Parow, A., Demopulos, C., Stoll, A. L., Lyoo, I. K., et al. (2004). Brain metabolic alterations in medication-free patients with bipolar disorder. *Archives General Psychiatry*, *61*, 450–458.

Dempsey, R. C., Gooding, P. A., & Jones, S. H. (2011). Positive and negative cognitive style correlates of the vulnerability to hypomania. *Journal of Clinical Psychology*, *67*, 673–690.

Dickstein, D. P., Axelson, D., Weissman, A. B., Yen, S., Hunt, J. I., Goldstein, B. I., et al. (2016). Cognitive flexibility and performance in children and adolescents with threshold and sub-threshold bipolar disorder. *European Child & Adolescent Psychiatry*, *25*, 625–638.

Do, E. K., & Mezuk, B. (2013). Comorbidity between hypomania and substance use disorders. *Journal of Affective Disorder*, *150*, 974–980.

Dodd, A. L., Mansell, W., Morrison, A. P., & Tai, S. (2011). Bipolar vulnerability and extreme appraisals of internal states: a computerized ratings study. *Clinical Psychology & Psychotherapy*, *18*, 387–396.

Fletcher, K., Parker, G., & Manicavasagar, V. (2013a). A qualitative investigation of hypomania and depression in bipolar II disorder. *Psychiatric Quarterly*, *84*, 455–474.

Fletcher, K., Parker, G. B., & Manicavasagar, V. (2013b). Coping profiles in bipolar disorder. *Comprehensive Psychiatry*, *54*, 1177–1184.

Fletcher, K., Parker, G., Paterson, A., & Synnott, H. (2013). High-risk behaviour in hypomanic states. *Journal of Affective Disorder, 150,* 50–56.

Grande, I., Berk, M., Birmaher, B., & Vieta, E. (2016). Bipolar disorder. *Lancet, 387,* 1561–1572.

Gutiérrez-Galve, L., Bruno, S., Wheeler-Kingshott, C. A., Summers, M., Cipolotti, L., & Ron, M. A. (2012). IQ and the fronto-temporal cortex in bipolar disorder. *Journal of the International Neuropsychological Society, 18,* 370–374.

Ha, T. H., Ha, K., Kim, J. H., & Choi, J. E. (2009). Regional brain gray matter abnormalities in patients with bipolar II disorder: a comparison study with bipolar I patients and healthy controls. *Neuroscience Letters, 456,* 44–48.

Ha, T. H., Her, J. Y., Kim, J. H., Chang, J. S., Cho, H. S., & Ha, K. (2011). Similarities and differences of white matter connectivity and water diffusivity in bipolar I and II disorder. *Neuroscience Letters, 505,* 150–154.

Hantouche, E. G., Angst, J., & Akiskal, H. S. (2003). Factor structure of hypomania: Interrelationships with cyclothymia and the soft bipolar spectrum. *Journal of Affective Disorder, 73,* 39–47.

Harada, M., Terao, T., Hatano, K., Kohno, K., Araki, Y., Mizokami, Y., et al. (2013). Hyperthymic temperament and brightness preference in healthy subjects: further evidence for involvement of left inferior orbitofrontal cortex in hyperthymic temperament. *Journal of Affective Disorder, 151,* 763–768.

Haznedar, M. M., Roversi, F., Pallanti, S., Baldini-Rossi, N., Schnur, D. B., Licalzi, E. M., et al. (2005). Fronto-thalamo-striatal gray and white matter volumes and anisotropy of their connections in bipolar spectrum illnesses. *Biological Psychiatry, 57,* 733–742.

Hosang, G. M., Cardno, A. G., Freeman, D., & Ronald, A. (2017). Characterization and structure of hypomania in a British nonclinical adolescent sample. *Journal of Affective Disorder, 207,* 228–235.

Hulvershorn, L. A., Karne, H., Gunn, A. D., Hartwick, S. L., Wang, Y., Hummer, T. A., et al. (2012). Neural activation during facial emotion processing in unmedicated bipolar depression, euthymia, and mania. *Biological Psychiatry, 71,* 603–610.

Kieseppä, T., Mäntylä, R., Tuulio-Henriksson, A., Luoma, K., Mantere, O., Ketokivi, M., et al. (2014). White matter hyperintensities and cognitive performance in adult patients with bipolar I, bipolar II, and major depressive disorders. *European Psychiatry, 29,* 226–232.

Kirkland, T., Gruber, J., & Cunningham, W. A. (2015). Comparing happiness and hypomania risk: a study of extraversion and neuroticism aspects. *PLoS ONE, 10,* e0132438.

Kozicky, J. M., Ha, T. H., Torres, I. J., Bond, D. J., Honer, W. G., Lam, R. W., et al. (2013). Relationship between frontostriatal morphology and executive function deficits in bipolar I disorder following a first manic episode: data from the Systemic Optimization Program for Early Mania (STOP-EM). *Bipolar Disorders, 15*(6), 657–668.

Lardi Robyn, C., Ghisletta, P., & Van der Linden, M. (2012). Self-defining memories and self-defining future projections in hypomania-prone individuals. *Consciousness and Cognition, 21,* 764–774.

Lee, K., Oh, H., Lee, E. H., Kim, J. H., Kim, J. H., & Hong, K. S. (2016). Investigation of the clinical utility of the hypomania checklist 32 (HCL-32) for the screening of bipolar disorders in the non-clinical adult population. *BMC Psychiatry, 16,* 124.

Lex, C., Hautzinger, M., & Meyer, T. D. (2011). Cognitive styles in hypomanic episodes of bipolar I disorder. *Bipolar Disorder, 13,* 355–364.

Li, C. T., Hsieh, J. C., Wang, S. J., Yang, B. H., Bai, Y. M., Lin, W. C., et al. (2012). Differential relations between fronto-limbic metabolism and executive function in patients with remitted bipolar I and bipolar II disorder. *Bipolar Disorder, 14*, 831–842.

Liu, J. X., Chen, Y. S., Hsieh, J. C., Su, T. P., Yeh, T. C., & Chen, L. F. (2010). Differences in white matter abnormalities between bipolar I and II disorders. *Journal of Affective Disorder, 127*, 309–315.

Malhi, G. S., Ivanovski, B., Hadzi-Pavlovic, D., Mitchell, P. B., Vieta, E., & Sachdev, P. (2007). Neuropsychological deficits and functional impairment in bipolar depression, hypomania and euthymia. *Bipolar Disorder, 9*, 114–125.

Malhi, G. S., Ivanovski, B., Wen, W., Lagopoulos, J., Moss, K., & Sachdev, P. (2007). Measuring mania metabolites: a longitudinal proton spectroscopy study of hypomania. *Acta Psychiatrica Scandinavica, 434*, 57–66.

Malhi, G. S., Lagopoulos, J., Sachdev, P., Mitchell, P. B., Ivanovski, B., & Parker, G. B. (2004). Cognitive generation of affect in hypomania: an fMRI study. *Bipolar Disorder, 6*, 271–285.

Maller, J. J., Anderson, R., Thomson, R. H., Rosenfeld, J. V., Daskalakis, Z. J., & Fitzgerald, P. B. (2015). Occipital bending (Yakovlevian torque) in bipolar depression. *Psychiatry Research, 231*, 8–14.

Maller, J. J., Thaveenthiran, P., Thomson, R. H., McQueen, S., & Fitzgerald, P. B. (2014). Volumetric, cortical thickness and white matter integrity alterations in bipolar disorder type I and II. *Journal of Affective Disorder, 169*, 118–127.

Martino, D. J., Igoa, A., Marengo, E., Scápola, M., & Strejilevich, S. A. (2011). Neurocognitive impairments and their relationship with psychosocial functioning in euthymic bipolar II disorder. *The Journal of Nervous and Mental Disease, 199*, 459–464.

Mason, L., O'Sullivan, N., Blackburn, M., Bentall, R., & El-Deredy, W. (2012). I want it now! Neural correlates of hypersensitivity to immediate reward in hypomania. *Biological Psychiatry, 71*, 530–537.

Mathew, I., Gardin, T. M., Tandon, N., Eack, S., Francis, A. N., Seidman, L. J., et al. (2014). Medial temporal lobe structures and hippocampal subfields in psychotic disorders: findings from the Bipolar-Schizophrenia Network on Intermediate Phenotypes (B-SNIP) study. *JAMA Psychiatry, 71*, 769–777.

McKenna, B. S., Sutherland, A. N., Legenkaya, A. P., & Eyler, L. T. (2014). Abnormalities of brain response during encoding into verbal working memory among euthymic patients with bipolar disorder. *Bipolar Disorder, 16*, 289–299.

Merikangas, K. R., & Lamers, F. (2012). The "true" prevalence of bipolar II disorder. *Current Opinion in Psychiatry, 25*, 19–23.

O'Sullivan, N., Szczepanowski, R., El-Deredy, W., Mason, L., & Bentall, R. P. (2011). fMRI evidence of a relationship between hypomania and both increased goal-sensitivity and positive outcome-expectancy bias. *Neuropsychologia, 49*, 2825–2835.

Pålsson, E., Figueras, C., Johansson, A. G., Ekman, C. J., Hultman, B., Östlind, J., et al. (2013). Neurocognitive function in bipolar disorder: a comparison between bipolar I and II disorder and matched controls. *BMC Psychiatry, 13*, 165.

Pompili, M., Innamorati, M., Mann, J. J., Oquendo, M. A., Lester, D., Del Casale, A., et al. (2008). Periventricular white matter hyperintensities as predictors of suicide attempts in bipolar disorders and unipolar depression. *Progress in Neuro-Psychopharmacology & Biological Psychiatry, 32*, 1501–1507.

Renaud, S., Corbalan, F., & Beaulieu, S. (2012). Differential diagnosis of bipolar affective disorder type II and borderline personality disorder: analysis of the affective dimension. *Comprehensive Psychiatry, 53*, 952–961.

Rey, G., Desseilles, M., Favre, S., Dayer, A., Piguet, C., Aubry, J. M., et al. (2014). Modulation of brain response to emotional conflict as a function of current mood in bipolar disorder: preliminary findings from a follow-up state-based fMRI study. *Psychiatry Research, 223*, 84–93.

Romero, E., Holtzman, J. N., Tannenhaus, L., Monchablon, R., Rago, C. M., Lolich, M., et al. (2016). Neuropsychological performance and affective temperaments in euthymic patients with bipolar disorder type II. *Psychiatry Research, 238*, 172–180.

Schenkel, L. S., Chamberlain, T. F., & Towne, T. L. (2014). Impaired theory of mind and psychosocial functioning among pediatric patients with type I versus type II bipolar disorder. *Psychiatry Research, 215*, 740–746.

Schenkel, L. S., West, A. E., Jacobs, R., Sweeney, J. A., & Pavuluri, M. N. (2012). Cognitive dysfunction is worse among pediatric patients with bipolar disorder type I than type II. *Journal of Child Psychology and Psychiatry, 53*, 775–781.

Scott, J., & Pope, M. (2003). Cognitive styles in individuals with bipolar disorders. *Psychological Medicine, 33*, 1081–1088.

Seal, K., Mansell, W., & Mannion, H. (2008). What lies between hypomania and bipolar disorder? A qualitative analysis of 12 non-treatment-seeking people with a history of hypomanic experiences and no history of major depression. *Psychology and Psychotherapy, 81*, 33–53.

Sparding, T., Silander, K., Pålsson, E., Östlind, J., Sellgren, C., Ekman, C. J., et al. (2015). Cognitive functioning in clinically stable patients with bipolar disorder I and II. *PLoS ONE, 10*, e0115562.

Spielberg, J. M., Beall, E. B., Hulvershorn, L. A., Altinay, M., Karne, H., & Anand, A. (2016). Resting state brain network disturbances related to hypomania and depression in medication-free bipolar disorder. *Neuropsychopharmacology, 41*, 3016–3024.

Stange, J. P., Sylvia, L. G., Magalhães, P. V., Frank, E., Otto, M. W., Miklowitz, D. J., et al. (2013). Extreme attributions predict transition from depression to mania or hypomania in bipolar disorder. *Journal of Psychiatric Research, 47*, 1329–1336.

Tighe, S. K., Reading, S. A., Rivkin, P., Caffo, B., Schweizer, B., Pearlson, G., et al. (2012). Total white matter hyperintensity volume in bipolar disorder patients and their healthy relatives. *Bipolar Disorder, 14*, 888–893.

Torrent, C., Martínez-Arán, A., Daban, C., Sánchez-Moreno, J., Comes, M., Goikolea, J. M., et al. (2006). Cognitive impairment in bipolar II disorder. *British Journal of Psychiatry, 189*, 254–259.

Townsend, J., & Altshuler, L. L. (2012). Emotion processing and regulation in bipolar disorder: a review. *Bipolar Disorder, 14*, 326–339.

Vinberg, M., Mikkelsen, R. L., Kirkegaard, T., Christensen, E. M., & Kessing, L. V. (2017). Differences in clinical presentation between bipolar I and II disorders in the early stages of bipolar disorder: a naturalistic study. *Journal of Affective Disorder, 208*, 521–527.

Wegbreit, E., Cushman, G. K., Puzia, M. E., Weissman, A. B., Kim, K. L., Laird, A. R., et al. (2014). Developmental meta-analyses of the functional neural correlates of bipolar disorder. *JAMA Psychiatry, 71*, 926–935.

Yang, H., Li, L., Peng, H., Liu, T., Young, A. H., Angst, J., et al. (2016). Alterations in regional homogeneity of resting-state brain activity in patients with major depressive disorder screening positive on the 32-item hypomania checklist (HCL-32). *Journal of Affective Disorder, 203*, 69–76.

Yasuno, F., Kudo, T., Matsuoka, K., Yamamoto, A., Takahashi, M., Nakagawara, J., et al. (2016). Interhemispheric functional disconnection because of abnormal corpus callosum integrity in bipolar disorder type II. *British Journal of Psychiatry Open, 2*, 335–340.

FURTHER READING

Akiskal, H. S., Hantouche, E. G., & Allilaire, J. F. (2003). Bipolar II with and without cyclothymic temperament: "dark" and "sunny" expressions of soft bipolarity. *Journal of Affective Disorder, 73,* 49–57.

Altamura, A. C., Buoli, M., Cesana, B., Dell'Osso, B., Tacchini, G., Albert, U., et al. (2010). Socio-demographic and clinical characterization of patients with bipolar disorder I vs II: A Nationwide Italian study. *European Archives of Psychiatry and Clinical Neuroscience Journal, 268,* 169–177.

Altamura, A. C., Mundo, E., Cattaneo, E., Pozzoli, S., Dell'osso, B., Gennarelli, M., et al. (2010). The MCP-1 gene (SCYA2) and mood disorders: preliminary results of a case-control association study. *Neuroimmunomodulation, 17,* 126–131.

Aminoff, S. R., Tesli, M., Bettella, F., Aas, M., Lagerberg, T. V., Djurovic, S., et al. (2015). Polygenic risk scores in bipolar disorder subgroups. *Journal Affective Disorder, 183,* 310–314.

Arnone, D., Abou-Saleh, M. T., & Barrick, T. R. (2016). Diffusion tensor imaging of the corpus callosum in addiction. *Neuropsychobiology, 54,* 107–113.

Benazzi, F. (2006). Gender differences in bipolar-II disorder. *European Archives of Psychiatry and Clinical Neuroscience Journal, 256,* 67–71.

Chang, C. C., Lu, R. B., Ma, K. H., Chang, H. A., Chen, C. L., Huang, C. C., et al. (2007). Association study of the norepinephrine transporter gene polymorphisms and bipolar disorder in Han Chinese population. *The World Journal of Biological Psychiatry, 8,* 188–195.

Cichon, S., Schmidt-Wolf, G., Schumacher, J., Müller, D. J., Hürter, M., Schulze, T. G., et al. (2001). A possible susceptibility locus for bipolar affective disorder in chromosomal region 10q25–q26. *Molecular Psychiatry, 6,* 342–349.

Clerici, M., Arosio, B., Mundo, E., Cattaneo, E., Pozzoli, S., Dell'osso, B., et al. (2009). Cytokine polymorphisms in the pathophysiology of mood disorders. *CNS Spectrums, 14,* 419–425.

Cullen, B., Ward, J., Graham, N. A., Deary, I. J., Pell, J. P., Smith, D. J., et al. (2016). Prevalence and correlates of cognitive impairment in euthymic adults with bipolar disorder: a systematic review. *Journal of Affective Disorder, 205,* 165–181.

D'Addario, C., Dell'Osso, B., Palazzo, M. C., Benatti, B., Lietti, L., Cattaneo, E., et al. (2012). Selective DNA methylation of BDNF promoter in bipolar disorder: differences among patients with BDI and BDII. *Neuropsychopharmacology, 37,* 1647–1655.

de Sousa, R. T., Uno, M., Zanetti, M. V., Shinjo, S. M., Busatto, G. F., Gattaz, W. F., et al. (2014). Leukocyte mitochondrial DNA copy number in bipolar disorder. *Progress in Neuro-Psychopharmacology & Biological Psychiatry, 48,* 32–35.

Delvecchio, G., Fossati, P., Boyer, P., Brambilla, P., Falkai, P., Gruber, O., et al. (2012). Common and distinct neural correlates of emotional processing in bipolar disorder and major depressive disorder: a voxel-based meta-analysis of functional magnetic resonance imaging studies. *European Neuropsychopharmacology, 22,* 100–113.

Dmitrzak-Weglarz, M., Rybakowski, J. K., Slopien, A., Czerski, P. M., Leszczynska-Rodziewicz, A., Kapelski, P., et al. (2006). Dopamine receptor D1 gene-48A/G polymorphism is associated with bipolar illness but not with schizophrenia in a polish population. *Neuropsychobiology, 53,* 46–50.

Foroud, T., Castelluccio, P. F., Koller, D. L., Edenberg, H. J., Miller, M., Bowman, E., et al. (2000). Suggestive evidence of a locus on chromosome 10p using the NIMH genetics initiative bipolar affective disorder pedigrees. *American Journal of Medical Genetics Part A, 96,* 18–23.

Hu, M. C., Lee, S. Y., Wang, T. Y., Chang, Y. H., Chen, S. L., Chen, S. H., et al. (2015). Interaction of DRD2TaqI, COMT, and ALDH2 genes associated with bipolar II disorder comorbid with anxiety disorders in Han Chinese in Taiwan. *Metabolic Brain Disease, 30*, 755–765.

Johansson, A. G., Nikamo, P., Schalling, M., & Landén, M. (2011). AKR1C4 gene variant associated with low euthymic serum progesterone and a history of mood irritability in males with bipolar disorder. *Journal of Affective Disorder, 133*, 346–351.

Johansson, A. G., Nikamo, P., Schalling, M., & Landén, M. (2012). Polymorphisms in AKR1C4 and HSD3B2 and differences in serum DHEAS and progesterone are associated with paranoid ideation during mania or hypomania in bipolar disorder. *European Neuropsychopharmacology, 22*, 632–640.

Mathieu, F., Dizier, M. H., Etain, B., Jamain, S., Rietschel, M., Maier, W., et al. (2010). European collaborative study of early-onset bipolar disorder: Evidence for genetic heterogeneity on 2q14 according to age at onset. *American Journal of Medical Genetics Part B: Neuropsychiatric Genetics, 153B*, 1425–1433.

McDonald, M. L., MacMullen, C., Liu, D. J., Leal, S. M., & Davis, R. L. (2012). Genetic association of cyclic AMP signaling genes with bipolar disorder. *Translational Psychiatry, 2*, e169.

Nöthen, M. M., Cichon, S., Rohleder, H., Hemmer, S., Franzek, E., Fritze, J., et al. (1999). Evaluation of linkage of bipolar affective disorder to chromosome 18 in a sample of 57 German families. *Molecular Psychiatry, 4*, 76–84.

Otani, K., Ujike, H., Tanaka, Y., Morita, Y., Katsu, T., Nomura, A., et al. (2005). The GABA type A receptor alpha5 subunit gene is associated with bipolar I disorder. *Neuroscience Letters, 381*, 108–113.

Palo, O. M., Soronen, P., Silander, K., Varilo, T., Tuononen, K., Kieseppä, T., et al. (2010). Identification of susceptibility loci at 7q31 and 9p13 for bipolar disorder in an isolated population. *American Journal of Medical Genetics Part B: Neuropsychiatric Genetics, 153B*, 723–735.

Phillips, M. L., & Kupfer, D. J. (2013). Bipolar disorder diagnosis: challenges and future directions. *Lancet, 381*, 1663–1671.

Rucci, P., Nimgaonkar, V. L., Mansour, H., Miniati, M., Masala, I., Fagiolini, A., et al. (2009). Gender moderates the relationship between mania spectrum and serotonin transporter polymorphisms in depression. *American Journal of Medical Genetics Part B: Neuropsychiatric Genetics, 150B*, 907–913.

Rybakowski, J. K., Skibinska, M., Leszczynska-Rodziewicz, A., Kaczmarek, L., & Hauser, J. (2009). Matrix metalloproteinase-9 gene and bipolar mood disorder. *Neuromolecular Medicine, 11*, 128–132.

Seok, J. W., Lee, K. H., Sohn, S., & Sohn, J. H. (2015). Neural substrates of risky decision making in individuals with internet addiction. *Australian & New Zealand Journal of Psychiatry, 49*, 923–932.

Spijker, A. T., Giltay, E. J., van Rossum, E. F., Manenschijn, L., DeRijk, R. H., Haffmans, J., et al. (2011). Glucocorticoid and mineralocorticoid receptor polymorphisms and clinical characteristics in bipolar disorder patients. *Psychoneuroendocrinology, 36*, 1460–1469.

Spijker, A. T., van Rossum, E. F., Hoencamp, E., Derijk, R. H., Haffmans, J., Blom, M., et al. (2009). Functional polymorphism of the glucocorticoid receptor gene associates with mania and hypomania in bipolar disorder. *Bipolar Disorder, 11*, 95–101.

Strakowski, S. M., Eliassen, J. C., Lamy, M., Cerullo, M. A., Allendorfer, J. B., Madore, M., et al. (2011). Functional magnetic resonance imaging brain activation in bipolar mania: evidence for disruption of the ventrolateral prefrontal-amygdala emotional pathway. *Biological Psychiatry, 69*, 381–388.

Wang, Z., Li, Z., Chen, J., Huang, J., Yuan, C., Hong, W., et al. (2012). Association of BDNF gene polymorphism with bipolar disorders in Han Chinese population. *Genes, Brain and Behavior*, *11*, 524–528.

Xu, C., Li, P. P., Cooke, R. G., Parikh, S. V., Wang, K., Kennedy, J. L., et al. (2009). TRPM2 variants and bipolar disorder risk: confirmation in a family-based association study. *Bipolar Disorder*, *11*, 1–10.

Xu, C., Macciardi, F., Li, P. P., Yoon, I. S., Cooke, R. G., Hughes, B., et al. (2006). Association of the putative susceptibility gene, transient receptor potential protein melastatin type 2, with bipolar disorder. *American Journal of Medical Genetics Part B: Neuropsychiatric Genetics*, *141B*, 36–43.

Early pharmacological interventions in youth

11

Dana Baker Kaplin*, Ekaterina Stepanova*,†, Bradley Grant*,†, Dejan B. Budimirovic*,†, Robert L. Findling†,‡

Division of Child and Adolescent Psychiatry, Department of Psychiatry and Behavioral Sciences, Johns Hopkins University School of Medicine, Baltimore, MD, United States* Children's Mental Health Center, Kennedy Krieger Institute, Baltimore, MD, United States† Division of Child and Adolescent Psychiatry, Department of Psychiatry and Behavioral Sciences, The Johns Hopkins Hospital, Bloomberg Children's Center, Johns Hopkins University School of Medicine, Baltimore, MD, United States‡

CHAPTER OUTLINE

INTRODUCTION

The diagnosis and treatment of bipolar disorder (BD) in adults is well-studied and researchers continue to elucidate psychopharmacological findings on effective treatments for this complex illness. Although there is a relative dearth of scientifically rigorous, controlled clinical trials on pediatric bipolar illness, there is growing evidence suggesting that bipolar disorder may begin not only prior to adulthood, but during pre-adolescence (Findling et al., 2005; Post et al., 2010).

In the United States, the overall prevalence of bipolar disorder in youth is approximately 1.8% according to a recent metaanalysis (Van Meter, Moreira, & Youngstrom, 2011).

This study examined prevalence across the continuum of bipolar disorders including bipolar disorder I (BDI), bipolar disorder II (BDII), cyclothymia (CYC), and bipolar disorder not otherwise specified (BD-NOS). Nevertheless, most clinical trials have focused on youth with a BDI diagnosis who more frequently experience symptoms of acute manic and mixed states (Leverich et al., 2007).

Pediatric bipolar disorder is a chronic illness and if untreated can be seriously debilitating and substantially reduce one's quality of life (Findling et al., 2010). Furthermore, longitudinal studies indicate that across all bipolar disorders, earlier age of onset is associated with greater severity of illness and with more serious negative outcomes in adulthood (Post et al., 2010) such as unemployment, school dropout, and substance use. A psychiatric diagnosis of BDI in youth is also a risk factor for both suicide attempts and suicide (Hauser, Galling, & Correll, 2013).

Current data indicate that pharmacologic intervention may be an effective treatment of BDI and BDII in youth. Initially, medications were almost exclusively studied in adult populations often with subsequent approval for youth by the U.S. Food and Drug Administration (FDA). However, there are now several medications with FDA-approved indications for use in pediatric populations.

The purpose of this chapter is to review what is known about the psychopharmacologic treatment of children and adolescents diagnosed with bipolar disorders. As BDI is the most studied regarding the pharmacotherapy of pediatric bipolar illness, this chapter will primarily focus on this population unless otherwise indicated.

ACUTE MONOTHERAPY FOR TREATMENT OF PEDIATRIC BIPOLAR I DISORDER (MIXED OR MANIC STATES)

Lithium carbonate (Li^+), antipsychotics and anticonvulsants appear to be the classes of medication that are most commonly used in treating youth suffering with BDI.

Table 1 details current FDA-approved monotherapies for the treatment of BDI in children and adolescents with acute manic or mixed episodes.

SELECTED STUDIES ON FDA-APPROVED ACUTE MONOTHERAPY FOR PEDIATRIC BDI (MIXED OR MANIC)

Lithium

Lithium carbonate (Li^+) was the first FDA-approved medication prescribed for treating BDI in adolescents ages 12 and older. Li^+ was initially studied and FDA-approved for the treatment of bipolar disorder in adults. There are, however, only a few rigorous, evidence-based treatment studies of Li^+ in pediatric populations.

In 2008, Findling and colleagues began a series of Li^+ studies funded by the National Institute of Child Health and Human Development (NICHD) called the Collaborative Lithium Trials (CoLT). The purpose of these initial studies was to determine the pharmacokinetics, dosing strategy and tolerability of Li^+ in pediatric participants ages 7–17 years (Findling et al., 2008).

Subsequently, in the first CoLT randomized, double-blind controlled trial that followed, Li^+ was superior in reducing symptoms of acute mania when compared with placebo in a group of children aged 12 years and older (Findling, Robb, et al., 2015).

Second-generation antipsychotics

More recently, second-generation antipsychotics (SGAs), also known as atypical antipsychotics, have been tested in randomized, double-blind clinical trials. Monotherapeutic treatment of pediatric BDI, with acute mixed and manic episodes, with SGAs seems to be highly effective in youth. However, the prescriber should be careful to monitor for side effects such as increased weight and hyperlipidemia, as these may be a limiting factor in treatment with certain SGAs.

SGAs that are currently FDA-approved for acute monotherapy of BDI with manic or mixed episodes include olanzapine, aripiprazole, asenapine, quetiapine, and risperidone. Evidence-based studies of these SGAs are detailed below and in Table 1.

Olanzapine is FDA-approved for treatment of acute mania in bipolar youth 13 years old or greater. In a 3-week, randomized, multisite, double-blind, placebo-controlled study of olanzapine in adolescents aged 13–17 with acute mixed and manic episodes, olanzapine monotherapy was superior to placebo in treating symptoms of acute mania (Tohen et al., 2007).

Table 1 FDA-approved monotherapy for the treatment of pediatric bipolar I disorder (BDI): acute manic or mixed episodes

Medication (class)	Study description	Dosing and side effects	Clinical significance of study	FDA-approved indication for treatment	Study reference(s)
Lithium	• Multicenter randomized, double-blind, placebo-controlled • 8-week study • Youth ages 7–17 • BDI—manic or mixed episodes	Starting dose of 300mg three times daily, with weekly increases of 300mg until therapeutic response or maximum blood level of 1.4mEq/L achieved or dose limiting side effect. Blood levels must be monitored. Findling et al. (2011) Common side effects include: sleepiness, muscle weakness, acne, and elevated thyroid-stimulating hormone (TSH) levels. Findling et al. (2008)	Lithium was superior to placebo in reducing manic symptoms Effect size was similar to that in historically reported in adult studies	Acute mania, in youth 12years or greater	Findling, Robb, et al. (2015)
Second generation antipsychotics					
Aripiprazole	• Randomized, double-blind, placebo-controlled • 4-week study • Youth ages 7–17 • BDI—manic or mixed episodes	*Target dose of 10mg/d* (initial dose for this group started at 2mg/d and was titrated up over a 5-day period until the target dose was reached) **or** *Target dose of 30mg/d* (initial dose for this group started at 10mg/d and was titrated up over a 7-day period until the target dose was reached) **or** Placebo once daily Common side effects include: extrapyramidal side effects and somnolence with higher rates in 30mg than in 10mg dose. Nausea and stomach upset were also common side effects	Aripiprazole was beneficial in treating BDI, acute manic or mixed states	Acute manic or mixed states, in youth 7–17years	Findling et al. (2009)

Asenapine	• Randomized, double-blind, placebo controlled • 3-week study • Youth ages 10–17 • BDI—manic or mixed episodes	*Target dose of 5mg twice daily (initial dose for this group was 2.5mg twice daily for 4 days until reaching target dose)* **or** *Target dose of 10mg twice daily (initial dose for this group was 2.5mg twice daily and titrated up over a 7-day period until reaching target dose)* **or** Placebo twice daily Common side effects: somnolence, sedation, hypoesthesia oral, paraesthesia oral, and increased appetite	Asenapine was superior in all doses compared to placebo in reducing manic or mixed symptoms	Acute manic or mixed states in youth ages 10–17 years	Findling, Landbloom, et al. (2015)
Olanzapine	• Randomized, double-blind, placebo controlled • 3-week study • Youth ages 13–17 • BDI—manic or mixed episodes	*Initial dose of either 2.5mg or 5mg once daily (dose could be titrated up to 20mg/d at the investigator's discretion based on participant response)* **or** Placebo once daily Substantial weight gain and lipid increases can be treatment limiting side effects. Other common side effect include: sedation and increase in appetite	Olanzapine was more effective than placebo in treating BDI with acute mania	Acute mania in youth ages 13–17	Tohen et al. (2007)
Quetiapine	• Randomized, double-blind, placebo controlled • 3-week study • Youth ages 10–17 • BDI—manic episodes	*Target doses were 400mg **or** 600mg once daily (initial dose for either target group started at 50mg/d and titrated up over a 3–7-day period until reaching target doses)* **or** Placebo once daily Generally well-tolerated Weight and lipids should be monitored	Quetiapine significantly more effective than placebo in reducing symptoms of acute mania	Acute mania in youth ages 10–17	Pathak et al. (2013)
Risperidone	• Randomized, double-blind • 3-week study • Youth ages 10–17 • BDI—manic or mixed episodes	*Initial dose was 0.25mg once daily (initial dose was titrated up to target ranges 0.5–2.5mg/d **or** 3–6mg/d)* **or** Placebo once daily Common side effects include: weight fatigue, dizziness, and dystonia	Risperidone significantly reduced acute manic and mixed symptoms	Acute manic or mixed states in youth ages 10–17 years	Haas et al. (2009)

Key: BDI, bipolar I disorder; mg/d, milligrams per day; mEq/L, milliequivalent per liter; TSH, elevated thyroid-stimulating hormone.

Findling et al. (2009) conducted a 4-week randomized, double-blind, placebo-controlled study of aripiprazole and found that it was effective in treating children and youth with BDI, acute manic or mixed episodes (Findling et al., 2009). Aripiprazole is FDA-approved for the treatment of BDI, acute manic or mixed states in youth aged 7–17. The same author with other colleagues conducted a 3-week randomized, double-blind, placebo-controlled trial of asenapine, which revealed that asenapine was superior to placebo at all doses. Asenapine is now FDA-approved for the treatment of BDI, acute manic or mixed episodes, in children and adolescents aged 10–17 (Findling, Landbloom, et al., 2015).

Findings from a recent randomized, double-blind, placebo-controlled study of quetiapine, showed it significantly reduced acute mania in youth aged 10–17 years with BDI. Quetiapine was generally well-tolerated in this 3-week study; however, monitoring weight and lipids was noted as a necessary precaution (Pathak et al., 2013).

In a 3-week randomized, double-blind, placebo-controlled, safety and efficacy study, risperidone significantly decreased symptoms of acute manic or mixed episodes experienced by youth with bipolar disorder (Haas et al., 2009). The study results also demonstrated that the benefit-risk profile for risperidone at daily dosing of 0.5–2.5 mg is better than in daily dosing of 3–6 mg. Risperidone is also FDA-approved for this indication in youth aged 10–17.

In another randomized, placebo-controlled efficacy and safety trial comparing divalproex sodium (DVPX) and risperidone monotherapy to placebo, Kowatch et al. (2015) found that risperidone was more effective when compared to placebo while DVPX was not more effective than placebo. This was a small, 6-week study in 46 children aged 3–7. Results from this study also suggested that particularly in preschool children with BDI, mixed or manic states, frequent monitoring for metabolic side effects is warranted (Kowatch et al., 2015).

SELECTED STUDIES WITH PROMISING RESULTS FOR ACUTE MONOTHERAPY OF PEDIATRIC BDI (MIXED OR MANIC)

Preliminary findings from a study of clozapine, an atypical antipsychotic, indicated that it may be effective in treating adolescents aged 12–17 with treatment-refractory BDI, manic or mixed episodes. If study results are confirmed, this may be a good option for treating adolescents who do not have a good clinical response to conventional treatment with other agents (Masi, Mucci, & Millepiedi, 2002).

Results from a double-blind, placebo-controlled trial conducted by DelBello et al. (2005) suggested that topiramate may be effective in treating pediatric BDI. However, this study was discontinued when adult bipolar disorder trials found topiramate not to be an effective treatment for that population. This study was therefore stopped before it was complete, and was statistically underpowered (Delbello et al., 2005).

In a 4-week randomized, double-blind, placebo-controlled trial followed by a 26-week open-label extension study, Findling, Cavus, et al. (2013) demonstrated that ziprasidone may be effective in treating youth with BDI, mixed or manic, aged 10–17 compared to placebo. The intent of the study was to examine short- and long-term safety and efficacy of ziprasidone in the treatment of youth with BDI. Results indicated ziprasidone was generally well-tolerated and, in contrast with most other SGAs, exhibited a neutral metabolic profile.

In a large prospective, open-label study of extended-release carbamazepine (ERC), Findling & Ginsberg (2014) showed that ERC may be beneficial in treating youth aged 10–17 with BDI acute manic or mixed episodes (Findling & Ginsberg, 2014). Definitive studies are needed to confirm these results.

Omega-3 fatty acids

Initial findings from an epidemiological study of omega-3 fatty acids showed a possible correlation between lower rates of bipolarity and greater consumption of omega-3 fatty acids-rich seafood (Noaghiul & Hibbeln, 2003).

Gracious et al. (2010) published findings from a randomized, placebo-controlled study using flax seed as a source of omega-3 fatty acids. The results showed a decrease in symptoms associated with pediatric bipolar disorder I and II. Additional evidence-based studies are needed to determine effective omega-3 dosing strategies in treating youth with bipolar disorder (Gracious et al., 2010).

In another study, the use of omega-3 in treating youth with bipolar disorder was conducted by Wozniak et al. (2015) in a small, randomized, placebo-controlled study of children aged 6–12, with a diagnosis of bipolar disorder I, II, or BD-NOS. This study examined treatment with combination therapy of omega-3 fatty acids and inositol. The authors observed that this combination may be beneficial in treating prepubertal children with bipolar spectrum disorders with acute, manic, mixed, or hypomanic symptoms (Wozniak et al., 2015). Larger controlled trials are needed to confirm these findings.

SELECTED STUDIES WITH NEGATIVE RESULTS FOR TREATMENT OF PEDIATRIC BDI (MIXED OR MANIC)

A safety and efficacy study of adjunctive gabapentin indicated that it was not an effective treatment for youth aged 16 and older and adults with BDI, manic, hypomanic, or mixed symptoms (Pande, Crockatt, Janney, Werth, & Tsaroucha, 2000). This study compared gabapentin to Li^+, DVPX, and Li^+ plus DVPX. Gabapentin was not more effective when compared to these other medications. Currently, there are no methodologically rigorous studies in pediatric-aged patients.

In a multicenter, randomized, double-blind safety and efficacy study on oxcarbazepine, Wagner et al. (2006) found that oxcarbazepine was not more effective than placebo in treating youth aged 7–18 with BDI.

ACUTE MONOTHERAPY FOR TREATMENT OF PEDIATRIC BIPOLAR I DISORDER (DEPRESSION)

There is a dearth of efficacy trials on the use of antidepressant treatment in youth with BDI with depressive symptoms (Cosgrove, Roybal, & Chang, 2013). Moreover, evidence from some studies suggests that treatment with antidepressants may induce mania in youth with depression, particularly in younger adolescents and those who have a parent with bipolar I disorder (Strawn Jr. et al., 2014). The observation that some youth may be particularly vulnerable to antidepressant-related mania may help explain why antidepressant monotherapy trials have not been conducted in youth who have already developed a bipolar disorder.

That said, there are now two medications that are FDA-approved for treating depression in youth, ages 10–17, with BDI as detailed below and in Table 2.

SELECTED STUDY ON FDA-APPROVED ACUTE MONOTHERAPY FOR PEDIATRIC BIPOLAR I DISORDER (DEPRESSION)

Detke, DelBello, Landry, and Usher (2015) published the results of a study using a combination of olanzapine and fluoxetine, known as combination OFC (brand name Symbyax), to treat acute depressive symptoms in youth. This was an 8-week randomized, double-blind, placebo-controlled study of youth aged 10–17 where OFC was significantly superior to placebo in reducing symptoms of depression in this population (Detke et al., 2015). Until recently, OFC was the only medication with FDA-approval for the treatment of bipolar depression in youth (see Table 2).

Lurasidone has been studied in youth and was recently FDA-approved for acute monotherapy in youth ages 10–17 with BDI depression. Well-studied in adult populations, lurasidone was initially approved as monotherapy for acute BDI depression in that population. In a study by Loebel and colleagues (Loebel et al., 2014), lurasidone in doses of 20–120 mg daily was found to be effective in decreasing depressive symptoms in adults.

Subsequently, Findling, Goldman, et al. (2015), in a multicenter, open-label pharmacokinetic and tolerability study in youth aged 6–17 with psychiatric disorders including BDI, demonstrated that lurasidone was better tolerated in doses less than 120 mg daily compared to higher doses.

In 2017, a pivotal study conducted by DelBello and colleagues found that lurasidone was more effective than placebo in reducing symptoms of depression in youth ages 10–17 with BDI (DelBello et al., 2017). These findings led to FDA-approval of lurasidone for this indication.

STUDIES WITH PROMISING RESULTS FOR ACUTE MONOTHERAPY IN PEDIATRIC BIPOLAR I DISORDER (DEPRESSION)

In a small, open-label efficacy and safety study of lithium for the treatment of youth with BDI with acute depression, Patel et al. (2006) found that Li^+ may be effective and safe in significantly reducing symptoms of depression in adolescents aged 12–18. The findings also indicate that Li^+ may be relatively well-tolerated (Patel et al., 2006).

SELECTED STUDIES WITH NEGATIVE RESULTS FOR MONOTHERAPY IN PEDIATRIC BIPOLAR I DISORDER (DEPRESSION)

In a pilot study of quetiapine, DelBello and colleagues found that quetiapine was not more effective than placebo in treating adolescents with bipolar depression

Table 2 FDA-approved therapy for the treatment of pediatric bipolar I disorder (BDI): acute depressive episodes

Medication (class)	Study description	Dosing and side effects	Clinical significance of Study	FDA-approved indication for treatment	Study reference(s)
Olanzapine/ Fluoxetine Combination (OFC) Brand name—Symbyax	• Randomized, double-blind, placebo controlled • 8-week study • Youth ages 10–17 • BDI–depressive episodes	Initial dose was 3 mg olanzapine/25 mg fluoxetine once daily, in one tablet (target doses were titrated up to 6/25 mg/d **or** 6/50 mg/d, 12/25 mg/d, **or** 12/50 mg/d over a 2-week period) **or** Placebo once daily Common side effects include: weight gain, increased appetite, hyperlipidemia, and somnolence	OFC was superior to placebo in reducing depressive symptoms	Acute depression in youth ages 10–17 years	Detke et al. (2015)
Monotherapy Lurasidone Brand name—Latuda	• Randomized, Double-blind, placebo controlled • 6-week Study • Youth ages 10–17 • BDI-depressive episodes	Initially was flexibly dosed from 20 mg/d to 80 mg/d, in one tablet **or** Placebo once daily Common side effects include: nausea and somnolence	Lurasidone was superior to placebo in significantly reducing depressive episodes	Acute depression in youth with BDI, ages 10–17 years	DelBello et al. (2017)

Key: BDI, bipolar I disorder; mg/d, milligrams per day.

(DelBello et al., 2009). It was suggested that the high placebo response rate may have had an impact on the study findings, which should be a consideration in designing future studies.

Another study of quetiapine extended-release (XR) by Findling, Pathak, Earley, Liu, & DelBello (2014) demonstrated that quetiapine XR was not more effective than placebo as a treatment for acute pediatric bipolar depression. This was a multicenter, double-blind, randomized trial in youth aged 10–17 (Findling, Pathak, et al., 2014).

MAINTENANCE THERAPY FOR TREATMENT OF PEDIATRIC BIPOLAR I DISORDER

Few scientifically rigorous maintenance studies on BDI in youth have been conducted to date. Pediatric bipolarity is, however, a chronic illness and clinical trials are very much needed to determine the safety and efficacy of pharmacologic treatments in pediatric patients with acute manic and mixed states, during the maintenance phase of the illness.

SELECTED PIVOTAL STUDIES OF MAINTENANCE TREATMENTS FOR PEDIATRIC BDI

A long-term efficacy study of aripiprazole for maintenance therapy in children with bipolar disorders was conducted by Findling et al. (2012). Participants enrolling in this double-blind, randomized placebo-controlled study were outpatients with BDI, manic or mixed states aged 4–9. The results of the study did show that in long-term treatment, aripiprazole was superior to placebo once these pediatric patients were stabilized with open-label aripiprazole (Findling et al., 2012).

This same group conducted a 30-week, double-blind, randomized study of aripiprazole as a maintenance treatment for youth aged 10–17 with BDI, manic or mixed episodes. Although completion rates were low, results suggest that aripiprazole was more effective than placebo (Findling, Correll, et al., 2013).

SELECTED STUDIES WITH PROMISING RESULTS FOR MAINTENANCE TREATMENT FOR PEDIATRIC BDI

Findling, Chang, et al. (2015), in a randomized, placebo-controlled withdrawal study comparing lamotrigine to placebo in youth aged 10–17, found that lamotrigine was not more beneficial than placebo as maintenance monotherapy. However, within this cohort, post-hoc analyses suggested possible benefits in youths aged 13–17, but not aged 10–12 (Findling, Chang, et al., 2015).

Findling et al. (2016), in a 50-week, open-label, maintenance, safety study of asenapine conducted in youth aged 10–17 with acute manic or mixed symptoms of BDI, found that asenapine was generally well-tolerated. A combination of somnolence, sedation, and hypersomnia was the most common treatment-emergent adverse event identified (Findling et al., 2016).

In a 26-week, open-label, continuation study of quetiapine, Findling and colleagues found that quetiapine flexibly dosed (400–800 mg/day) in youth with BDI and schizophrenia was generally well-tolerated and safe. The authors noted that monitoring both weight gain and lipids is clinically indicated when treating either disorder with quetiapine (Findling, Pathak, Earley, Liu, & DelBello, 2013).

Both asenapine and quetiapine are considered generally safe and may be effective maintenance treatment of BDI in youth ages 10–17.

A HEAD-TO-HEAD STUDY FOR MAINTENANCE TREATMENT OF PEDIATRIC BDI

To date, only one study compared DVPX to Li^+ in an 18-month, double-blind trial examining maintenance monotherapeutic treatment of youth with BDI, acute manic and mixed episodes (Findling et al., 2005). Once youth, aged 5–17, were clinically stabilized using combination therapy DVPX and Li^+, they were treated with either DVPX or Li^+ monotherapy for up to 76 weeks. Results indicated that DVPX was not superior to Li^+ for maintenance monotherapy in bipolar youth stabilized with combination DVPX and Li^+.

COMBINATION THERAPY FOR TREATMENT OF PEDIATRIC BIPOLAR DISORDER I

Although youth suffering with symptoms of BDI acute manic or mixed episodes on average benefit from monotherapeutic treatments, many do not become completely well.

For this reason, researchers have more recently begun studying combination pharmacotherapy as an additional treatment strategy for pediatric bipolar disorder. Some research examining combination therapies has been promising.

SELECTED STUDIES OF ACUTE COMBINATION TREATMENTS FOR PEDIATRIC BDI

In a controlled trial conducted by Delbello, Schwiers, Rosenberg, and Strakowski (2002), DVPX plus quetiapine was compared to DVPX plus placebo. The combination DVPX plus quetiapine was more effective in treating youth with acute manic and mixed episodes. Sedation in the quetiapine group was, however, a significant side effect.

Findling et al. (2003) conducted an open-label trial of Li^+ and DVPX. This study showed that combination DVPX and Li^+ may be an effective treatment for youth with BDI with acute symptoms of mania and depression.

SELECTED STUDY WITH PROMISING RESULTS FOR COMBINATION TREATMENT OF PEDIATRIC BDI

A 12-month, open-label safety and efficacy trial on risperidone augmentation of Li^+ was conducted by Pavuluri et al. (2006). This study examined youth aged 4–17 years with preschool-onset bipolar disorder with manic or mixed episodes that had an insufficient response to Li^+ monotherapy. After 8 weeks of Li^+ monotherapy, participants who did not respond adequately or who relapsed received augmentation with risperidone. Study results suggested that combination treatment of Li^+ and risperidone was effective, well-tolerated, and generally safe in this population (Pavuluri et al., 2006).

A SELECTED STUDY WITH NEGATIVE RESULTS FOR COMBINATION TREATMENT OF PEDIATRIC BDI

A study on the long-term efficacy of Li^+, in combination with mood stabilization therapy, in treating refractory mania was conducted with youth aged 7–17. Findings suggest that although maintenance treatment with Li^+ may be beneficial for youth who respond well to acute treatment with Li^+, partial responders to acute Li^+ may not see improved symptoms even with adjunct medications (Findling, Kafantaris, et al., 2013).

COMORBIDITY

Bipolar disorder in youth can be challenging to diagnose and treat. This is, in part, because of high rates of comorbidity with other psychiatric illnesses (Kowatch, Youngstrom, Danielyan, & Findling, 2005; Marangoni, De Chiara, & Faedda, 2015; Van Meter, Burke, Kowatch, Findling, & Youngstrom, 2016), including attention-deficit/hyperactivity disorder (ADHD), anxiety, oppositional defiant disorder (ODD), substance use disorders, and more (see Fig. 1). Overlapping

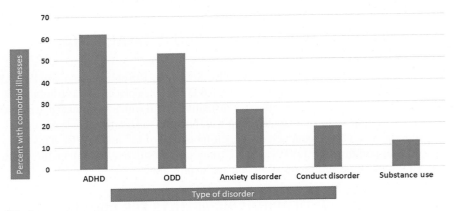

FIG. 1

Common psychiatric comorbidities in pediatric bipolar disorder I.
Goldstein, B. I., Sassi, R., & Diler, R. S. (2012). Pharmacologic treatment of bipolar disorder in children and adolescents. *Child and Adolescent Psychiatric Clinics of North America, 21*(4), 911–939. Doi: https://doi.org/10.1016/j.chc.2012.07.004.

symptoms of these comorbidities require rigorous clinical assessment to result in an accurate diagnosis (Goldstein, Sassi, & Diler, 2012). Unfortunately, comorbid ADHD and comorbid substance use are the only conditions that have been studied in controlled clinical trials.

SELECTED PIVOTAL STUDIES OF COMORBIDITY IN YOUTH WITH ACUTE PEDIATRIC BDI

In a double-blind, placebo-controlled study of Li^+ in youth with bipolar disorders and substance dependency, Li^+ was found to be of some potential benefit (Geller, Cooper, Sun, et al., 1998).

Scheffer, Kowatch, Carmody, and Rush (2005) conducted a safety and efficacy study and found that youth aged 6–17 with BDI or BDII and comorbid ADHD were effectively and safely treated with mixed amphetamine salts (MAS) subsequent to stabilizing treatment for manic symptoms with DVPX. The initial 8-week study of DVPX alone was not effective in treating ADHD with bipolar disorder (Scheffer et al., 2005).

In a 4-week, double-blind, placebo-controlled trial of youth aged 5–17 with bipolar disorder and comorbid ADHD, a study suggested that methylphenidate (MPH) may be an effective, short-term treatment for euthymic youth with bipolar disorder and ADHD (Findling, Short, et al., 2007).

Findling et al. (2005) also conducted a randomized clinical trial comparing DVPX and Li^+ for maintenance treatment in youth aged 5–17 with BDI. A significant number of participants received adjunct psychostimulants for comorbid ADHD across study and treatment groups. The authors found that presence of stimulants was not a risk factor in relapse, suggesting the possibility of relative safety of MPH beyond acute exposure (Findling et al., 2005).

A COMORBIDITY COURSE AND OUTCOME STUDY, ACUTE PEDIATRIC BDI

Studies have found that untreated comorbid conditions may negatively impact the course of one's bipolar illness by increasing the time it takes to recover from a mood episode or shortening the time between mood episodes (Yen et al., 2016). Early identification and treatment of comorbid illnesses are therefore important to the success of treating bipolar disorder in youth.

GENETICALLY AT RISK

The treatment of children and youth who may be genetically at higher risk for developing bipolar disorders has not been well-studied.

A promising randomized, double-blind, placebo-controlled study, the results of which are currently under review for publication, examined symptomatic high-risk youth aged 5–17 diagnosed with cyclothymia or BD-NOS. Participants were defined as being genetically at risk if they had one parent diagnosed with bipolar disorder or had a first- or second-degree relative diagnosed with a mood disorder that did not previously respond to psychotherapeutic intervention. Participants were randomly assigned to either aripiprazole or placebo. In this 12-week study, aripiprazole was significantly more effective than placebo in decreasing symptoms of mania in youth diagnosed with cyclothymia or BD-NOS (Findling et al., 2017).

A double-blind, placebo-controlled trial that included youth aged 5–17 who were diagnosed with either BD-NOS or cyclothymia, and had at least one parent diagnosed with bipolar disorder, was also conducted by Findling, Frazier, et al. (2007). Study participants were randomly assigned to either monotherapy with DVPX or with placebo over the course of 5 years. This trial indicated that monotherapy with DVPX did not clinically improve symptoms of BD-NOS or cyclothymia in youth with genetic risk for developing bipolar disorder (Findling, Frazier, et al., 2007).

In a double-blind, placebo-controlled study of Li^+ in youth with prepubertal major depressive disorder (PMDD) and a family history of BDI or mania or major depressive disorder (MDD), treatment was not found to be better than placebo in predicting bipolar disorder in this pediatric population (Geller et al., 1998).

CONCLUSION

As a direct result of double-blind, placebo controlled studies, pharmacologic treatments for bipolar disorder in pediatric populations have recently expanded and give physicians some evidence-based options in identifying effective treatments for their patients. However, there is still a significant need for evidence-based efficacy and safety studies of medications, particularly in treatment options for maintenance therapy and for comorbid conditions. A need also exists for identifying effective treatment options for BPI with depressive episodes. Given the chronic and harmful impact of bipolar disorder on children and adolescents, rigorous research that may ultimately lead to the delay of onset or progression of pediatric bipolarity is greatly needed.

REFERENCES

Cosgrove, V. E., Roybal, D., & Chang, K. D. (2013). Bipolar depression in pediatric populations: epidemiology and management. *Paediatric Drugs*, *15*(2), 83–91. https://doi.org/10.1007/s40272-013-0022-8.

DelBello, M. P., Chang, K., Welge, J. A., Adler, C. M., Rana, M., Howe, M., et al. (2009). A double-blind, placebo-controlled pilot study of quetiapine for depressed adolescents with bipolar disorder. *Bipolar Disorders*, *11*(5), 483–493. https://doi.org/10.1111/j.1399-5618.2009.00728.x.

Delbello, M. P., Findling, R. L., Kushner, S., Wang, D., Olson, W. H., Capece, J. A., et al. (2005). A pilot controlled trial of topiramate for mania in children and adolescents with bipolar disorder. *Journal of the American Academy of Child and Adolescent Psychiatry*, *44*(6), 539–547. https://doi.org/10.1097/01.chi.0000159151.75345.20.

DelBello, M. P., Goldman, R., Phillips, D., Deng, L., Cucchiaro, J., & Loebel, A. (2017). Efficacy and safety of Lurasidone in children and adolescents with bipolar I depression: a double-blind, placebo-controlled study. *Journal of the American Academy of Child and Adolescent Psychiatry*, *56*(12), 1015–1025. https://doi.org/10.1016j.jaac.2017.10.006.

Delbello, M. P., Schwiers, M. L., Rosenberg, H. L., & Strakowski, S. M. (2002). A double-blind, randomized, placebo-controlled study of quetiapine as adjunctive treatment for adolescent mania. *Journal of the American Academy of Child and Adolescent Psychiatry*, *41*(10), 1216–1223. https://doi.org/10.1097/00004583-200210000-00011.

Detke, H. C., DelBello, M. P., Landry, J., & Usher, R. W. (2015). Olanzapine/fluoxetine combination in children and adolescents with bipolar I depression: a randomized, double-blind, placebo-controlled trial. *Journal of the American Academy of Child and Adolescent Psychiatry*, *54*(3), 217–224. https://doi.org/10.1016/j.jaac.2014.12.012.

Findling, R. L., Cavus, I., Pappadopulos, E., Vanderburg, D. G., Schwartz, J. H., Gundapaneni, B. K., et al. (2013). Efficacy, long-term safety, and tolerability of ziprasidone in children and adolescents with bipolar disorder. *Journal of Child and Adolescent Psychopharmacology*, *23*(8), 545–557. https://doi.org/10.1089/cap.2012.0029.

Findling, R. L., Chang, K., Robb, A., Foster, V. J., Horrigan, J., Krishen, A., et al. (2015). Adjunctive maintenance Lamotrigine for pediatric bipolar I disorder: a placebo-controlled, randomized withdrawal study. *Journal of the American Academy of Child and Adolescent Psychiatry*, *54*(12), 1020–1031. e1023 https://doi.org/10.1016/j.jaac.2015.09.017.

Findling, R. L., Correll, C. U., Nyilas, M., Forbes, R. A., McQuade, R. D., Jin, N., et al. (2013). Aripiprazole for the treatment of pediatric bipolar I disorder: a 30-week, randomized, placebo-controlled study. *Bipolar Disorders*, *15*(2), 138–149. https://doi.org/10.1111/bdi.12042.

Findling, R. L., Frazier, J. A., Kafantaris, V., Kowatch, R., McClellan, J., Pavuluri, M., et al. (2008). The collaborative lithium trials (CoLT): Specific aims, methods, and implementation. *Child and Adolescent Psychiatry and Mental Health*, *2*(1), 21. https://doi.org/10.1186/1753-2000-2-21.

Findling, R. L., Frazier, T. W., Youngstrom, E. A., McNamara, N. K., Stansbrey, R. J., Gracious, B. L., et al. (2007). Double-blind, placebo-controlled trial of divalproex monotherapy in the treatment of symptomatic youth at high risk for developing bipolar disorder. *The Journal of Clinical Psychiatry*, *68*(5), 781–788.

Findling, R. L., & Ginsberg, L. D. (2014). The safety and effectiveness of open-label extended-release carbamazepine in the treatment of children and adolescents with bipolar I disorder suffering from a manic or mixed episode. *Neuropsychiatric Disease and Treatment*, *10*, 1589–1597. https://doi.org/10.2147/NDT.S68951.

Findling, R. L., Goldman, R., Chiu, Y. Y., Silva, R., Jin, F., Pikalov, A., et al. (2015). Pharmacokinetics and tolerability of Lurasidone in children and adolescents with psychiatric disorders. *Clinical Therapeutics*, *37*(12), 2788–2797. https://doi.org/10.1016/j.clinthera.2015.11.001.

Findling, R. L., Kafantaris, V., Pavuluri, M., McNamara, N. K., Frazier, J. A., Sikich, L., et al. (2013). Post-acute effectiveness of lithium in pediatric bipolar I disorder. *Journal of Child and Adolescent Psychopharmacology*, *23*(2), 80–90. https://doi.org/10.1089/cap.2012.0063.

Findling, R. L., Kafantaris, V., Pavuluri, M., McNamara, N. K., McClellan, J., Frazier, J. A., et al. (2011). Dosing strategies for lithium monotherapy in children and adolescents with bipolar I disorder. *J Child Adolesc Psychopharmacol.*, *21*(3), 195–205. https://doi.org/10.1089/cap.2010.0084.

Findling, R. L., Landbloom, R. L., Mackle, M., Wu, X., Snow-Adami, L., Chang, K., et al. (2016). Long-term safety of Asenapine in pediatric patients diagnosed with bipolar I disorder: a 50-week open-label, flexible-dose trial. *Paediatric Drugs*, *18*(5), 367–378. https://doi.org/10.1007/s40272-016-0184-2.

Findling, R. L., Landbloom, R. L., Szegedi, A., Koppenhaver, J., Braat, S., Zhu, Q., et al. (2015). Asenapine for the acute treatment of pediatric manic or mixed episode of bipolar I disorder. *Journal of the American Academy of Child and Adolescent Psychiatry*, *54*(12), 1032–1041. https://doi.org/10.1016/j.jaac.2015.09.007.

Findling, R. L., McNamara, N. K., Gracious, B. L., Youngstrom, E. A., Stansbrey, R. J., Reed, M. D., et al. (2003). Combination lithium and divalproex sodium in pediatric bipolarity. *Journal of the American Academy of Child and Adolescent Psychiatry*, *42*(8), 895–901. https://doi.org/10.1097/01.chi.0000046893.27264.53.

Findling, R. L., McNamara, N. K., Youngstrom, E. A., Stansbrey, R., Gracious, B. L., Reed, M. D., et al. (2005). Double-blind 18-month trial of lithium versus divalproex maintenance treatment in pediatric bipolar disorder. *Journal of the American Academy of Child and Adolescent Psychiatry*, *44*(5), 409–417. https://doi.org/10.1097/01.chi.0000155981.83865.ea.

Findling, R. L., Nyilas, M., Forbes, R. A., McQuade, R. D., Jin, N., Iwamoto, T., et al. (2009). Acute treatment of pediatric bipolar I disorder, manic or mixed episode, with aripiprazole: a randomized, double-blind, placebo-controlled study. *The Journal of Clinical Psychiatry*, *70*(10), 1441–1451. https://doi.org/10.4088/JCP.09m05164yel.

Findling, R. L., Pathak, S., Earley, W. R., Liu, S., & DelBello, M. (2013). Safety, tolerability, and efficacy of quetiapine in youth with schizophrenia or bipolar I disorder: a 26-week, open-label, continuation study. *Journal of Child and Adolescent Psychopharmacology*, *23*(7), 490–501. https://doi.org/10.1089/cap.2012.0092.

Findling, R. L., Pathak, S., Earley, W. R., Liu, S., & DelBello, M. P. (2014). Efficacy and safety of extended-release quetiapine fumarate in youth with bipolar depression: an 8 week, double-blind, placebo-controlled trial. *Journal of Child and Adolescent Psychopharmacology*, *24*(6), 325–335. https://doi.org/10.1089/cap.2013.0105.

Findling, R. L., Robb, A., McNamara, N. K., Pavuluri, M. N., Kafantaris, V., Scheffer, R., et al. (2015). Lithium in the acute treatment of bipolar I disorder: a double-blind, placebo-controlled study. *Pediatrics*, *136*(5), 885–894. https://doi.org/10.1542/peds.2015-0743.

Findling, R. L., Short, E. J., McNamara, N. K., Demeter, C. A., Stansbrey, R. J., Gracious, B. L., et al. (2007). Methylphenidate in the treatment of children and adolescents with bipolar disorder and attention-deficit/hyperactivity disorder. *Journal of the American Academy of Child and Adolescent Psychiatry*, *46*(11), 1445–1453. https://doi.org/10.1097/chi.0b013e31814b8d3b.

Findling, R. L., Youngstrom, E. A., Fristad, M. A., Birmaher, B., Kowatch, R. A., Arnold, L. E., et al. (2010). Characteristics of children with elevated symptoms of mania: the Longitudinal Assessment of Manic Symptoms (LAMS) study. *The Journal of Clinical Psychiatry*, *71*(12), 1664–1672. https://doi.org/10.4088/JCP.09m05859yel.

Findling, R. L., Youngstrom, E. A., McNamara, N. K., Stansbrey, R. J., Wynbrandt, J. L., Adegbite, C., et al. (2012). Double-blind, randomized, placebo-controlled long-term maintenance study of aripiprazole in children with bipolar disorder. *The Journal of Clinical Psychiatry*, *73*(1), 57–63. https://doi.org/10.4088/JCP.11m07104.

Findling, R. L., Youngstrom, E. A., Rowles, B. M., Deyling, E., Lingler, J., Stansbrey, R. J., et al. (2017). A double-blind and placebo-controlled trial of aripiprazole in symptomatic youths at genetic high risk for bipolar disorder. *Journal of Child and Adolescent Psychopharmacology*, *27*(10), 864–874. https://doi.org/10.1089/cap.2016.0160.

Geller, B., Cooper, T. B., Sun, K., Zimerman, B., Frazier, J., Williams, M., et al. (1998). Double-blind and placebo-controlled study of lithium for adolescent bipolar disorders with secondary substance dependency. *Journal of the American Academy of Child & Adolescent Psychiatry*, *37*(2), 171–178. https://doi.org/10.1097/00004583-199802000-00009.

Geller, B., Cooper, T. B., Zimerman, B., Frazier, J., Williams, M., Heath, J., et al. (1998). Lithium for prepubertal depressed children with family history predictors of future bipolarity: a double-blind, placebo-controlled study. *Journal of Affective Disorders*, *51*(2), 165–175.

Goldstein, B. I., Sassi, R., & Diler, R. S. (2012). Pharmacologic treatment of bipolar disorder in children and adolescents. *Child and Adolescent Psychiatric Clinics of North America*, *21*(4), 911–939. https://doi.org/10.1016/j.chc.2012.07.004.

Gracious, B. L., Chirieac, M. C., Costescu, S., Finucane, T. L., Youngstrom, E. A., & Hibbeln, J. R. (2010). Randomized, placebo-controlled trial of flax oil in pediatric bipolar disorder. *Bipolar Disorders*, *12*(2), 142–154. https://doi.org/10.1111/j.1399-5618.2010.00799.x.

Haas, M., Delbello, M. P., Pandina, G., Kushner, S., Van Hove, I., Augustyns, I., et al. (2009). Risperidone for the treatment of acute mania in children and adolescents with bipolar disorder: a randomized, double-blind, placebo-controlled study. *Bipolar Disorders*, *11*(7), 687–700. https://doi.org/10.1111/j.1399-5618.2009.00750.x.

Hauser, M., Galling, B., & Correll, C. U. (2013). Suicidal ideation and suicide attempts in children and adolescents with bipolar disorder: a systematic review of prevalence and incidence rates, correlates, and targeted interventions. *Bipolar Disorders*, *15*(5), 507–523. https://doi.org/10.1111/bdi.12094.

Kowatch, R. A., Scheffer, R. E., Monroe, E., Delgado, S., Altaye, M., & Lagory, D. (2015). Placebo-controlled trial of valproic acid versus risperidone in children 3–7 years of age with bipolar I disorder. *Journal of Child and Adolescent Psychopharmacology*, *25*(4), 306–313. https://doi.org/10.1089/cap.2014.0166.

Kowatch, R. A., Youngstrom, E. A., Danielyan, A., & Findling, R. L. (2005). Review and meta-analysis of the phenomenology and clinical characteristics of mania in children and adolescents. *Bipolar Disorders*, *7*(6), 483–496. https://doi.org/10.1111/j.1399-5618.2005.00261.x.

Leverich, G. S., Post, R. M., Keck, P. E., Jr., Altshuler, L. L., Frye, M. A., Kupka, R. W., et al. (2007). The poor prognosis of childhood-onset bipolar disorder. *Journal of Pediatrics*, *150*(5), 485–490. https://doi.org/10.1016/j.jpeds.2006.10.070.

Loebel, A., Cucchiaro, J., Silva, R., Kroger, H., Hsu, J., Sarma, K., et al. (2014). Lurasidone monotherapy in the treatment of bipolar I depression: a randomized, double-blind, placebo-controlled study. *The American Journal of Psychiatry*, *171*(2), 160–168. https://doi.org/10.1176/appi.ajp.2013.13070984.

Marangoni, C., De Chiara, L., & Faedda, G. L. (2015). Bipolar disorder and ADHD: comorbidity and diagnostic distinctions. *Current Psychiatry Reports*, *17*(8), 604. https://doi.org/10.1007/s11920-015-0604-y.

Masi, G., Mucci, M., & Millepiedi, S. (2002). Clozapine in adolescent inpatients with acute mania. *Journal of Child and Adolescent Psychopharmacology*, *12*(2), 93–99. https://doi.org/10.1089/104454602760219135.

Noaghiul, S., & Hibbeln, J. R. (2003). Cross-national comparisons of seafood consumption and rates of bipolar disorders. *The American Journal of Psychiatry*, *160*(12), 2222–2227.

Pande, A. C., Crockatt, J. G., Janney, C. A., Werth, J. L., & Tsaroucha, G. (2000). Gabapentin in bipolar disorder: a placebo-controlled trial of adjunctive therapy. Gabapentin bipolar disorder study group. *Bipolar Disorders*, *2*(3 Pt 2), 249–255.

Patel, N. C., DelBello, M. P., Bryan, H. S., Adler, C. M., Kowatch, R. A., Stanford, K., et al. (2006). Open-label lithium for the treatment of adolescents with bipolar depression. *Journal of the American Academy of Child and Adolescent Psychiatry*, *45*(3), 289–297. https://doi.org/10.1097/01.chi.0000194569.70912.a7.

Pathak, S., Findling, R. L., Earley, W. R., Acevedo, L. D., Stankowski, J., & Delbello, M. P. (2013). Efficacy and safety of quetiapine in children and adolescents with mania associated with bipolar I disorder: a 3-week, double-blind, placebo-controlled trial. *The Journal of Clinical Psychiatry*, *74*(1), e100–109. https://doi.org/10.4088/JCP.11m07424.

Pavuluri, M. N., Henry, D. B., Carbray, J. A., Sampson, G. A., Naylor, M. W., & Janicak, P. G. (2006). A one-year open-label trial of risperidone augmentation in lithium nonresponder youth with preschool-onset bipolar disorder. *Journal of Child and Adolescent Psychopharmacology*, *16*(3), 336–350. https://doi.org/10.1089/cap.2006.16.336.

Post, R. M., Leverich, G. S., Kupka, R. W., Keck, P. E., Jr., McElroy, S. L., Altshuler, L. L., et al. (2010). Early-onset bipolar disorder and treatment delay are risk factors for poor outcome in adulthood. *The Journal of Clinical Psychiatry*, *71*(7), 864–872. https://doi.org/10.4088/JCP.08m04994yel.

Scheffer, R. E., Kowatch, R. A., Carmody, T., & Rush, A. J. (2005). Randomized, placebo-controlled trial of mixed amphetamine salts for symptoms of comorbid ADHD in pediatric bipolar disorder after mood stabilization with divalproex sodium. *The American Journal of Psychiatry*, *162*(1), 58–64. https://doi.org/10.1176/appi.ajp.162.1.58.

Strawn, A. C., Jr., RK, M. N., Welge, J. A., Bitter, S. M., Mills, N. P., Barzman, D. H., et al. (2014). Antidepressant tolerability in anxious and depressed youth at high risk for bipolar disorder: a prospective naturalistic treatment study. *Bipolar Disorder*, *16*(5), 523–530. https://doi.org/10.1111/bdi.12113.

Tohen, M., Kryzhanovskaya, L., Carlson, G., Delbello, M., Wozniak, J., Kowatch, R., et al. (2007). Olanzapine versus placebo in the treatment of adolescents with bipolar mania. *The American Journal of Psychiatry*, *164*(10), 1547–1556. https://doi.org/10.1176/appi.ajp.2007.06111932.

Van Meter, A. R., Burke, C., Kowatch, R. A., Findling, R. L., & Youngstrom, E. A. (2016). Ten-year updated meta-analysis of the clinical characteristics of pediatric mania and hypomania. *Bipolar Disorders*, *18*(1), 19–32. https://doi.org/10.1111/bdi.12358.

Van Meter, A. R., Moreira, A. L., & Youngstrom, E. A. (2011). Meta-analysis of epidemiologic studies of pediatric bipolar disorder. *The Journal of Clinical Psychiatry*, *72*(9), 1250–1256. https://doi.org/10.4088/JCP.10m06290.

Wagner, K. D., Kowatch, R. A., Emslie, G. J., Findling, R. L., Wilens, T. E., McCague, K., et al. (2006). A double-blind, randomized, placebo-controlled trial of oxcarbazepine in the treatment of bipolar disorder in children and adolescents. *American Journal of Psychiatry*, *163*(7), 1179–1186. https://doi.org/10.1176/appi.ajp.163.7.1179.

Wozniak, J., Faraone, S. V., Chan, J., Tarko, L., Hernandez, M., Davis, J., et al. (2015). A randomized clinical trial of high eicosapentaenoic acid omega-3 fatty acids and inositol as monotherapy and in combination in the treatment of pediatric bipolar spectrum disorders: a pilot study. *The Journal of Clinical Psychiatry*, *76*(11), 1548–1555. https://doi.org/10.4088/JCP.14m09267.

Yen, S., Stout, R., Hower, H., Killam, M. A., Weinstock, L. M., Topor, D. R., et al. (2016). The influence of comorbid disorders on the episodicity of bipolar disorder in youth. *Acta Psychiatrica Scandinavica*, *133*(4), 324–334. https://doi.org/10.1111/acps.12514.

Psychological interventions in offspring of parents with bipolar disorder

12

Rudolf Uher[*,†], **Barbara Pavlova**[*,†]

Department of Psychiatry, Dalhousie University, Halifax, NS, Canada [*] *Nova Scotia Health Authority, Halifax, NS, Canada* [†]

CHAPTER OUTLINE

INTRODUCTION

Sons and daughters of parents with bipolar disorder are at high risk for developing major mood or psychotic disorders themselves (Rasic, Hajek, Alda, & Uher, 2014). By early adulthood, 1 in 10 will develop bipolar disorder, 2 in 10 will develop major depression, and 1 in 20 will develop schizophrenia or another psychotic disorder (Rasic et al., 2014). The very high risk indicates a need for preemptive early interventions for offspring of parents with bipolar disorder (Maziade, 2017). While parents living with bipolar disorder may express their wish for monitoring and intervention for their children (Cumby, Neville, Garnham, Pavlova, & Uher, 2018), existing health systems do not offer services to children of affected parents. In this chapter we will examine what is known about the selection, feasibility, acceptability, and efficacy of preemptive early interventions for offspring of parents living with bipolar

Bipolar Disorder Vulnerability. https://doi.org/10.1016/B978-0-12-812347-8.00012-9

247

disorder, with a specific focus on psychological interventions. We will describe specific psychological interventions that are being tested and offer suggestions for the next steps in designing and delivering interventions to offspring of affected parents.

NEED FOR INTERVENTIONS IN OFFSPRING OF AFFECTED PARENTS

The need for early preemptive interventions stems from the observation of high risk of mood and psychotic disorders in offspring of affected parents (Rasic et al., 2014) and from the concerns expressed by parents living with bipolar disorder (Cumby et al., 2018). Relatively little is known about the needs and wishes of the offspring themselves. A small qualitative study suggests that adolescent offspring who do not suffer from symptoms themselves prefer nonmedical support and are more concerned about other family members than about themselves (Davison & Scott, 2017). The perceived stigma of psychiatric services constitutes a barrier to accepting interventions (Davison & Scott, 2017). We lack systematic data on how a family history of bipolar disorder in a parent affects treatment-seeking in offspring who develop symptoms of mood or other disorders. Anecdotal observations suggest that parents who are in contact with services for their own bipolar disorder are more likely to instigate assessment and treatment for their sons and daughters with symptoms. Overall, there appears to be tension between the concerns of family members and the reluctance of youth to accept interventions if they are offered. Therefore, any planning for interventions has to consider the timing, acceptability, risk of stigmatization, and selection of individuals who are likely both to accept and to benefit from an intervention.

CLINICAL HIGH-RISK AND FAMILY HIGH-RISK APPROACHES TO EARLY IDENTIFICATION AND INTERVENTION

Indicated prevention seeks to reduce the likelihood of illness in individuals who have been identified as being at increased risk. Over the last two decades, the identification of risk has primarily focused on clinical assessments of individuals who are actively seeking treatment for mental health problems. These assessments identify individuals who have symptoms resembling a major mental illness, but fall short of meeting the diagnostic criteria. Such clinical high-risk criteria have been formulated for schizophrenia (Miller et al., 2003; Yung et al., 2005) and for bipolar disorder (Bechdolf et al., 2012) (Box 1).

Individuals who fulfill the clinical-high risk criteria have a 15%–30% likelihood of developing a definite psychotic or bipolar disorder over the 2–5 years following assessment (Bechdolf et al., 2014; Fusar-Poli et al., 2017). The likelihood of developing a severe mental illness may be reduced through psychological interventions in individuals at high clinical risk, as has been demonstrated for psychotic disorders (Hutton

BOX 1 CLINICAL BIPOLAR AT-RISK CRITERIA

Inclusion criteria

Aged between 15 and 25 years and fulfills criteria of at least one of three groups within the last 12 months:

Group 1: Subthreshold mania.
Group 2: Depression and cyclothymic features.
Group 3: Depression and genetic risk.

Definitions

Subthreshold mania: Two-to-three consecutive days of abnormally and persistently elevated, expansive, or irritable mood accompanied by two or more of: (1) inflated self-esteem/grandiosity; (2) decreased need for sleep; (3) unusual talkativeness of pressure of speech; (4) flight of ideas/racing thoughts; (5) distractibility; (6) increase goal-directed activity/agitation.

Depression: At least 1 week of depressed mood, or loss of interest or pleasure accompanied by two or more of: (1) significant weight loss; (2) insomnia or hypersomnia nearly every day; (3) psychomotor retardation or agitation; (4) loss of energy, fatigue; (5) worthlessness or excessive/inappropriate guilt; (6) diminished ability to think or concentrate; (7) recurrent thoughts of death or suicidal ideation.

Cyclothymic features: Numerous episodes with subthreshold manic symptoms not meeting criteria for subthreshold mania and numerous episodes with depressive symptoms. E.g. subthreshold mania symptoms lasting only for 4h within a 24-h period and at least 4 cumulative lifetime days meeting the criteria.

Genetic risk: First-degree relative with bipolar disorder.

Exclusion criteria

(a) History of a treated or untreated manic episode lasting 4 days or longer.
(b) History of a treated or untreated psychosis lasting 7 days or longer.
(c) Past treatment with a mood stabilizer for 6 weeks or longer.
(d) Past treatment with a therapeutic dose[1] of an antipsychotic for 3 weeks or longer.
(e) IQ below the normal range.
(f) Organic brain disorder.

[1]Equivalent to 15 mg of haloperidol per week or more.

& Taylor, 2014). However, the impact of the clinical high-risk early interventions on population health is limited because only a small fraction of individuals who are going to develop severe mental illness present with symptoms that qualify for clinical high risk (van Os & Guloksuz, 2017), and those who do are already impaired (Fusar-Poli et al., 2015). Therefore, other approaches to risk identification are needed that can find individuals at risk for mental illness earlier, when there is still time to normalize the developmental trajectory through timely interventions. The family high-risk approach is attractive, because family history can be established early in life and is strongly predictive of risk over a long period of time, thus enabling interventions that can be delivered before impairing symptoms develop. The family high-risk approach may also facilitate accessing at-risk individuals early if an affected family member is already in touch with mental health services. The clinical high-risk and family high-risk approaches are not exclusive and may be complementary. Clinical high-risk criteria typically take into account family history (Box 1). In the presence of positive family history of severe mental illness in close relatives, even milder psychopathology may have strong predictive value. Therefore, a combination of family history and clinical assessment may represent a more comprehensive and effective approach to risk identification than either would alone, and we cover both approaches in this chapter. The key distinction between early interventions that stem primarily from the family high-risk perspective and those being developed from the clinical high-risk perspective may be whether nontreatment-seeking individuals are approached for interventions. The challenge of approaching nontreatment-seeking individuals will likely shape the nature and focus of preemptive interventions for offspring at familial high risk.

WHAT TYPE OF INTERVENTION?

Pharmacological, nutritional, and psychological interventions have all been considered as potential primary and secondary prevention of bipolar disorder and other types of mental illness. The selection of interventions to be offered to offspring of affected parents may need to take into account the developmental and clinical stage, safety, acceptability, and potential efficacy of interventions. In younger children and adolescents at low stages of clinical risk, only safe interventions that do not cause adverse effects or burden or stigmatize the individual would be acceptable. The use of psychotropic medication may not be an acceptable option in the early stages. In young offspring of parents with bipolar disorder, the tolerability of antidepressant medication is relatively poor (Strawn et al., 2014). In a study of adolescent offspring of parents with bipolar disorder, more than half of those who were prescribed an antidepressant for depression or anxiety discontinued the medication because of adverse effects including irritability and activation (Strawn et al., 2014). Use of atypical antipsychotics is associated with large increase in body weight and risk of diabetes in children and adolescents (Correll et al., 2009). In addition, youth are typically reluctant to use long-term medication even if it does not cause major adverse effects.

Nutritional interventions are relatively safe and nonstigmatizing, and may be more acceptable to youth and families than psychotropic medication. However, the beneficial effects of neutriceuticals on mental health tend to be small or nonreplicable (McGorry et al., 2017).

Psychological interventions are time-limited, relatively free of adverse effects, and may be more acceptable to youth and their families. Evidence that substantial beneficial effects last for a number of years after completion (Brent et al., 2015; Saavedra, Silverman, Morgan-Lopez, & Kurtines, 2010) makes psychological interventions a particularly attractive choice in prevention. Consequently, the majority of studies of early interventions in offspring of parents with bipolar disorder have focused on psychological interventions. There is a large number of psychological interventions that may potentially be beneficial. They differ in who is the primary client (offspring, parent, or family), the mode of delivery (individual vs group; face-to-face vs remote technology-facilitated delivery), and intensity (number and frequency of sessions). In the next section, we will outline the psychological interventions that have been applied or are being tested in offspring of parents with bipolar disorder.

WHEN TO INTERVENE?

One of the key questions concerns the timing of interventions. The first major mood episode, typically a depressive one, tends to occur in the late teens, with a modal age at onset of 16 years. The age at onset of first mood episode is lower in offspring of parents with bipolar disorder than in the general population. Therefore, a truly preventive intervention has to target children or adolescents. Accordingly, most existing intervention studies have focused on offspring aged 9–17 years. Children aged 9 and older have the capacity to engage in psychological intervention and learn beneficial skills (Garber, Frankel, & Herrington, 2016). It is possible that interventions at an even earlier age may be beneficial, but they would require a very long follow-up to establish efficacy. To our knowledge, no previous study tested interventions in children of parents with bipolar disorder younger than 8 years of age. However, parent-focused interventions for mothers with depression have beneficially influenced the well-being of their offspring in the toddler and preschool age range (Goodman & Garber, 2017), suggesting that interventions in early childhood may also need to be considered for offspring of parents with bipolar disorder.

The first onset depressive episode in offspring of parents with bipolar disorder may offer another window of opportunity for intervention that may aim to reduce the risk of depression recurrence and/or development of (hypo)manic episodes. Such interventions would target youth aged 14–25 who develop mood symptoms but have not experienced a hypomanic or manic episode. Those who have subthreshold hypomanic symptoms in addition to depressive episodes may be at very high risk for developing bipolar disorder over a relatively short time period of 1–2 years (Bechdolf et al., 2010; Fiedorowicz et al., 2011; Nadkarni & Fristad, 2010).

The combination of family history (affected parent) with depressive episode and subthreshold hypomanic symptoms may identify a group at ultra-high risk for developing bipolar disorder and in definite need for secondary-prevention interventions. Clearly the interventions appropriate at this later stage would differ from interventions in asymptomatic or mildly symptomatic offspring of affected parents. Staging models have been developed to match interventions to the stage of risk or disorder development in treatment-seeking individuals at clinical high risk. Fig. 1 presents a staging model adapted to offspring of parents with bipolar disorder.

Compared to staging in individuals without family history of major mental illness, milder "antecedent" symptoms that may not reach a diagnostic threshold for a specific disorder may need to be considered as targets for early interventions, because they strongly predict onsets of major mood disorders in the context of positive family history. Affective lability, anxiety, psychotic symptoms, and basic symptoms have been identified as antecedents that precede and predict the onset of major mood and psychotic disorders in long-term prospective studies, and are used as indications for psychological preemptive early interventions within the Families Overcoming Risks and Building Opportunities for Well-being (FORBOW) cohort. Therefore, in Fig. 1, the antecedent stage, marked as 1/0, is highlighted as distinct from stage 0 (no symptoms) and stage 1 (distress disorder). In addition, stage 2 (first episode) is

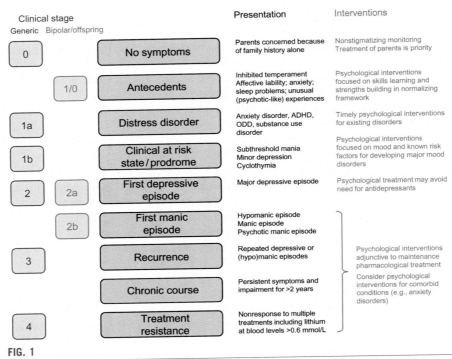

FIG. 1

Stages of risk and interventions, adapted to offspring of parents with bipolar disorder.

separated into onset of first major depressive episode (2a) and onset of first (hypo) manic episode (2b), which typically occur several years apart and require different intervention options. Psychological interventions matching each stage of risk are proposed in the right-hand column of Fig. 1.

PUBLISHED STUDIES OF PSYCHOLOGICAL INTERVENTIONS IN OFFSPRING OF AFFECTED PARENTS

Three specific psychological interventions have been tested in small samples of offspring of parents with bipolar disorder: family-focused therapy, interpersonal and social rhythm therapy, and mindfulness-based cognitive therapy (Table 1).

Miklowitz and colleagues first developed a *family-focused therapy* (FFT) for youth with bipolar disorder and then adapted it for use with symptomatic offspring of parents with bipolar disorder. The adaptation of FFT has been described in a series of 13 symptomatic offspring of parents with bipolar disorder who were affected with major depressive disorder with or without subthreshold hypomanic symptoms (Miklowitz et al., 2011). The FFT-high risk consists of three modules that are delivered in 12 weekly family sessions. They focus on identifying triggers of mood changes and developing a mood management plan, effective communication, and problem solving. The FFT high-risk intervention has been compared to a minimal intervention control condition (brief psychoeducation) in a randomized controlled trial of 40 offspring aged 9–18, who had major depressive disorder with or without subthreshold manic symptoms or cyclothymia (Miklowitz et al., 2013). In addition to FFT, the study participants had access to pharmacotherapy by study psychiatrists and 75% participants received psychotropic medication. Over a 12-month follow-up, the youth who had received FFT spent more time without subthreshold mood symptoms than those who had received brief educational control intervention (Miklowitz et al., 2013). However, FFT did not reduce the risk of mood episodes and the only onset of bipolar disorder over the follow-up occurred in a participant who had received FFT. Clearly the sample size and length of follow-up were not sufficient to establish whether the FFT intervention can actually prevent the onset of bipolar disorder or reduce the risk of recurrence of major depressive episodes. A larger randomized controlled trial is currently underway that aims to compare FFT against an "enhanced care" control condition in 150 youth aged 9–17, who will be followed for up to 4 years (Miklowitz, 2011).

Goldstein and colleagues have adapted the *interpersonal and social rhythm therapy* (IPSRT), an established treatment for bipolar disorder, to use with adolescents who are at high risk for developing bipolar disorder because they have a first-degree relative (most often a parent) with bipolar disorder. IPSRT focuses on regularizing patterns of sleep and social activities, but also includes psychoeducation about the risk for bipolar disorder, exploring the young person's feeling about being at risk, and mourning the "lost healthy family member" and creating a care plan for management of emergent mood symptoms. IPSRT is delivered in 12 individual sessions over 6 months. The development and initial experiences with IPSRT have been reported in 13 adolescents at risk for bipolar disorder (Goldstein et al., 2014). The authors screened 85 interested families, of whom 28 (33%) attended an initial assessment; 57 declined participation in this initial assessment, commonly citing unwillingness of the adolescent to attend sessions. They identified 19 eligible adolescents, of whom 13 (68%) attended at least one intervention session. Those who engaged attended on average 7.7 sessions (range 2–16), found the intervention helpful, and made meaningful changes to their sleep, including less oversleeping on weekends (Goldstein et al., 2014). A shortened 8-session version of the IPSRT intervention has been compared to no intervention in a randomized controlled trial with 42 youth and 6-month follow-up (Goldstein, 2017). Preliminary data suggest improved acceptability, positive effects on sleep, and reduction in subthreshold hypomanic symptoms. With no onsets of major mood episodes, the sample size and duration of follow-up are not sufficient to provide information on whether IPSRT can prevent mood disorders.

Recently, another type of psychological intervention has been applied in a small group of offspring of parents with bipolar disorder: *mindfulness-based cognitive therapy* (MBCT). MBCT has shown long-term efficacy in reducing anxiety and preventing recurrence of major depression in adults. Cotton et al. (2016) have

Table 1 Trials of psychological intervention in offspring of parents with bipolar disorder.

Intervention	Control condition	Mode of delivery	Number of sessions	
			Intervention	Control
Family-focused therapy (FFT)	Education	Family	12	2
Interpersonal and social rhythm therapy (IPSRT)	No intervention	Individual	8	0
Mindfulness-based cognitive therapy (MBCT)	No control group	Individual	12	0
Early specific cognitive-behavioral therapy BEsT (be) for(e) bipolar	Unstructured group meetings	Group	14	14
Cognitive-behavioral therapy—Regulation (CBT-R)		Individual	24	
Cognitive behavioral therapy (BART)	Treatment as usual	Individual	25	0
Multimodal antecedent-focused cognitive-behavioral training (maCBT), Skills for Wellness (SWELL)	No intervention	Individual	16	0

Number of participants		Age (years)	Eligibility	Length of follow-up	Reported results	Reference
Intervention	**Control**					
21 (+75 projected in a second trial)	19 (+75 projected in a second trial)	9–17	First degree relative with bipolar disorder, diagnosis of a mood disorder and current mood symptoms	12 months	Intervention group had faster recovery from existing depressive symptoms and more symptom-free weeks over follow-up	Miklowitz et al. (2013)
21	21	12–18	First degree relative with bipolar disorder	6 months	Improvements in subthreshold hypomanic symptoms and sleep	Goldstein (2017)
10	0	9–17	Biological parent with bipolar I disorder and diagnosis of an anxiety disorder	12 weeks	Reduction in anxiety, improvement in emotional regulation	Cotton et al. (2016)
38 enrolled (projected 50)	37 enrolled (projected 50)	15–30	First degree relative with bipolar disorder and clinical bipolar at risk criteria	18 months	Results not yet available	Pfennig et al. (2014)
15 (projected)			First degree relative with bipolar disorder		Results not yet available	Vallarino et al. (2015)
38 (projected)	38 (projected)	16–25	Clinical bipolar at risk criteria	12 months	Results not yet available	Parker (2015)
24 enrolled (projected 150)	24 enrolled (projected 150)	9–21	One or more antecedent: affective lability, anxiety, psychotic symptoms, basic symptoms	3–10 years	Results not yet available	Uher (2013)

applied MBCT to treatment of anxiety in a series of 11 offspring (9 daughters and 2 sons) of parents with bipolar disorder. Over 12 weeks, 9 of 10 treatment completers responded with substantial reduction of anxiety to the extent of meeting criteria for remission of their anxiety disorders (Cotton et al., 2016). There was no comparison group and the reported follow-up did not extend beyond the 12 weeks of treatment. Therefore, there are no data on whether MBCT may reduce the long-term risk of major mood disorder. This is being tested in an ongoing larger trial (DelBello, 2014).

Additional information can be gleaned indirectly from results of intervention in other family high-risk groups, including offspring of parents with depression (Garber et al., 2009). In 2003–06, Garber and colleagues compared a group cognitive-behavioral therapy (CBT) intervention against usual care in a four-center randomized controlled trial that included 316 symptomatic adolescent offspring (aged 13–17) of parents with depression (Garber et al., 2009). The intervention reduced the risk of depression onset by one-third (21% vs 33%) and the beneficial effects of the intervention persisted for at least 6 years (Beardslee et al., 2013; Brent et al., 2015). This study strongly suggests that a psychological intervention can prevent an onset of mood disorder and raises hopes that bipolar disorder could also be prevented.

ONGOING STUDIES OF PSYCHOLOGICAL INTERVENTIONS IN OFFSPRING OF AFFECTED PARENTS

Although little published evidence is available to date, early interventions for offspring of parents with bipolar disorder are a current topic of interest and there are at least four ongoing studies of novel psychological interventions (Table 1).

Pfennig et al. (2014) have developed an intervention based on *cognitive-behavioral therapy* (CBT) for youth related to individuals with major mood disorders. The BEsT (be)for(e) Bipolar intervention, also described as *early specific CBT* (esCBT), focuses on psychoeducation about emotions and bipolar disorder, stress-management, problem solving, and mindfulness. It is delivered in 14 weekly sessions in groups of 4–5 youths. The BEsT (be)for(e) Bipolar is currently being compared to an unstructured attention control condition in a multicenter randomized controlled trial that aims to enroll 100 youth aged 15–30 who fulfill clinical high-risk criteria for bipolar disorder (Box 1) and also have a first-degree relative with a major depressive disorder, bipolar disorder, or schizoaffective disorder (Pfennig et al., 2014). It is expected that the combination of clinical and family history criteria should lead to a selected sample of individuals at extremely high risk for developing bipolar disorder. The investigators plan to follow the participants for 18 months. The BEsT (be)for(e) Bipolar intervention aims to reduce mood symptoms, improve functioning, and reduce the risk of onset of bipolar disorder (Pfennig et al., 2014).

Scott and colleagues have developed another adaptation of cognitive-behavioral therapy for offspring of parents with bipolar disorder. The *CBT-Regulation* (CBT-R) follows a three-phase model and targets developmental processes that are associated with risk of mood disorders, including rumination, sleep/circadian rhythmicity, and physical activity. CBT-R is delivered in 24 individual sessions over 6 months and is currently being tested in a pilot study (Vallarino et al., 2015).

Parker and colleagues have also adapted CBT for use in high-risk youth, including offspring of parents with bipolar disorder, following a model they previously applied in the prevention of psychosis (Morrison et al., 2004, 2012). The CBT intervention is offered as up to 25 individual sessions of CBT over the course of 6 months. It is currently being compared to "treatment as usual" in the *Bipolar At-Risk Trial* (BART), a randomized controlled trial that plans to enroll 76 help-seeking youth (38 in the intervention group and 38 in the control group) aged 16–25 who experience "mood swings," meet the clinical bipolar at-risk criteria (Box 1), and may also have first-degree relatives with bipolar disorder (Parker, 2015). The investigators aim to reduce the risk of onset of bipolar disorder.

A fourth CBT-based preventive intervention has been developed with the broader aim of preventing major mood and psychotic disorders, including major depressive disorder, bipolar disorder, and schizophrenia. Skills for Wellness (SWELL) is a *multimodal antecedent-focused cognitive-behavioral training* (maCBT) intervention developed based on experience with youth at risk and the needs of families living with mental illness (Uher, 2013). SWELL contains core modules (e.g., problem solving, present moment focus) and specific modules that are selected based on what type of early antecedent is present (mood lability, anxiety, or unusual experiences). SWELL is delivered in 8–16 individual sessions (each lasting 50–60 min) with a "coach," who is a clinical psychologist trained in cognitive-behavioral therapy. Each module progresses from formulation of a current problem to skill development that is designed to counter future challenges in similar areas in addition to addressing current difficulties. A normative focus on skill-learning and coaching makes the intervention acceptable to nontreatment-seeking youth and counters self-stigmatization. The length of the course depends on how many modules are applicable and how long it takes the participant to acquire and practice the target skills. Most participants require between 12 and 16 sessions. The acceptability and efficacy of SWELL are being tested in an ongoing randomized trial (Uher, 2013) that is embedded within a cohort of youth that includes sons and daughters of parents with bipolar disorder, depression, and schizophrenia, the Families Overcoming Risks and Building Opportunities for Wellbeing (FORBOW; Uher et al., 2014). Youth aged 9–21 years are eligible if they have one or more "antecedents" that are known to predict increase risk of major mood and psychotic disorders: affective lability, anxiety, psychotic symptoms, and basic symptoms. A proportion of eligible youth are randomly selected to be offered the SWELL intervention while all remain participants of the FORBOW cohort. The trial-within-cohort (TwiC) design enables externally valid comparison of intervention offer against naturalistic follow-up, which is arguably the most appropriate comparison condition for a preventive intervention (Relton, Torgerson, O'Cathain, & Nicholl, 2010). The aims include reducing antecedent psychopathology and lowering the risk of onset of major mood or psychotic disorders in youth who receive

SWELL. Initial experience suggests that the SWELL intervention is acceptable to nontreatment-seeking youth and their families, with over 80% of youth to whom intervention was offered completing the intervention.

Overall, there are more intervention trials currently in progress than have ever been completed to date. Most of the studies include relatively small numbers of participants and the planned follow-ups range from 6 to 18 months. These studies will provide data on the change in current symptoms. Only one study, SWELL, is designed to address the effect of psychological interventions in preventing onsets of major mood and psychotic illness, including bipolar disorder. Progress to date suggests that this study will take a decade to complete. Since four ongoing trials are testing adaptations of CBT, integration of data across the four, and potentially further additional studies, may bring us closer to estimating the preventive potential of psychological early interventions in offspring of individuals with bipolar disorder.

CONTENT OF INTERVENTIONS

Although the various psychological interventions follow different theoretical models, there is substantial convergence and overlap in their therapeutic techniques. Table 2 lists the core modules and techniques of the six interventions for which sufficient detail is available.

Problem solving, a generally beneficial skill with established modes of delivery and good outcomes, is explicitly included in three of the interventions (FFT, esCBT, maCBT). Regular *sleep and activity* is a well-known factor implicated in the onsets of mood episodes in bipolar disorder, and it is a primary focus of two interventions (IPSRT, CBT-R) and an optional module in a third one (maCBT). *Mindfulness* and/ or attention training toward *external focus* is also used in three of the interventions (MBCT, esCBT, maCBT). *Rumination*, a known predictor of depression, is targeted by CBT-R and maCBT. Learning clear and efficient ways of *communication* is a focus of FFT and an optional module in maCBT. The partial overlap in modules and techniques among the currently trialed interventions may allow identifying the effective elements through joint analysis of results across studies of the different interventions.

HOW ACCEPTABLE AND FEASIBLE ARE PSYCHOLOGICAL INTERVENTIONS FOR OFFSPRING AND THEIR FAMILIES?

No matter how effective an intervention may potentially be, it can only succeed if it is taken up by a large proportion of the target group. Therefore, acceptability is the prerequisite of efficacy and it needs to be a first focus of any intervention program. Interventions that are offered but not taken up will not be beneficial, but may also be harmful because the unacceptable offer may incite self-stigmatization and reduce the

Table 2 Content of psychological interventions for offspring of parents with bipolar disorder

Intervention	Components
Family-focused therapy (FFT)	(1) Education about bipolar disorder (2) Communication (3) Problem solving
Interpersonal and social rhythm therapy (IPSRT)	(1) Education about bipolar disorder (2) Sleep and activity regularization (3) Addressing interpersonal problems (coming to terms with being at risk, mourning a lost healthy family member)
Mindfulness-based cognitive therapy (MBCT)	Mindfulness skills training
Early specific cognitive-behavioral therapy BEsT (be)for(e) Bipolar	Education about bipolar disorder Stress management Problem solving Mindfulness
Cognitive-behavioral therapy—Regulation (CBT-R)	(1) Rumination (2) Sleep (3) Activity

Multimodal antecedent-focused cognitive-behavioral training (maCBT), Skills for wellness (SWELL)	(1) Core: problem solving, mindfulness/external focus
	(2) Antecedent focused:
	  • affective stability: distress tolerance, exposure to emotion, problem solving, communication
	  • anxiety: overcoming avoidance, behavioral experiments, external focus, realistic thinking.
	  • unusual experiences: normalizing, reality testing, communication
	(3) Optional: sleep, psychoactive substances, communication, rumination, perfectionism

likelihood of engaging with other programs or services in the future. The opportunity cost of making an unacceptable offer may be hard to quantify, but it is probably not negligible (Sekhon, Cartwright, & Francis, 2017) and may include reduced willingness to accept help in the future (Lovell et al., 2017). While there is no established cut-off of what proportion of eligible individuals need to start or complete an intervention for it to have impact at population level, it is clear that it should be a large majority. Examining data on efficacy from the initial trial may inform us how we can design and deliver acceptable interventions in the future. In this respect, unsuccessful trials may also be informative.

Researchers in the Netherlands have attempted to test the preventive effects of combined CBT and family therapy on offspring of parents with depression and anxiety (Nauta et al., 2012). Prior to starting the trial, the investigators estimated that they would need to screen 554 children to include 204 in the intervention trial. After 30 months, the trial was abandoned due to lack of recruitment: after screening 13,000 and finding 1300 eligible participants, only 26 enrolled in the trial (Festen et al., 2014). The investigators examined the reasons for nonparticipation through interviews with 24 potential participants who quoted parent burden, shame, and stigma among the barriers to participation. The interviewed parents were concerned about their offspring symptoms, but chose not to discuss their own problems with their offspring and expressed a preference for a parent-focused intervention. This suggests that intervention aimed at nontreatment-seeking youth may be more acceptable if they are framed in a normalizing nonclinical way and if they avoid the need to discuss the parent's illness. However, research design and framing of the intervention offer may also play a major role. The SWELL trial uses a cohort-within-trial design with a two-stage person-centered consent process and it uses nonmedical terminology ("couching" rather than "therapy," "skills" rather than "risk"). To date, over 95% of eligible participants accepted the intervention offered and over 85% completed the SWELL intervention. This contrasting experience suggests that research design and framing also plays a major role and that psychological early interventions can be made acceptable to nontreatment-seeking youth and their families.

CAN PSYCHOLOGICAL INTERVENTIONS PREVENT BIPOLAR DISORDER?

The primary motivation for high-risk research is to prevent the onset of bipolar disorder in (some) individuals at risk. The summary of completed and ongoing intervention studies in high-risk offspring (Table 1) shows that we do not have enough evidence to make even an initial guess on whether a psychological intervention can prevent bipolar disorder. The published studies are too small and the follow-up period too short to analyze onsets as an outcome. Consequently, short-term changes in milder levels of psychopathology have been used as proxy outcomes. Given the strong continuity of psychopathology over time, it is reasonable to extrapolate that short-term reduction in subthreshold symptoms may make a later onset of more severe disorder less likely, but it is not equivalent to demonstrating preventive efficacy. Three ongoing studies are planning to enroll larger samples and complete longer follow-ups and include prevention of bipolar onset among their aims. The similarity of interventions suggests a potential for combining results from the three ongoing studies to obtain a more robust answer than any one study could provide on its own. The amount of work ahead of the investigators is enormous and the need for long-term follow-up means that data will not be available until several years after the interventions are completed. However, the efforts currently ongoing in Germany, England, Canada, and possibly elsewhere give us strong hope that within 5–10 years, we may know whether youth-focused CBT can prevent bipolar disorder.

CONCLUSION

Recently completed and ongoing research brings suggestions on what may be the next steps in psychological interventions for offspring of parents with bipolar disorder. The overarching topics are transdiagnostic approach and staging. A synthesis of family studies shows that offspring of parents with bipolar disorder are at increased risk for a broad range of unfavorable outcomes including bipolar disorder, major depressive disorder, and schizophrenia, and, vice versa, offspring of parents with depression and schizophrenia are also at increased risk of bipolar disorder (Rasic et al., 2014). It is becoming evident that risk of mental illness can be predicted early in life, but the prediction may not be specific to the ultimate diagnosis until the prodrome onsets. The need for earlier interventions before functional impairment occurs has led to calls for early, broad-based, and transdiagnostic risk identification and intervention strategies (Keshavan, DeLisi, & Seidman, 2011; McGorry & Nelson, 2016). A recent review concluded that broad-based transdiagnostic interventions have outcomes as good as interventions that have been specifically developed for bipolar disorder, and suggested that future interventions may be stage-specific rather than disorder-specific (Vallarino et al., 2015). Targets for early interventions have been identified, including sleep difficulties, ruminative thinking style, and affective lability (Uher et al., 2014; Vallarino et al., 2015), which are nonspecifically associated with risk for bipolar disorder as well as other disorders. At this stage, when there are more ongoing studies than completed studies, it may be hard to predict the shape of the next generation of early interventions. However, it is obvious that there is increased recognition of the needs of offspring of parents with mental illness (Maziade, 2017) and commitment to large, long-term trials to test potential preventive effects of interventions in reducing the risk of onset of severe mental illness. As emerging evidence identifies the key components that shape an effective intervention, the high-risk offspring research may move into the implementation phase when early multimodal intervention for offspring may be integrated with services for their parents and family.

REFERENCES

Beardslee, W. R., Brent, D. A., Weersing, V. R., Clarke, G. N., Porta, G., Hollon, S. D., et al. (2013). Prevention of depression in at-risk adolescents: longer-term effects. *JAMA Psychiatry*, *70*, 1161–1170.

Bechdolf, A., Nelson, B., Cotton, S. M., Chanen, A., Thompson, A., Kettle, J., et al. (2010). A preliminary evaluation of the validity of at-risk criteria for bipolar disorders in help-seeking adolescents and young adults. *Journal of Affective Disorders*, *127*, 316–320.

Bechdolf, A., Ratheesh, A., Cotton, S. M., Nelson, B., Chanen, A. M., Betts, J., et al. (2014). The predictive validity of bipolar at-risk (prodromal) criteria in help-seeking adolescents and young adults: a prospective study. *Bipolar Disorders*, *16*, 493–504.

Bechdolf, A., Ratheesh, A., Wood, S. J., Tecic, T., Conus, P., Nelson, B., et al. (2012). Rationale and first results of developing at-risk (prodromal) criteria for bipolar disorder. *Current Pharmaceutical Design*, *18*, 358–375.

Brent, D. A., Brunwasser, S. M., Hollon, S. D., Weersing, V. R., Clarke, G. N., Dickerson, J. F., et al. (2015). Effect of a cognitive-behavioral prevention program on depression 6 years after implementation among at-risk adolescents: a randomized clinical trial. *JAMA Psychiatry, 72*, 1110–1118.

Correll, C. U., Manu, P., Olshanskiy, V., Napolitano, B., Kane, J. M., & Malhotra, A. K. (2009). Cardiometabolic risk of second-generation antipsychotic medications during first-time use in children and adolescents. *JAMA, 302*, 1765–1773.

Cotton, S., Luberto, C. M., Sears, R. W., Strawn, J. R., Stahl, L., Wasson, R. S., et al. (2016). Mindfulness-based cognitive therapy for youth with anxiety disorders at risk for bipolar disorder: a pilot trial. *Early Intervention in Psychiatry, 10*, 426–434.

Cumby, J., Neville, S., Garnham, J., Pavlova, B., & Uher, R. (2018). The needs of parents with severe mental illness and their offspring. Submitted for publication.

Davison, J., & Scott, J. (2017). Should we intervene at stage 0? A qualitative study of attitudes of asymptomatic youth at increased risk of developing bipolar disorders and parents with established disease. *Early Intervention in Psychiatry*.

DelBello, M. (2014). *Mindfulness based cognitive therapy for youth with anxiety at risk for bipolar.* https://clinicaltrials.gov/ct2/show/NCT02090595.

Festen, H., Schipper, K., de Vries, S. O., Reichart, C. G., Abma, T. A., & Nauta, M. H. (2014). Parents' perceptions on offspring risk and prevention of anxiety and depression: a qualitative study. *BMC Psychology, 2*, 17.

Fiedorowicz, J. G., Endicott, J., Leon, A. C., Solomon, D. A., Keller, M. B., & Coryell, W. H. (2011). Subthreshold hypomanic symptoms in progression from unipolar major depression to bipolar disorder. *The American Journal of Psychiatry, 168*, 40–48.

Fusar-Poli, P., Rocchetti, M., Sardella, A., Avila, A., Brandizzi, M., Caverzasi, E., et al. (2015). Disorder, not just state of risk: meta-analysis of functioning and quality of life in people at high risk of psychosis. *The British Journal of Psychiatry, 207*, 198–206.

Fusar-Poli, P., Rutigliano, G., Stahl, D., Davies, C., De, M. A., Ramella-Cravaro, V., et al. (2017). Long-term validity of the at risk mental state (ARMS) for predicting psychotic and non-psychotic mental disorders. *European Psychiatry, 42*, 49–54.

Garber, J., Clarke, G. N., Weersing, V. R., Beardslee, W. R., Brent, D. A., Gladstone, T. R., et al. (2009). Prevention of depression in at-risk adolescents: a randomized controlled trial. *JAMA, 301*, 2215–2224.

Garber, J., Frankel, S. A., & Herrington, C. G. (2016). Developmental demands of cognitive behavioral therapy for depression in children and adolescents: cognitive, social, and emotional processes. *Annual Review of Clinical Psychology, 12*, 181–216.

Goldstein, T. R. (2017). *Early assessment and intervention for adolescents at risk for bipolar disorder.* https://clinicaltrials.gov/ct2/show/NCT03203707.

Goldstein, T. R., Fersch-Podrat, R., Axelson, D. A., Gilbert, A., Hlastala, S. A., Birmaher, B., et al. (2014). Early intervention for adolescents at high risk for the development of bipolar disorder: pilot study of interpersonal and social rhythm therapy (IPSRT). *Psychotherapy, 51*, 180–189.

Goodman, S. H., & Garber, J. (2017). Evidence-based interventions for depressed mothers and their young children. *Child Development, 88*, 368–377.

Hutton, P., & Taylor, P. J. (2014). Cognitive behavioural therapy for psychosis prevention: a systematic review and meta-analysis. *Psychological Medicine, 44*, 449–468.

Keshavan, M. S., DeLisi, L. E., & Seidman, L. J. (2011). Early and broadly defined psychosis risk mental states. *Schizophrenia Research, 126*, 1–10.

Lovell, K., Bower, P., Gellatly, J., Byford, S., Bee, P., McMillan, D., et al. (2017). Low-intensity cognitive-behaviour therapy interventions for obsessive-compulsive disorder compared to waiting list for therapist-led cognitive-behaviour therapy: 3-arm randomised controlled trial of clinical effectiveness. *PLoS Medicine, 14*, e1002337.

Maziade, M. (2017). At risk for serious mental illness—screening children of patients with mood disorders or Schizophrenia. *The New England Journal of Medicine, 376*, 910–912.

McGorry, P., & Nelson, B. (2016). Why we need a Transdiagnostic staging approach to emerging psychopathology, early diagnosis, and treatment. *JAMA Psychiatry, 73*, 191–192.

McGorry, P. D., Nelson, B., Markulev, C., Yuen, H. P., Schafer, M. R., Mossaheb, N., et al. (2017). Effect of omega-3 polyunsaturated fatty acids in young people at ultrahigh risk for psychotic disorders: the NEURAPRO randomized clinical trial. *JAMA Psychiatry, 74*, 19–27.

Miklowitz, D. J. (2011). *Early intervention for youth at risk for bipolar disorder*. https://clinicaltrials.gov/ct2/show/NCT01483391.

Miklowitz, D. J., Chang, K. D., Taylor, D. O., George, E. L., Singh, M. K., Schneck, C. D., et al. (2011). Early psychosocial intervention for youth at risk for bipolar I or II disorder: a one-year treatment development trial. *Bipolar Disorders, 13*, 67–75.

Miklowitz, D. J., Schneck, C. D., Singh, M. K., Taylor, D. O., George, E. L., Cosgrove, V. E., et al. (2013). Early intervention for symptomatic youth at risk for bipolar disorder: a randomized trial of family-focused therapy. *Journal of the American Academy of Child and Adolescent Psychiatry, 52*, 121–131.

Miller, T. J., McGlashan, T. H., Rosen, J. L., Cadenhead, K., Cannon, T., Ventura, J., et al. (2003). Prodromal assessment with the structured interview for prodromal syndromes and the scale of prodromal symptoms: predictive validity, interrater reliability, and training to reliability. *Schizophrenia Bulletin, 29*, 703–715.

Morrison, A. P., French, P., Stewart, S. L., Birchwood, M., Fowler, D., Gumley, A. I., et al. (2012). Early detection and intervention evaluation for people at risk of psychosis: multisite randomised controlled trial. *BMJ, 344*, e2233.

Morrison, A. P., French, P., Walford, L., Lewis, S. W., Kilcommons, A., Green, J., et al. (2004). Cognitive therapy for the prevention of psychosis in people at ultra-high risk: randomised controlled trial. *British Journal of Psychiatry, 185*, 291–297.

Nadkarni, R. B., & Fristad, M. A. (2010). Clinical course of children with a depressive spectrum disorder and transient manic symptoms. *Bipolar Disorders, 12*, 494–503.

Nauta, M. H., Festen, H., Reichart, C. G., Nolen, W. A., Stant, A. D., Bockting, C. L., et al. (2012). Preventing mood and anxiety disorders in youth: a multi-centre RCT in the high risk offspring of depressed and anxious patients. *BMC Psychiatry, 12*, 31.

Parker, S. (2015). *Bipolar at risk trial (BART)*. ISRCTN Registry. http://www.isrctn.com/ISRCTN10773067.

Pfennig, A., Leopold, K., Bechdolf, A., Correll, C. U., Holtmann, M., Lambert, M., et al. (2014). Early specific cognitive-behavioural psychotherapy in subjects at high risk for bipolar disorders: study protocol for a randomised controlled trial. *Trials, 15*, 161.

Rasic, D., Hajek, T., Alda, M., & Uher, R. (2014). Risk of mental illness in offspring of parents with schizophrenia, bipolar disorder, and major depressive disorder: a meta-analysis of family high-risk studies. *Schizophrenia Bulletin, 40*, 28–38.

Relton, C., Torgerson, D., O'Cathain, A., & Nicholl, J. (2010). Rethinking pragmatic randomised controlled trials: introducing the "cohort multiple randomised controlled trial" design. *BMJ, 340*, c1066.

Saavedra, L. M., Silverman, W. K., Morgan-Lopez, A. A., & Kurtines, W. M. (2010). Cognitive behavioral treatment for childhood anxiety disorders: long-term effects on anxiety and secondary disorders in young adulthood. *Journal of Child Psychology and Psychiatry, 51,* 924–934.

Sekhon, M., Cartwright, M., & Francis, J. J. (2017). Acceptability of healthcare interventions: an overview of reviews and development of a theoretical framework. *BMC Health Services Research, 17,* 88.

Strawn, J. R., Adler, C. M., McNamara, R. K., Welge, J. A., Bitter, S. M., Mills, N. P., et al. (2014). Antidepressant tolerability in anxious and depressed youth at high risk for bipolar disorder: a prospective naturalistic treatment study. *Bipolar Disorders, 16,* 523–530.

Uher, R. (2013). *Skills for Wellness (SWELL)*. https://clinicaltrials.gov/ct2/show/NCT01980147.

Uher, R., Cumby, J., MacKenzie, L. E., Morash-Conway, J., Glover, J. M., Aylott, A., et al. (2014). A familial risk enriched cohort as a platform for testing early interventions to prevent severe mental illness. *BMC Psychiatry, 14,* 344.

Vallarino, M., Henry, C., Etain, B., Gehue, L. J., Macneil, C., Scott, E. M., et al. (2015). An evidence map of psychosocial interventions for the earliest stages of bipolar disorder. *Lancet Psychiatry, 2,* 548–563.

van Os, J., & Guloksuz, S. (2017). A critique of the "ultra-high risk" and "transition" paradigm. *World Psychiatry, 16,* 200–206.

Yung, A. R., Yuen, H. P., McGorry, P. D., Phillips, L. J., Kelly, D., Dell'Olio, M., et al. (2005). Mapping the onset of psychosis: the comprehensive assessment of at-risk mental states. *The Australian and New Zealand Journal of Psychiatry, 39,* 964–971.

Summary and integration of current findings: A model for bipolar disorder development

13

Consuelo Walss-Bass*, Jair C. Soares‡, Paolo Brambilla*,†

Translational Psychiatry Program, Department of Psychiatry and Behavioral Sciences, University of Texas Health Sciences Center at Houston, Houston, TX, United States*
Department of Neurosciences and Mental Health, Fondazione IRCCS Ca' Granda Ospedale Maggiore Policlinico, University of Milan, Milan, Italy†
Center of Excellence on Mood Disorders, Department of Psychiatry and Behavioral Sciences, McGovern Medical School, University of Texas Health Science Center at Houston (UTHealth), Houston, TX, United States‡

Bipolar disorder patients with a childhood or adolescent onset have a worse prospective course of mood symptoms and higher rates of comorbid illness compared to those with adult onset (Arango, Fraguas, & Parellada, 2014; Chang, 2009). The poor prognosis in patients with early onset is due, in part, to the long gap between onset of illness and first pharmacological treatment. This has raised the interest in identifying what is known as the "bipolar prodrome" (Chapter 1). As discussed herein, longitudinal clinical studies show that early anxiety, excessive alertness, and emotional sensitivity may represent early signs that evolve into mood and adjustment problems, and may eventually develop into major depression and/or hypomania/mania. These studies are evidence that bipolar prodromal symptoms manifest years before the onset of bipolar disorder. Identification of these clinical signs before onset of the disorder may aid in the identification of vulnerable subjects.

Most of the literature available to date on the pathophysiological mechanisms of bipolar disorder focuses on studies conducted with chronically ill adult individuals. Therefore, it is not clear whether any abnormalities found in patients precede the onset of illness, emerge during early illness development, or follow bipolar disorder onset. First-degree relatives of bipolar disorder patients present a higher polygenic load of risk variants than control individuals. Particularly, offspring of bipolar disorder parents have a fourfold increased risk of developing bipolar disorder compared to offspring of healthy parents. Therefore high-risk family members represent an ideal population for the study of biomarkers of risk, as the effects of illness and psychotropic medications, which confound studies of affected individuals, are not

Bipolar Disorder Vulnerability. https://doi.org/10.1016/B978-0-12-812347-8.00013-0

present (Chapter 5). This book highlights studies in pediatric and high-risk populations using diverse translational methodologies including genetic studies (Chapters 4 and 5), functional and structural neuroimaging (Chapter 7), and neuropsychological testing (Chapter 8) that point toward structural, neurochemical, and functional abnormalities in pediatric and high-risk subjects. Altogether, these findings suggest that these intermediate phenotypes are useful for identification of subjects at risk for bipolar disorder, may predict BD onset, and therefore will have diagnostic value and allow for specific targeting of interventions to the groups at highest risk.

Studies of molecular trajectories in youth with bipolar disorder and unaffected subjects at high risk for developing bipolar disorder, as well as the examination of interaction between genomic and environmental influences that are shaping behavior at these critical periods, are presented in this work. Of great importance, several lines of evidence show the harmful effect of early stress, particularly childhood maltreatment, on the course of bipolar disease, leading to earlier age of onset and greater symptom severity compared to patients without such history (Chapter 3). Indeed, as discussed in this work, high-risk offspring are more likely to have experienced episodic and chronic interpersonal stress, by virtue of their family environment. Accordingly, high-risk offspring who experienced high interpersonal chronic stress display a larger cortisol rise following awakening than those reporting low interpersonal chronic stress (Ostiguy et al., 2011). Low levels of structure provided by parents have been predictive of an elevated cortisol response following awakening and during a laboratory psychosocial stressor (Ellenbogen & Hodgins, 2009). Altogether, studies suggest an important role of family environment in modulating hypothalamus-pituitary-adrenal (HPA) axis activity in youth at high risk. Impaired stress resilience and a dysfunctional HPA axis may play key roles in bipolar disorder and its risk in offspring of bipolar parents (Duffy et al., 2016). The glucocorticoid receptor (GR) is a cytosolic transcription factor activated by cortisol, and is responsible for the negative feedback of the HPA axis that reduces cortisol levels after the cessation of stress. Therefore, a dysfunctional GR signaling pathway, which has been reported in adult BD patients and is suggested by our recent study in youth (Fries et al., 2014, 2017), could underlie the increased levels of cortisol seen in high-risk subjects. As discussed in Chapter 6, glucocorticoids are considered possible mediators of neuroprogression toward mania, because in excess levels they lead to oxidative stress and inflammation, causing neuronal damage (Grande et al., 2012). Importantly, prospective studies suggest that abnormalities in the HPA axis can predict the onset of an affective disorder in offspring of BD parents (Ellenbogen et al., 2011). HPA axis dysfunction, along with a dysfunction in the immune system, has been proposed as one of the main biological mechanisms underlying the risk for BD in offspring of parents with bipolar disorder (Duffy et al., 2012, 2016). HPA axis dysfunction may in turn lead to dysregulation of brain regions associated with emotion including the ventrolateral prefrontal cortex (VLPFC), anterior cingulate cortex (ACC), thalamus, striatum, and hippocampus (Chapter 6). Abnormal emotion regulation may then give rise to symptoms seen in bipolar disorder (Fig. 1). Indeed, the summary of imaging studies presented here suggests that brain regions previously implicated in bipolar disorder have a potential maladaptive response to stress that precedes the onset of psychiatric symptoms.

As a whole, results from the different lines of study presented in this work point toward imbalances in different biological mechanisms, including neurodevelopment,

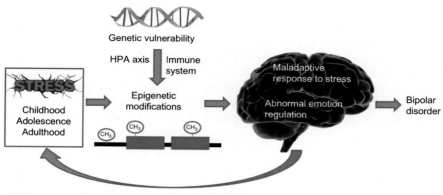

FIG. 1

Model for bipolar disorder development. Gene-environment interactions early in life may lead to a maladaptive response to stress and brain abnormalities that are worsened by further exposure to stress, eventually leading to manifestation of bipolar disorder.

HPA axis, emotion regulation, and immune signaling, in development of BD. Individually, alterations in any one of these pathways may partially increase the risk of bipolar disorder, but no one of them can by itself explain the high degree of heritability of bipolar disorder. In fact, the proposed theories are not mutually exclusive, but rather are highly interconnected. Inherited genetic alterations in one specific biological pathway may lead to an inability to respond appropriately to a given environmental insult early in life, causing biological alterations and epigenetic modifications that in combination may lead to brain functional abnormalities. These early brain abnormalities could initiate a feedback loop of increased sensitivity to stress, such that brain abnormalities would increase in severity with exposure to further stress, in time leading to a global systemic dysfunction and disease manifestation (Fig. 1). An overall understanding of biological and environmental factors influencing disease manifestation could lead to an early detection of brain and behavioral abnormalities, and, importantly, lead to identification of novel treatments.

In regard to therapeutic treatments, pharmacologic interventions have already proven to be successful in treatment of BD in pediatric populations, thus providing evidence-based options for clinicians in determining the most effective treatments for young patients already suffering from the disorder (Chapter 11). However, treatments that prevent full onset of bipolar disorder are greatly needed, to avoid the chronic and harmful impact of the disorder to patients and families. The studies discussed herein are highly promising for providing new targets for development of novel pharmacologic approaches to achieve this goal.

Recent findings from ongoing studies involving psychological interventions offer encouraging evidence suggesting these interventions may be effective in prevention of bipolar disorder in youth at risk, as substantial beneficial effects appear to last for a number of years after treatment completion, and the treatment modalities may reduce negative side effects, burden, and risk of stigmatization in young children and adolescents (Chapter 12).

As emerging discoveries from research in high-risk subjects reveal the underlying mechanisms leading to the development of bipolar disorder, these studies will establish an empirical basis for the application of novel therapeutics and multimodal interventions in an early phase of illness during which specific treatments could more effectively alter the disease's course, ultimately reducing and possibly avoiding the lasting damage caused by this illness to patients and their families.

REFERENCES

Arango, C., Fraguas, D., & Parellada, M. (2014). Differential neurodevelopmental trajectories in patients with early-onset bipolar and schizophrenia disorders. *Schizophrenia Bulletin*, *40*(Suppl 2), S138–146.

Chang, K. D. (2009). Diagnosing bipolar disorder in children and adolescents. *The Journal of Clinical Psychiatry*, *70*(11), e41.

Duffy, A., et al. (2012). Biological indicators of illness risk in offspring of bipolar parents: targeting the hypothalamic-pituitary-adrenal axis and immune system. *Early Intervention in Psychiatry*, *6*(2), 128–137.

Duffy, A., et al. (2016). Candidate risks indicators for bipolar disorder: early intervention opportunities in high-risk youth. *The International Journal of Neuropsychopharmacology*, *19*(1).

Ellenbogen, M. A., & Hodgins, S. (2009). Structure provided by parents in middle childhood predicts cortisol reactivity in adolescence among the offspring of parents with bipolar disorder and controls. *Psychoneuroendocrinology*, *34*(5), 773–785.

Ellenbogen, M. A., et al. (2011). Elevated daytime cortisol levels: a biomarker of subsequent major affective disorder? *Journal of Affective Disorders*, *132*(1–2), 265–269.

Fries, G. R., et al. (2014). Hypothalamic-pituitary-adrenal axis dysfunction and illness progression in bipolar disorder. *The International Journal of Neuropsychopharmacology*, *18*(1).

Fries, G. R., et al. (2017). Integrated transcriptome and methylome analysis in youth at high risk for bipolar disorder: a preliminary analysis. *Translational Psychiatry*, *7*(3), e1059.

Grande, I., et al. (2012). Mediators of allostasis and systemic toxicity in bipolar disorder. *Physiology & Behavior*, *106*(1), 46–50.

Ostiguy, C. S., et al. (2011). Sensitivity to stress among the offspring of parents with bipolar disorder: a study of daytime cortisol levels. *Psychological Medicine*, *41*(11), 2447–2457.

Index

Note: Page numbers followed by *f* indicate figures, *t* indicate tables, and *b* indicate boxes.

269